Lecture Notes in Mathematics

Edited by A. Dold and B. Eckmann

919

Séminaire Pierre Lelong - Henri Skoda (Analyse) Années 1980/81

et

Colloque de Wimereux, Mai 1981 "Les fonctions plurisousharmoniques en dimension finie ou infinie", organisé en l'honneur de Pierre Lelong

Edité par Pierre Lelong et Henri Skoda

Springer-Verlag
Berlin Heidelberg New York 1982

Editeurs
Pierre Lelong
Henri Skoda
Université de Paris VI, Mathématiques
Tour 45-46 -5éme étage,
4 Place Jussieu, 75005 Paris, France

AMS Subject Classifications(1980): 32-XX

ISBN 3-540-11482-3 Springer-Verlag Berlin Heidelberg New York
ISBN 0-387-11482-3 Springer-Verlag New York Heidelberg Berlin

CIP-Kurztitelaufnahme der Deutschen Bibliothek
Séminaire Pierre Lelong, Henri Skoda (Analyse): Séminaire Pierre Lelong, Henri Skoda (Analyse): années ... - Berlin; Heidelberg; New York: Springer
1980/81. ... Et Colloque de Wimereux, Mai 1981, „Les Fonctions Plurisousharmoniques en Dimension Finie ou Infinie": organisé en l'honneur de Pierre Lelong. - 1982.
(Lecture notes in mathematics; Vol. 919)
ISBN 3-540-11482-3 (Berlin, Heidelberg, New York)
ISBN 0-387-11482-3 (New York, Heidelberg, Berlin)
NE: Colloque Les Fonctions Plurisousharmoniques en Dimension Finie ou Infinie <1981, Wimereux>; GT

Printing and binding: Beltz Offsetdruck, Hemsbach/Bergstr.
2141/3140-543210

Introduction au Séminaire P.LELONG,H.SKODA 1980-1981.

Ce volume est divisé en deux parties.

La première partie se compose des exposés faits au séminaire P.Lelong,H.Skoda durant les années 1980 et 1981, qui n'ont pas déjà été publiés ailleurs. La deuxième partie reprend certains des exposés du Colloque de Wimereux , de Mai 1981, sous le titre : "Les fonctions plurisousharmoniques en dimension finie ou infinie", organisé en l'honneur de Pierre Lelong.

On a reproduit ici parmi les allocutions prononcées à cette occasion celle de G.Coeuré et la réponse de P.Lelong.

Indiquons brièvement les sujets traités dans les exposés des deux parties en les regroupant suivant leurs thèmes scientifiques. Il s'agit toujours de travaux apportant des résultats nouveaux dans le domaine de l'Analyse complexe, domaine qui continue à se développer rapidement.
On retrouve dans ce volume des thèmes classiques qui sont déjà apparus plusieurs fois dans ce séminaire.
1/ En premier lieu apparaissent les notions capacitaires et la théorie des fonctions plurisousharmoniques.

L'exposé de R.E.Molzon et B.Shiffman introduit des notions de capacité, de diamètre transfini et de constante de Tchebycheff sur $\mathbb{P}_{n-1}$ en liaison avec la théorie quantitative des ensembles analytiques dans $\mathbb{C}^n$ et apporte des résultats de comparaison entre ces différentes notions.

E.Bedford étend aux espaces analytiques complexes sa théorie de l'opérateur de Monge-Ampère $(dd^c)^n$ et la notion de capacité associée. Comme application, il étend le théorème de Josephson sur les ensembles localement pluripolaires à certains espaces analytiques (éventuellement sans fonctions holomorphes).

B.Gaveau et J.Lawrynowicz définissent à l'aide de la théorie des jeux de Von Neumann une intégrale de Dirichlet puis des notions capacitaires invariantes par isomorphisme analytique. Leur approche semble assez différente de celle qui utilise l'équation de Monge-Ampère complexe. Ils donnent une application à la physique théorique des particules élémentaires.

2/ C.O.Kiselman étudie le nombre de Lelong de la restriction d'une fonction plurisousharmonique aux différents germes de sous-variétés analytiques passant par l'origine.
Il montre que ce nombre est indépendant de la sous-variété sauf pour des sous-variétés appartenant à un ensemble exceptionnel. L'ensemble des germes de sous-variété étant de dimension infinie, l'étude des fonctions plurisousharmoniques sur des espaces de dimension infinie apparaît ici comme particulièrement naturelle. Une question posée à cette occasion à Wimereux par C.O.Kiselman a suscité une réponse de P.Lelong qui donne un nouveau moyen de calculer ce nombre et en déduit un lemme de Schwarz en dimension infinie.

M.Blel approfondit dans son article les propriétés de sommabilité locale de $\exp(-V)$ où V est le potentiel canonique plurisousharmonique associé par H.Skoda à un ensemble analytique X. Par une série de contre-exemples, il montre qu'il n'y a pas de liaison simple entre les propriétés algébriques des singularités de l'ensemble X (par exemple être une intersection complète) et les singularités du potentiel V.

V.Avanissian utilise la complexification et la notion de cellule d'harmonicité, introduite pour la première fois par Aronszajn et P.Lelong, pour l'étude des fonctions harmoniques d'ordre infini et donne des résultats sur celles d'entre elles qui sont arithmétiques, c'est-à-dire telles que $f(\mathbb{N}^n) \subset \mathbb{N}$.

3/ Un autre thème classique du séminaire est celui de l'analyse harmonique et de la synthèse spectrale. Il est bien connu depuis les travaux d'Ehrenpreis, B.Malgrange et V.P.Palamodov, que ces questions se ramènent à l'étude des variétés d'interpolation, c'est-à-dire au problème de l'extension d'une fonction holomorphe

avec croissance à partir d'une sous-variété V de $\mathbb{C}^n$. C.A.Berenstein et B.A.Taylor d'une part, J.-P.Demailly d'autre part montrent que si les mineurs de la matrice jacobienne de l'application holomorphe définissant V , ne décroissent pas trop vite, alors V est d'interpolation. En fait, ils construisent une bonne rétraction holomorphe sur V dans un voisinage de V , permettant d'utiliser ensuite les méthodes semi-globales pour l'opérateur $\bar{\partial}$.
Plus généralement, J.-P.Demailly s'intéresse au problème du scindage holomorphe avec croissance d'une suite exacte de fibrés holomorphes au-dessus d'une variété de Stein . Le problème précédent correspond au cas de la suite exacte définissant le fibré normal à V . Il obtient des résultats à peu près optimaux, dans le même esprit que ceux obtenus par H.Skoda dans les précédents séminaires pour les morphismes de fibrés vectoriels semi-positifs.

- Dans un article de la même veine, il introduit une troisième notion de positivité pour les fibrés vectoriels holomorphes, la positivité forte, complétant les notions de positivité au sens de P.A.Griffiths et S.Nakano.
Il compare ces différentes notions et transpose les résultats obtenus à la positivité des courants. Il obtient ainsi d'intéressantes relations entre les positivités faibles et fortes des courants.

4/ L'article de B.Barlet concerne les opérateurs différentiels et l'étude des singularités. Il observe que l'intégration d'une forme différentielle C^∞ à support compact sur les fibres d'une application holomorphe d'un espace analytique X à valeurs dans $\mathbb{C}$ ne fournit pas en général une fonction C^∞ . Il décrit les singularités de cette dernière fonction à l'aide d'un développement asymptotique dont les coefficients sont des courants sur X . De plus, les nombres rationnels qui interviennent dans ce développement sont reliés à la résolution des singularités de X et aux racines du polynôme de Bernstein-Sato.

5/ Le thème de la théorie fine des fonctions holomorphes est représenté dans ce volume par les articles de B.Gaveau et M.Range.

B.Gaveau décrit des conditions nécessaires quantitatives vérifiées par la courbure scalaire des diviseurs d'une fonction holomorphe dans une classe de Hardy de la boule de $\mathbb{C}^n$. Des conditions de ce type, distinctes de la condition de Blaschke et de celle de P.Malliavin sont en effet activement recherchées. Un résultat similaire est obtenu pour les diviseurs dans $\mathbb{C}^n$.

M.Range, dans un article de synthèse, fait le point sur la théorie des estimations hölderiennes pour $\bar{\partial}$ dans un domaine pseudoconvexe. Cette théorie n'est encore satisfaisante que dans le cas strictement pseudoconvexe et dans le cas d'un ouvert convexe de $\mathbb{C}^2$ à frontière analytique réelle. Il étudie également la régu larité höldérienne des dérivées d'ordre k des solutions explicites pour $\bar{\partial}$ en bidegré (p,q) avec $q \geqslant 1$. Le résultat nouveau, correspondant à $q > 1$, est obtenu par une modification convenable de la solution de Henkin.

6/ Le thème des applications à la Physique théorique, quelque peu oublié dans le précédent séminaire, réappparait ici.

V.S.Vladimirov présente dans son article un panorama actuel des liens entre la Physique Théorique et les fonctions holomorphes de plusieurs variables. Il montre tour à tour le rôle joué en Physique Théorique par le théorème du "edge of the wedge" , le théorème de l'enveloppe C-convexe, le théorème de covariance finie et l'étude des fonctions holomorphes de partie réelle > 0 définies dans un tube.

B.Gaveau et G.Laville montrent le lien entre les fonctions propres d'un certain hamiltonien et les fonctions holomorphes vérifiant les équations de Cauchy-Riemann tangentielles sur le groupe de Heisenberg. Ce lien est établit par l'intermédiaire d'une transformée de Fourier partielle sur le groupe de Heisenberg.

B.Gaveau et J.Lawrynowicz dans leur travail déjà cité donnent également une application à la Physique Théorique.

Nous remercions les auteurs qui nous ont confié leurs textes , Madame Orion qui a préparé de nombreux manuscrits et la librairie Springer qui édite ce séminaire et qui contribue ainsi à la diffusion rapide de résultats nouveaux.

Pierre LELONG - Henri SKODA

TABLE DES MATIÈRES

I SÉMINAIRE d'ANALYSE (PARIS).

1 BERENSTEIN (C.A.) et TAYLOR (B.A.). - On the geometry of interpolating varieties 1

2 BLEL (M.). - Fonctions plurisousharmoniques et idéal définissant un ensemble analytique 26

3 DEMAILLY (J.-P.). - Relations entre les différentes notions de fibrés et de courants positifs 56

4 " - Scindage holomorphe d'un morphisme de fibrés vectoriels semi-positifs avec estimations L^2 77

5 GAVEAU (B.). - Intégrales de courbure et potentiels sur les hypersurfaces analytiques de $\mathbb{C}^n$ 108

6 GAVEAU (B.) et LAVILLE (G.). - Fonctions holomorphes et particule chargée dans un champ magnétique uniforme 123

7 GAVEAU (B.) et ŁAWRYNOWICZ (J.). - Intégrale de Dirichlet sur une variété complexe I 131

8 LELONG (P.). - Calcul du nombre densité $\nu(x,f)$ et lemme de Schwarz pour les fonctions plurisousharmoniques dans un espace vectoriel topologique 167

9 RANGE (M.). - Boundary regularity for the Cauchy-Riemann complex 177

II COLLOQUE de WIMEREUX, Mai 1981.

Allocution prononcée par G. COEURÉ en l'honneur du Professeur P.LELONG. 187

Réponse de P.LELONG........ 190

1 AVANISSIAN (V.). - Sur les fonctions harmoniques d'ordre quelconque et leur prolongement analytique dans $\mathbb{C}^n$ 192

2 BARLET (D.). - Développements asymptotiques des fonctions obtenues par intégration sur les fibres 282

3 BEDFORD (E.). - "The operator $(dd^c)^n$ on complex spaces 294

4 KISELMAN (Ch.O.). - Stabilité du nombre de LELONG par restriction à une sous-variété 324

5 MOLZON (R.E.) et SHIFFMAN (B.). - Capacity, Tchebycheff constant, and transfinite hyperdiameter on complex projective space 337

6 VLADIMIROV (V.S.). - Several complex variables in Mathematical physics 358

Exposés faits à WIMEREUX dont les résultats paraîtront ailleurs :

DIEDERICH (K.), DINEEN (S.), GRAUERT (H.), HARVEY (R.), NACHBIN (L.), SIBONY (N.), SICIAK (J.), SIU (Y.T.), STOLL (W.).

Séminaire P.LELONG,H.SKODA
(Analyse)
20e et 21e année, 1980-1981.

ON THE GEOMETRY OF INTERPOLATING VARIETIES

Carlos A. Berenstein and B. A. Taylor

1. The problem we want to consider here is the following. Let V be an analytic variety in $\mathbb{C}^n$ and p a plurisubharmonic weight function. What are the necessary and sufficient conditions on V such that every analytic function φ on V satisfying an estimate of the form

$$(1) \qquad |\varphi(z)| \leq A \exp(Bp(z)), \qquad \text{for all } z \in V,$$

has an extension to an entire function Φ satisfying the same kind of estimate for all $z \in \mathbb{C}^n$, namely

$$(2) \qquad |\Phi(z)| \leq A' \exp(B'p(z)).$$

We will then say that V is an *interpolating variety* (for the weight p). Actually one has to impose some minimal conditions on V and p in order that the problem does not become trivial (e.g., no functions on V satisfy (1) except for the constants), and in applications to harmonic analysis one also needs to consider the problem of multiplicities. Rather than burden the reader with unnecessary details we refer to our paper [5], where essentially the opposite problem is considered, i.e. given V find the necessary and sufficient conditions on φ such that it has an extension Φ satisfying (2). We also refer to [4] and [5] for an explanation of the relevance of this kind of questions to mean-periodicity and related subjects in harmonic analysis.

To simplify the exposition we will assume throughout that $p(z) = |z|$ and point out whether the results below hold for more general weight functions. Hence, $A_p := \{\Phi$ entire functions in $\mathbb{C}^n$ satisfying (2)$\}$ = space of functions of exponential type, which is a space of considerable interest by itself [10], [4]. We remind the reader though, that even simple looking variations of this weight $|z|$ like $p(z) = |\operatorname{Im} z| + \log(1+|z|)$ often lead to considerable extra difficulties. In this second example, $A_p = \hat{\mathcal{E}}'(\mathbb{R}^n)$ = space of Fourier transforms of distributions in $\mathbb{R}^n$ with compact support.

Consider first in more detail the case $n = 1$ where the above problem is settled (at least for $p(z) = |z|$). If $V = V(f) := \{z \in \mathbb{C} : f(z) = 0\}$ (counted with multiplicities) for some $f \in A_p$, then $V = \{(z_k, m_k) : z_k$ zero of f of multiplicity $m_k\}$, any analytic function on V is just a doubly indexed sequence $\{a_{kj}\}$ $(0 \leq j < m_k)$ and we have a restriction map

$$\rho : A(\mathbb{C}) \longrightarrow A(V)$$

from the space of entire functions to the analytic functions on V, given by

(3) $$\rho(\varphi) = \left\{\frac{\varphi^{(j)}(z_k)}{j!}\right\}.$$

In this case $\rho(A_p) \subseteq A_p(V) := \{\{a_{kj}\} : |a_{kj}| \leq A\exp(B|z_k|)\}$. And the interpolation problem is simply put: Is the map $\rho : A_p \to A_p(V)$ onto? In [3] we showed that the necessary and sufficient condition for V to be interpolating (for the weight $|z|$) is

(4) $$\left|\frac{f^{(m_k)}(z_k)}{m_k!}\right| \geq \varepsilon \exp(-C|z_k|)$$

for some ε, $C > 0$. In particular, if all the points z_k in V are simple, this condition reduces to

(5) $$|f'(z_k)| \geq \varepsilon \exp(-C|z_k|),$$

a result due to A. F. Leont'ev. (Similar statements hold for arbitrary weights p, see [3].)

Recall that the function f defining V in the case $V = V(f)$ is unique up to an exponential factor and hence (4) is really a condition on V. More generally, one assumes only that $V \subseteq V(f)$, $f \in A_p$. As it is well known [19], this condition $V \subseteq V(f)$ is equivalent to the geometric condition

(6) $$n(r) := \sum_{|z_k|\leq r} m_k = O(r) \qquad \text{as } r \to \infty,$$

which measures the (0-dimensional) area of $V \cap B(0,r)$, $B(a,r) := \{z \in \mathbb{C} : |z-a| \leq r\}$. (Similarly one can describe geometrically the statement $V = V(f)$, $f \in A_p$, $p(z) = |z|$, see [10].) Under this weaker assumption (6) one can describe V as the set of common zeroes of several functions in A_p and correspondingly a statement similar to (4) holds, see [3, Theorem 4]. This analytic condition (4) has been translated into geometric terms by W. A. Squires [26]. He shows that V is an interpolating variety (for $p(z) = |z|$) if and only if for some fixed constants $A,B > 0$ one has

$$(7) \qquad \int_0^{|z_k|} n(z_k,t)\frac{dt}{t} \le A|z_k| + B \qquad \text{for every } z_k \in V,$$

where

$$(8) \qquad n(z_k,r) = \sum_{0<|z_j-z_k|\le r} m_j$$

is the "area" of $V \cap (B(z_k,r)\setminus\{z_k\})$. It was shown by Gelfond that every exponential-polynomial satisfies (7) (see e.g. [30] where sharp estimates for the constants A and B are given). Let us also point out that the corresponding geometric characterization for $p(z) = |\mathrm{Im}\, z| + \log(1+|z|)$ is not known although (4) still holds when $V = V(f)$, f slowly decreasing, see [3]. While the technical condition of f being slowly decreasing is a generic condition for this weight, [26] shows that some interpolating varieties cannot be defined by slowly decreasing f.

We discuss also other natural definitions of interpolating varieties in [3], [5]. In fact, there is a whole scale of such interpolating problems and the corresponsing geometric characterizations are not always completely known. In relation to this let us finish this secion by pointing out a small misprint in formula (23), [3, p. 121], it should read

$$\sum_{\ell=0}^{m_k-1} |a_{k,\ell}|r^\ell \le A\exp(Bp(z_k;r)), \qquad k = 1,2,\ldots .$$

2. In the case of several variables we can give sufficient conditions for interpolation when V is a smooth manifold of codimension k, say

$$(9) \quad V = V(f_1,\ldots,f_N) = \{z \in \mathbb{C}^n : f_1(z) = \cdots = f_N(z) = 0\},\ f_1,\ldots,f_N \in A_p,$$

and

(10) at each point $z \in V$, the rank of the jacobian matrix $Df = \left\| \frac{\partial f_i}{\partial z_j} \right\|$ is k.

Skoda has shown in [23] that if V is a subvariety of $\mathbb{C}^n$ of pure dimension $n-k$, then (for $p(z) = |z|$) (9) is a consequence of the area condition

(11) $$\frac{\sigma(r)}{r^{2(n-k)}} = O(r) \qquad \text{as } r \to \infty,$$

where $\sigma(r)$ denotes the $(2(n-k)$-dimensional) area of $V \cap V(0,r)$ (compare with (6)). For $k = 1$ one can give a sharpening of (11) ensuring that $N = 1$ in (9) (this is included in the earlier work [18] of Lelong). Hence, for $k = 1$, one can see that both (9) and (10) are a consequence of this sharper area condition when V is a manifold.

It is also well known that algebraic varieties are characterized by $\sigma(r)r^{-2(n-k)}$ being bounded [27].

The following condition analogous to (5) was stated without proof as sufficient for interpolation in [5]. We denote by $\Delta_{I,J}$ the $k \times k$ minors of the matrix Df, $I = (i_1,\ldots,i_k)$, $J = (j_1,\ldots,j_k)$, $1 \le i_1 < \cdots < i_k \le N$, $1 \le j_1 < \cdots < j_k \le n$.

THEOREM 1. The smooth manifold V defined by (9) and (10) is interpolating if for some ε, $C > 0$ the minors $\Delta_{I,J}$ satisfy

(12) $$\sum |\Delta_{I,J}(z)| \ge \varepsilon \exp(-Cp(z)), \qquad \text{for all } z \in V.$$

The sum being taken over all $k \times k$ minors of Df.

Neither the statement nor the proof of this theorem depend on $p(z) = |z|$, it only requires that p satisfies the conditions on [15]. Namely: $p(z) \ge 0$, $\log(1+|z|) = O(p(z))$, and there exist four constants $c_1,\ldots,c_4$ such that $|z| \le \exp(-c_1p(z)-c_2)$ implies that $p(w) \le c_3p(z) + c_4$.

PROOF. The idea of the proof is very simple. One has to extend the function φ analytic on V and satisfying (1), to a sufficiently large neighborhood of V preserving the growth conditions and then apply the following Semilocal Interpolation Theorem whose proof can be found in [5], [17]. Denote by $|f(z)| := (\sum |f_j(z)|^2)^{1/2}$ and

(13) $$S(\delta,D,f) := \{z \in \mathbb{C}^n : |f(z)| < \delta \exp(-Dp(z))\},$$

where $\delta, D > 0$. (Actually only the components of $S(\delta,D,g)$ which contain points of V are of any interest.)

SEMILOCAL INTERPOLATION THEOREM. If $\tilde{\varphi}$ is an analytic function in $S = S(\delta,D,f)$ satisfying the estimate

$$|\tilde{\varphi}(z)| \le A_1 \exp(B_1p(z)), \qquad \text{for all } z \in S,$$

then there is an entire function $\Phi \in A_p$ such that $\Phi|_V = \tilde{\varphi}|_V$.

The way we accomplish the extension from V to S is by showing that for a convenient choice of δ, D there is a holomorphic retraction

$$\pi : S \longrightarrow V$$

such that along the fibers of π the weight function does not change much, namely if $\pi(z) = \pi(z')$ then

$$p(z) \leq D_1 p(z') + D_2$$

for some constants $D_1, D_2 > 0$. Then the function $\tilde{\varphi}(z) := \varphi(\pi(z))$ will satisfy the hypothesis of the Semilocal Interpolation Theorem and we will be done. Actually, one can construct π only on the union of those components of S containing points of V, but then one can choose $\tilde{\varphi} = 0$ on the remaining components, which gets around this difficulty. Henceforth we will disregard any of the latter components.

From hypothesis (12) and the fact that $\frac{\partial f_i}{\partial z_j} \in A_p$ it follows that there are $\varepsilon_1, c_1, \varepsilon_2, c_2 > 0$ such that

$$\text{(14)} \qquad \sum |\Delta_{I,J}| \geq \varepsilon_1 \exp(-c_1 p(z)) \qquad \text{for } z \in S(\varepsilon_2, c_2, f).$$

(It is here where we have to throw out the components with no points in V.) Hence, by a theorem of Hörmander [15] there exist $\varepsilon_3, c_3 > 0$ and functions $\alpha_{I,J} \in A_p(S(\varepsilon_3, c_3, f))$ (with the obvious notation) such that

$$\text{(15)} \qquad \sum \alpha_{I,J} \Delta_{I,J} = 1 \quad \text{on } S(\varepsilon_3, c_3, f).$$

For $Z = \sum a_i \frac{\partial}{\partial z_i}$ a tangent vector to $\mathbb{C}^n$ at $z \in S(\varepsilon_3, c_3, f)$ define

$$\text{(16)} \qquad \eta(Z) = \eta(Z,z) := \sum_{I,J} \alpha_{I,J}(z) X(Z,I,J),$$

where the vector field $X(Z,I,J)$ is defined as follows. Let $a_\ell(Z,I,J)$ be the determinant of the $k \times k$ matrix obtained by replacing the ℓth column of $\left\| \frac{\partial f_i}{\partial z_j} \right\|_{i \in I, j \in J}$ by the column vector $\| Z(f_i) \|_{i \in I}$ and

$$\text{(17)} \qquad X(Z,I,J) = \sum_{j \in J} a_j(Z,I,J) \frac{\partial}{\partial z_j}.$$

Claim. The vector fields $X(Z,I,J)$ have the following properties.

(18) $X(Z,I,J)$ is linear in Z,

(19) $X(Z,I,J)(f_\ell) = \Delta_{I,J}(z)Z(f_\ell)$ at all points $z \in V$,

and

(20) η gives a splitting of T "normal" to $T(V)$,

where

T = tangent bundle to C^n along $V \equiv$ pullback of the trivial bundle $\mathbb{C}^n \times \mathbb{C}^n$ on $\mathbb{C}^n$ by the inclusion map $V \hookrightarrow \mathbb{C}^n$ and, $T(V)$ is the tangent bundle of V.

Property (18) being clear we proceed to prove (19). Consider the $(k+1) \times (k+1)$ augmented matrix

$$A = \left\| \begin{array}{c|c} Z(f_\ell) & \dfrac{\partial f_\ell}{z_j} \\ \hline Z(f_i) & \dfrac{\partial f_i}{\partial z_j} \end{array} \right\|_{i\in I,\, j\in J} .$$

At points $z \in V$, only k of the vectors $\nabla f_1,\dots,\nabla f_N$ $(\nabla f_i = (\frac{\partial f_i}{\partial z_1},\dots,\frac{\partial f_i}{\partial z_n}))$ can be linearly independent since V has codimension k, hence the matrix A is singular. Thus

$$0 = \det A = Z(f_\ell)\Delta_{I,J} - \sum_{j\in J} a_j(Z,I,J)\frac{\partial f_\ell}{\partial z_j} ,$$

which proves (19).

If $Z \in T(V)$ then $Z(f_i) = 0$, hence $\eta(Z) = 0$. Moreover, for any $Z \in T$, $\eta(Z)(f_i) = Z(f_i)$ for all i since

$$\eta(Z)(f_i) = \sum \alpha_{I,J}X(Z,I,J)(f_i) \qquad \text{(by (19))}$$

$$= \sum \alpha_{I,J}\Delta_{I,J}Z(f_i) = Z(f_i) \quad \text{(by (15))}.$$

Hence, $Z - \eta(Z)$ is tangent to V and $Z = \eta(Z) + (Z-\eta(Z))$ gives the splitting

(21) $T = \text{range}\,\eta \oplus T(V) = N \oplus T(V)$

N is then a "normal" bundle to V.

There is an associated holomorphic map

$$\nu : N \longrightarrow \mathbb{C}^n$$

given by

$$(22) \qquad \nu(z, \textstyle\sum a_j \frac{\partial}{\partial z_j}) = z + (a_1,\ldots,a_n),$$

where $z \in V$, $\sum a_j \frac{\partial}{\partial z_j} \in$ range η and $(a_1,\ldots,a_n) \in \mathbb{C}^n$. For $\delta, D > 0$ let

$$\Omega(\delta,D) := \{(z,Z) \in N : |z| < \delta \exp(-Dp(z))\}.$$

Claim. For $\delta, D, \delta_0, D_0 > 0$ convenient one has:

(23) ν is a one-to-one map on $\Omega(\delta_0, D_0)$ and

(24) $$\nu(\Omega(\delta_0,D_0)) \supseteq S(\delta,D,f).$$

Note that once this claim has been proved the map ν^{-1} followed by the projection of N onto V is the retraction map π we are looking for. (V is just the zero section of N and $\nu(z,0) = z$.)

PROOF OF THE CLAIM. Fix a point $q_0 \in V$. We can choose I, J, and positive constants ε_4, ε_5, c_4, c_5 such that

(25) $|\Delta_{I,J}(z)| \geq \varepsilon_4 \exp(-c_4 p(q_0))$ for $|z-q_0| \leq \varepsilon_5 \exp(-c_5 p(q_0))$.

We may also assume $I = J = \{1,2,..,k\}$. From (25), via the implicit function theorem, we can choose $\varepsilon_6, c_6 > 0$ such that

(26) $f_1,\ldots,f_k,z_{k+1},\ldots,z_n$ give local coordinates on the open set B,

$B = B(q_0,\varepsilon_6,c_6) = \{\zeta \in \mathbb{C}^n : |q_0-\zeta| < \varepsilon_6 \exp(-c_6 p(q_0))\}$, and

(27) $z_{k+1},\ldots,z_n$ give local coordinates on V.

There are sections $Z_1,\ldots,Z_k$ of T over B such that

$$(28) \qquad Z_i(f_j) = \delta_{ij}, \qquad 1 \leq i,j \leq k.$$

Namely, set $Z_i = \sum_{j=1}^{k} \beta_{ij} \frac{\partial}{\partial z_j}$, where $\|\beta_{ij}\| = \left\|\frac{\partial f_i}{\partial z_j}\right\|^{-1}$ $(1 \le i \le k)$,

Let us take now their normal components, that is,

$$X_i = \eta(Z_i), \qquad i = 1,\dots,k. \tag{29}$$

Then $X_1,\dots,X_k \in N$ and $X_i(f_j) = Z_i(f_j) = \delta_{ij}$. Hence, they are linearly independent and provide a local holomorphic frame for N and corresponding local coordinates $\zeta_1,\dots,\zeta_k$, $z_{k+1},\dots,z_n$ for N. If $z_1,\dots,z_k$ are the functions of $z_{k+1},\dots,z_n$ which define V and we write

$$X_i = \sum_{j=1}^{k} a_{ij} \frac{\partial}{\partial z_i} + \sum_{j=k+1}^{n} b_{ij} \frac{\partial}{\partial z_j} \qquad (1 \le i \le k),$$

then the map ν can be written explicitly as

$$\nu(\zeta_1 \cdots \zeta_k, z_{k+1} \cdots z_n) = (z_1,\dots,z_n) + \sum_{i=1}^{k} \zeta_i (a_{i,1},\dots,a_{i,k}, b_{i,k+1},\dots,b_{i,n}).$$

Now, from the explicit formulas for a_{ij} and b_{ij} it is clear that these functions are bounded by $A_0 \exp(B_0 p(q_0))$ on B. Further we can assume that the tangent plane at q_0 is given by $z_1 = \cdots = z_k = 0$, hence $\frac{\partial f_i}{\partial z_j}(q_0) = 0$ for $1 \le i \le k$, $k+1 \le j \le n$ and, at that point q_0, we have

$$1 = \det\|Z_i(f_j)\| = \det\|X_i(f_j)\| = \det\left\|\sum_{\ell=1}^{k} a_{i\ell} \frac{\partial f_j}{\partial z_\ell}\right\| \tag{30}$$

$$= \det\|a_{ij}\| \cdot \det\left\|\frac{\partial f_i}{\partial z_j}\right\| .$$

On the other hand, in the $\zeta_1,\dots,\zeta_k, z_{k+1},\dots,z_n$ coordinates, the Jacobian of the map ν is given by

$$\frac{\partial \nu}{\partial(\zeta, z)} = \det\left\|\begin{array}{c|c} a_{ij} & b_{i,\ell} \\ \hline 0 & I \end{array}\right\| = \det\|a_{ij}\|$$

So all entries and their derivatives are bounded by $A_1 \exp(B_1 p(q_0))$ for $z \in B$ and $|\zeta| \le 1$. Now, (30) gives a lower bound for this Jacobian at the point $(q_0, 0)$, hence we can apply the implicit function theorem and get that ν is a biholomorphic map on

$$(31)\qquad |\zeta| < \varepsilon_7 \exp(-c_7 p(q_0)), \quad |z'-z_0'| < \varepsilon_7 \exp(-c_7 p(q_0)),$$

$$z' = (z_{k+1},\dots,z_n), \quad z_0' = z'(q_0),$$

and its image contains a ball $B(q_0,\varepsilon_8,c_8)$. The set described by (31) contains a subset of the form

$$(B(q_0,\varepsilon_9,c_9) \cap V) \times \{|Z| < \varepsilon_{10} \exp(-c_{10}p(z))\}$$

by (26)-(29), and on this subset the image of ν contains a ball $B(q_0,\varepsilon_{11},c_{11})$. We define δ_0 and D_0 in (24) by

$$2\delta_0 < \varepsilon_9,\varepsilon_{10} \quad \text{and} \quad D > c_9,c_{10}.$$

We can show now that ν is globally one-to-one. In fact, if $\nu(w_1,W_1) = \nu(w_2,W_2)$ then

$$|w_1-w_2| < 2\delta_0 \exp(-D_0 p(w_1)) < \varepsilon_9 \exp(-c_9 p(w_1)),$$

and, since ν is locally one-to-one in that region, it follows $(w_1,W_1) = (w_2,W_2)$. Finally, by (14) it is clear that $\bigcup_{q_0\in V} B(q_0,\varepsilon_{11},c_{11})$ contains $S(\delta,D,f)$ for $\delta,D > 0$ convenient. This concludes the proof of the claim.

The proof of the theorem follows by observing that the required small change of the weight p on the fibres of π is a consequence of the above construction and the restrictions imposed above on the weight function p. □

A few simple examples of application of this theorem to (smooth) hypersurfaces are the following.

Example 1. Let f be a polynomial and assume $V = \{f = 0\}$ is a smooth submanifold of $\mathbb{C}^n$. Then V is interpolating for any weight p. Namely, the polynomials $f, \frac{\partial f}{\partial z_1},\dots,\frac{\partial f_n}{\partial z_n}$ have no common zeroes, hence by Hilbert's Nullenstellensatz [29] there exist polynomials $g_0,g_1,\dots,g_n$ such that

$$g_0 f + g_1 \frac{\partial f}{\partial z_1} + \cdots + g_n \frac{\partial f}{\partial z_n} = 1.$$

This identity implies $|\nabla f| \geq \varepsilon(1+|z|)^{-m}$ on V for some $\varepsilon > 0$, $m \geq 0$.

Here $\nabla f = (\frac{\partial f}{\partial z_1}, \ldots, \frac{\partial f}{\partial z_n})$ is the gradient of f and $|\nabla f|$ its length. Since we are assuming that all weights p satisfy the condition $\log(1+|z|) = O(p(z))$ the above assertion holds. Similarly for algebraic submanifolds of $\mathbb{C}^n$.

We should mention here that interpolation problems for algebraic varieties in $\mathbb{C}^n$ are completely understood. Ehrenpreis in [10], and then Palamodov [21], Liess [20] and Björck [6] have given proofs of the following statement. Given an algebraic variety V (with multiplicities), there are finitely many algebraic varieties V_j, $\cup V_j = V$ and differential operators with polynomial coefficients $Q_j(D)$ such that the necessary and sufficient condition for an analytic function φ on V to have an extension $\Phi \in A_p(\mathbb{C}^n)$ is that for all j it satisfies the estimates

$$|Q_j(D)\varphi(z)| \leq Ae^{Bp(z)}, \qquad z \in V_j.$$

When one can take all $Q_j(D)$ = identity operator, the variety is interpolating. A particular case occurs when $V = \{f = 0\}$, f polynomial without multiple factors [5, 21]. Since in the above definition of interpolating varieties, the multiplicities are not taken into account, one can always assume that f has no multiple factors and hence every algebraic hypervariety is interpolating.

<u>Example 2</u>. Let $z = (z', z_n)$, $z' \in \mathbb{C}^{n-1}$, $g \in A_{p_0}(\mathbb{C}^{n-1})$ and $p(z) \geq p_0(z') + \log(1+|z_n|)$ then

$$f(z) = z_n - g(z')$$

belongs to the space $A_p(\mathbb{C}^n)$ and $V = \{f = 0\}$ is an interpolating hypersurface since $|\nabla f| \geq 1$ throughout.

We remark that when the conditions of Theorem 1 are satisfied for a weight p, they are automatically satisfied for any weight $q \geq p$. It follows that V is an interpolating variety also for $A_q(\mathbb{C}^n)$. In the case of one variable it is easy to see that if a variety V is interpolating for a weight p, it is also interpolating for any weight $q \geq p$. We don't know whether this is the case for $n > 1$.

<u>Problem 1</u>. Iet V be a interpolating variety for $A_p(\mathbb{C}^n)$ $(n > 1)$ and q another weight satisfying $q \geq p$. Is V interpolating for $A_q(\mathbb{C}^n)$?

What happens in an interpolating manifold V for weights smaller than p may be very hard to decide. There is the following beautiful example due to Demailly [8].

Example 3. Let $V \subsetneq \mathbb{C}^2$ be defined by the equation

$$e^{z_1} + e^{z_2} = 1.$$

V clearly satisfies the hypothesis of Theorem 1 for $p(z) = |z|$. On the other hand [8] shows that if φ is an analytic function in V satisfying, for some integer m and $c > 0$,

$$|\varphi(z)| \leq c(1+|z|)^m \qquad (z \in V)$$

then φ is the restriction of a polynomial of degree $\leq m$. In particular the only bounded functions on V are the constants.

Finally, the method of proof of Theorem 1 is based in the fact that V is a holomorphic retract of a neighborhood. This can only occur if V is actually a smooth submanifold of $\mathbb{C}^n$ [13]. We also would like to point out that our method is related to Skoda's work [24]. Professor Skoda has kindly informed us that J. P. Demailly has generalized our proof of the existence of a holomorphic retraction in that context [9].

3. We will discuss the question of the necessity of condition (12) for interpolating submanifolds of $\mathbb{C}^n$ and some geometric conditions related to this condition in the next section. It is clear that the main difficulty to prove the necessity of any kind of condition is the construction of sufficiently many test functions. In the case of one variable this was solved very easily due to the fact that zero-one functions were at our disposal (see e.g. [3, p. 124]).

In this section we will consider a particular kind of hypervarieties (not necessarily smooth) for which sufficient conditions for interpolation analogous to (12) can be found and one example where we can find a condition that is both necessary and sufficient.

Recall that to simplify the exposition we are assuming throughout that the weight $p(z) = |z|$. In [2] and [5, Section 7] we discussed the hypervarieties $V = \{f = 0\}$ defined by a pseudopolynomial $(z=(s,w) \in \mathbb{C}^{n-1} \times \mathbb{C})$

$$f(z) = f(s,w) := \sum_{j=0}^{m} f_j(s)w^{m-j}, \tag{32}$$

where the f_j are entire functions of exponential type in $\mathbb{C}^{n-1}$ normalized by the condition

$$(33) \qquad \sum_{j=0}^{m} |f_j(s)| \geq a \exp(-b|s|)$$

for some $a,b > 0$. The function f being a polynomial in w has a well defined discriminant whose square $\Delta(s)$ is a polynomial in the coefficients f_j [29], hence $\Delta(s)$ is also an entire function of exponential type. It has been shown in [5, Section 9], [2, Chapter 4] that the following condition is sufficient for V to be an interpolating variety

(34) There are constants $c,d > 0$ and $q \geq 0$ such that for every $0 < \varepsilon < 1$ we have

$$\max_{B(s,\varepsilon)} |\Delta(s)| \geq c\varepsilon^{-q} \exp(-d|s|).$$

If the f_j are exponential polynomials this condition is satisfied as soon as $\Delta(s) \not\equiv 0$ [33]. This condition says that *on the average* two points in V cannot get too close in the euclidean metric unless they are comparably close to each other by measuring distances along the variety. We will see later that this condition is related to (12).

Let us discuss now an example where a more subtle version of (34) is both necessary and sufficient. Let $f(s,w)$ be defined in $\mathbb{C}^2$ by

$$(35) \qquad f(s,w) = (w-g(s))(w-h(s)),$$

and we assume $F(s) = h(s) - g(s) \not\equiv 0$. Both g, h are entire functions of exponential type in $\mathbb{C}$. Consider the condition:

(36) The map $\alpha(s) \longmapsto F(s)\alpha(s)$ of the space

$\{\alpha$ entire in $\mathbb{C} : |\alpha(s)| \leq A \exp B(|s|+|g(s)|)$ for some $A,B \geq 0\}$

into itself, has closed range.

THEOREM 2. Let $p(z) = |z| = (|s|^2+|w|^2)^{1/2}$ and $V = \{z \in \mathbb{C}^2 : f(s,w) = 0\}$. Then V is an interpolation variety for $A_p(\mathbb{C}^2)$ if and only if (36) holds.

PROOF. Write $q(s) = |s| + |g(s)|$, which is essentially $p(z)$ when $w - g(s) = 0$. Assume the interpolation property holds. Let $\alpha(s)$ be an entire function with $|\alpha(s)F(s)| \leq Ae^{Bq(s)}$. Define an analytic function on V by

$$\varphi(s,w) = \begin{cases} \alpha(s)F(s) & \text{if } w = g(s) \\ 0 & \text{if } w = h(s)\,. \end{cases}$$

Since φ satisfies the correct growth conditions there is an extension $\Phi \in A_p(\mathbb{C}^2)$ with $\Phi|V = \varphi$.

From the mean value theorem we have that if $|w_1-w| \le 1$, $|w_2-w| \le 1$,

$$|\Phi(s,w_1) - \Phi(s,w_2)| \le A_1|w_1-w_2|e^{B_1(|s|+|w|)}.$$

Thus if $|F(s)| \le 1/2$, we have

$$\begin{aligned} |\alpha(s)F(s)| &= |\varphi(s,g(s)) - \varphi(s,h(s))| \\ &= |\Phi(s,g(s)) - \Phi(s,h(s))| \le A_1|g(s)-h(s)|e^{B_1(|s|+|g(s)|)}. \end{aligned}$$

Hence,

$$|\alpha(s)| \le A_1 e^{B_1(|s|+|g(s)|)}.$$

But, if $|F(s)| \ge 1/2$, the division estimate is clear. This proves (36).

To prove the converse. First, we observe that $V_1 = \{w-h(s) = 0\}$ is an interpolation variety, e.g. by Theorem 1 or by the above mentioned results in this section. Hence, given φ on V, we can extend $\varphi|V_1$ to a global function Φ_1 and consider $\varphi - \Phi_1$. Therefore, we can always assume without any loss of generality that $\varphi = 0$ on V_1. But, if this condition is satisfied there is an entire function α such that

$$\varphi\Big|_{w=g(s)} = \varphi(s,g(s)) = \alpha(s)F(s).$$

Since φ satisfies the required growth condition

$$|\varphi(z)| \le Ae^{Bp(z)}, \qquad z \in V$$

we conclude from (36) that

$$|\alpha(s)| \le A_1 e^{B_1|q(s)|}.$$

Then, as $V_2 = \{w-g(s) = 0\}$ is also an interpolation variety we can extend α, considered as a function in V_2 to an entire function $A(s,w)$ of exponential type. Set

$$\Phi(s,w) = A(s,w)(w-h(s)),$$

clearly $\Phi \in A_p(\mathbb{C}^2)$. Furthermore

$$\Phi = 0 = \varphi \quad \text{on the set} \quad V_1 = \{h(s) - w = 0\},$$

and, on the set $V_2 = \{g(s) - w = 0\}$ we have

$$\Phi = (g(s)-h(s))A(s,g(s)) = F(s)\alpha(s) = \varphi.$$

Hence Φ is the required extension. □

To conclude this section we would like to show that there are in fact varieties for which (36) does not hold. Taking $g(s) = \cos s$, the answer is given by the following.

THEOREM 3. Let $q(s) = |s| + \exp|\operatorname{Im} s|$, and $A_q = \{\alpha$ entire functions on $\mathbb{C} : |\alpha(s)| \leq Ae^{Bq(s)}$ for some $A,B \geq 0\}$. Then there exists an entire function F, of exponential type, such that the map $A_q \to A_q$ given by $\alpha \longmapsto F\alpha$ does not have closed range. That is, there is an entire function β such that $\beta F \in A_q$ but $\beta \notin A_q$.

PROOF. Let $\{x_n\}$, $\{y_n\}$ be sequences of positive numbers converging to ∞, with the property

$$x_n < y_n < x_{n+1} < y_{n+1}, \qquad y_n < 2x_n.$$

Set $\ell_n = y_n - x_n$ and, for a given $\varepsilon > 0$, define $\Omega_n = \Omega_n(\varepsilon)$ by

(37) $\Omega_n = \{s = \sigma + i\tau : x_n < \sigma < y_n,\ e^{|\tau|} < 1 + \varepsilon(|s|\log(3+|s|))\}.$

We then define in Ω_n, h_n = unique harmonic function on Ω_n with $h_n = q$ on $\partial\Omega_n$. Then, set

(38)
$$U(s) = \begin{cases} h_n(s), & s \in \Omega_n \\ q(s), & s \notin \cup\Omega_n. \end{cases}$$

Clearly $U(s)$ is subharmonic in $\mathbb{C}$. We want to know some properties of this auxiliary function U; in particular, that $U >> q$ at many points in Ω_n. For this, we will need to assume something about the shape of Ω_n, that ℓ_n is large compared to the height of Ω_n. The key is the following simple estimate.

LEMMA 1. Let $M_n = \min\{q(s) : s \in \overline{\Omega}_n$ and $\operatorname{Re} s = \sigma = x_n$ or $\sigma = y_n\}$, and $w_n = 2 \max\{\tau : s = \sigma + i\tau \in \overline{\Omega}_n$. Then, if $\ell_n > w_n$ we have

$$U\left(\frac{x_n+y_n}{2}\right) > \frac{M_n}{2}. \tag{39}$$

PROOF OF THE LEMMA. Recall that on Ω_n we have $U = h_n$. Enclose Ω_n by the rectangle R_n with vertical sides on $\sigma = x_n$, $\sigma = y_n$ and horizontal sides on $\tau = \pm\frac{W_n}{2}$. Let $\tilde{h}_n$ be the harmonic function on R_n with $\tilde{h}_n = M_n$ on the horizontal sides and $\tilde{h}_n = 0$ on the vertical sides. We then have

$$\tilde{h}_n \le h_n \quad \text{on} \quad \partial\Omega_n .$$

So, by the maximum principle $\tilde{h}_n \le h_n$ in Ω_n. Since $\ell_n > w_n$, we clearly have

$$\tilde{h}_n\left(\frac{x_n+y_n}{2}\right) >> \frac{1}{2} M_n$$

and (29) follows.

COROLLARY. $U(s) \ne O(q(s))$ if $\ell_n = y_n - x_n > x_n^{\alpha}$ for some $\alpha > 0$.

PROOF. Since for large n, $w_n \approx \log y_n \approx \log x_n$ $(y_n < x_n)$, we can apply the lemma when $\ell_n > x_n^{\alpha}$. Let $s_n = \frac{x_n+y_n}{2}$. Then $q(s_n) = s_n$ but

$$U(s_n) > \frac{M_n}{2} = \frac{1}{2}\, q(x_n+i\tau_n), \quad \text{where}$$

$$e^{\tau_n} = 1 + \varepsilon(|x_n+i\tau_n| \log(3+|x_n+i\tau_n|)).$$

So for n large, we have

$$q(x_n+i\tau_n) \ge \frac{\varepsilon}{2} x_n \log x_n .$$

Using once more that $y_n < 2x_n$ we see that

$$U(s_n) \ge \frac{\varepsilon}{4} x_n \log x_n \quad \text{and} \quad q(s_n) \le \frac{3}{2} x_n .$$

So $U(s) \neq O(q(s))$. □

LEMMA 2. Let $\{a_n\}$ be the set of Gaussian integers located in the union of the squares of center $s_n = 2^n$ and half-side $2^{n/2}$. Then there is an entire function F of exponential type with zero set precisely $\{\pm a_n\}$, and it satisfies

$$(40) \qquad |F(s)| \leq e^{-\delta|s|\log|s|} \quad \text{when} \quad |s-2^n| \leq \frac{1}{4}\, 2^{n/2},$$

for some $\delta > 0$.

We omit the proof since it only involves standard estimates of Hadamard products. We have already used properties of this function elsewhere [3, Example 15].

We can now continue with the proof of Theorem 3. Set

$$\delta_n = \frac{x_n+y_n}{2} = 2^n,$$

with

$$y_n - x_n = (x_n)^{1/3}.$$

Choose $\varepsilon > 0$ so small that the discs $|s-s_n| < \frac{1}{4}\sqrt{s_n}$ contain $\Omega_n = \Omega_n(\varepsilon)$. Let $\tilde{q}(s) = U(s)$. We claim that

(41) There is $\beta \in A_{\tilde{q}} \backslash A_q$ such that $\beta F \in A_q$.

Namely, using interpolation we can find an entire function β such that,

$$(42) \qquad \beta(s_n) = \exp(\varepsilon_1 \tilde{q}(s_n)) \quad \text{and} \quad |\beta(s)| \leq c\exp(\tilde{q}(s))$$

(see [3]). If we choose $\varepsilon < \delta, \delta$ as in (40), we will have

$$|\beta(s)F(s)| \leq c_1 e^{B|s|} \quad \text{in} \quad U\Omega_n$$

since $\tilde{q}(s) = h_n(s) \leq A|s_n| + \varepsilon|s|\log|s|$ on Ω_n. From (38) it follows

$$\beta F \in A_q.$$

The first part of (42) and the Corollary to Lemma 1 shows $\beta \notin A_q$. □

4. For the case of discrete varieties $V = \{a_k\} \subsetneq \mathbb{C}^n$, condition (12) corresponds to the case where all the points are counted with multiplicity one and the following condition is then necessary and sufficient for interpolation [5]

$$(43) \qquad |J(a_k)| \geq \varepsilon e^{-cp(a_k)},$$

where $J(z)$ the Jacobian of the defining equations, $V = \{f_1 = \cdots = f_n = 0\}$, $f_j \in A_p$. For $n = 1$, as mentioned, we can prove the corresponding theorem without any restriction on the multiplicities [3]. This leads to the following

Problem 2. Given V interpolating submanifold of $\mathbb{C}^n$ for A_p defined by (9) and (10) is condition (12) then satisfied?

Even for hyper-surfaces and $p(z) = |z|$ this is unknown, let us rephrase it

Problem 3. Let $p(z) = |z|$, $f \in A_p$ such that $\nabla f(z) \neq 0$ whenever $f(z) = 0$. Suppose that $V = \{z \in \mathbb{C}^n = 0\}$ is an interpolating variety for A_p, does there exist constants $\varepsilon, c > 0$ such that

$$(44) \qquad |\nabla f(z)| \geq \varepsilon \exp(-c|z|) \quad \text{whenever} \quad f(z) = 0?$$

We will discuss here some of the geometric conditions on V imposed by (44) and by the interpolation condition. To simplify matters we will assume $n = 2$. Observe then that V is just an open Riemann surface embedded in $\mathbb{C}^2$, which we will consider with the induced metric. The associated distance in V will be denoted by d.

First, note that (44) is really an intrinsic property of V, since if g is any other function of exponential type vanishing exactly on V and such that the multiplicity of the zeroes of g is one (possibly with the exception of a lower dimensional set), then g is the product of f by an exponential function $\exp(\alpha z_1 + \beta z_2 + \gamma)$, for some $\alpha, \beta, \gamma \in \mathbb{C}$. Hence (44) also holds for g.

As a consequence of the construction in Theorem 1 of the open set $S(\delta, D, f)$ of which V is a holomorphic retract one has the following relation between the "external distance" between two points in V (called *diastasis* in [7]) and the distance d measured along the manifold V.

(45) There are constants $c_1, c_2, c_3 > 0$, $0 < \varepsilon_1 < 1$ such that if two points $z, w \in V$ satisfy $|z-w| \leq \varepsilon_1 e^{-c_1|z|}$, then

$$|z-w| \le d(z,w) \le c_3 e^{c_4|z|}|z-w|.$$

Of course, the first inequality is always true since d is given by the induced metric. Calabi has shown that the diastasis between different points is in fact an intrinsic property of V [7]. Namely, given any isometric imbedding of V into $\mathbb{C}^N$ ($N = \infty$ is also allowed) then the diastasis $|z-w|$ between two points of V remains the same. This condition means that the manifold V cannot come back on itself too often and corresponds to a similar condition in the case of one variable. It is also related to the fact that the necessary and sufficient conditions for extending a given function φ can be expressed in terms of the restriction of this function to all lines through the origin [5, Theorem 7.1]. If one can guarantee that the estimate (1) for φ gives an estimate for the tangential derivaties of φ then (45) implies an estimate on the divided difference $\frac{\varphi(z)-\varphi(w)}{z-w}$ when z and w are in the same line.

The question of the estimation of the derivatives of φ is related both to interpolation and to the curvature of the manifold V as we will see shortly.

We recall that V is a one dimensional complex manifold and all curvatures coincide. Hence using the definition given in [31] or reducing oneself locally to the case of a graph and using [28, p. 823] one obtains the following formula for the curvature K of V, V = $\{f = 0\}$

$$K = -2\,\frac{\left|f_{z_2}^2 f_{z_1z_1} - 2f_{z_1}f_{z_2}f_{z_1z_2} + f_{z_1}^2 f_{z_2z_2}\right|^2}{|\nabla f|^6}. \tag{46}$$

As an immediate consequence, we have the following.

PROPOSITION 1. If f is a function of exponential type and satisfies (44), the curvature of the submanifold V of $\mathbb{C}^2$ given by $V = \{z \in \mathbb{C}^2 : f(z) = 0\}$ satisfies

$$0 \le -K(z) \le Ae^{B|z|} \tag{47}$$

for some constants $A,B > 0$.

PROPOSITION 2. If $\zeta \in V$ is a point where $\left|\frac{\partial f}{\partial z_2}(\zeta)\right| = |\nabla f(\zeta)|$ and $|K(z)| \le R$ for $z \in V \cap B(\zeta;1)$, then V can be described by the equation $z_2 = g(z_1)$ over a disk of radius at least $(8R)^{-1/2}$ (we recall $B(\zeta;1) = \{z \in \mathbb{C}^2 : |z-\zeta| < 1\}$.)

PROOF. We can assume $\zeta = 0$ for simplicity. The tangent line T to V at 0 is then described by $T = \{z \in \mathbb{C}^2 : z_2 = 0\}$ since by hypothesis $\frac{\partial f}{\partial z_1}(0) = 0$. Hence, near 0, V can be described by $z_2 = g(z_1)$. Let Δ be the largest disk in T of center 0 such that the graph of g is contained in $V \cap B(0;1)$. Denote its radius by r. It follows that in Δ, $\frac{\partial f}{\partial z_2}(z) \neq 0$, where $z = (z_1, g(z_1))$. Let

$$\Omega = \{z_1 \in \Delta : \left|\frac{\partial f}{\partial z_2}(z)\right| \geq \frac{1}{2}|\nabla f(z)|\}.$$

Either $\Omega = \Delta$ or $\partial\Omega \cap \Delta \neq \emptyset$. Choose $w \in \partial\Delta$ in the first case and $w \in \partial\Omega \cap \Delta$ in the second. Integrating over the straight line segment from 0 to w we obtain (with $z = (tw, g(tw))$ over this segment)

$$\frac{1}{4} \leq \left|\frac{\frac{\partial f}{\partial z_2}(w,g(w))}{\nabla f(w,g(w))}\right|^2 = \int_0^1 \frac{d}{dt}\left\{\frac{\left|\frac{\partial f}{\partial z_2}\right|^2}{|\nabla f|^2}\right\} dt$$

$$= 2 \operatorname{Re} \int_0^1 \frac{d}{dz_1}\left\{\frac{\left|\frac{\partial f}{\partial z_2}\right|^2}{|\nabla f|^2}\right\} \frac{dz_1}{dt}\, dt$$

$$\leq 2|w| \int_0^1 \left|\frac{d}{z_1}\left\{\frac{\left|\frac{\partial f}{\partial z_2}\right|^2}{|\nabla f|^2}\right\}\right| dt.$$

Since $|w| \leq r$ we only need to estimate the integrand in terms of R. We have on V,

$$\frac{d}{dz_1} \frac{\left|\frac{\partial f}{\partial z_2}\right|^2}{|\nabla f|^2} = -\frac{\overline{f}_{z_2} \overline{f}_{z_1}}{f_{z_2}} \frac{\left[f_{z_2}^2 f_{z_1 z_1} + f_{z_1}^2 f_{z_2 z_2} - 2 f_{z_1} f_{z_2} f_{z_1 z_2}\right]}{|\nabla f|^4}.$$

Hence, using (46) and the hypothesis of the proposition we obtain

$$\left|\frac{d}{dz_1} \frac{\left|\frac{\partial f}{\partial z_2}\right|^2}{|\nabla f|^2}\right| \leq \sqrt{\frac{|K(z)|}{2}} \leq \sqrt{\frac{R}{2}}\ .$$

This estimate concludes the proof of Proposition 2. □

Proposition 2 can be written in a way that shows its similarity with the well known Ahlfors' lemma on the hyperbolic metric [1] where one places and estimate on K from *above*, while here we have an estimate from *below*. In our special case one has the following theorem.

THEOREM 4. Let V be a Riemann surface given as above by $V = \{f = 0\}$, $f \in A_p(\mathbb{C}^2)$, $p(z) = |z|$. Then there exist constants $A,B > 0$ such that for all $z \in V$.

$$-Ae^{B|z|} \leq K(z) \leq 0$$

if and only if there exist $\varepsilon, C > 0$ such that for each $z \in V$ there is an analytic map of the unit disk $\Delta = \{s \in \mathbb{C} : |s| < 1\}$

$$\varphi : \Delta \longrightarrow V \cap B(z;1), \qquad \varphi(0) = z,$$

and

$$|\varphi'(0)| \geq \varepsilon e^{-c|z|}.$$

Further, φ can be taken to be schlicht.

If V is actually an algebraic manifold, it follows from Example 1 that $|K(z)| = O((1+|z|)^m)$ for some integer m. Vitter has shown that if V is non-singular at infinity then actually $|K(z)| = O((1+|z|)^{-2})$ [28]. It is the fact that the curvature might grow at infinity which does not allow us to use below the refined analytic methods form [11], [22], [32].

To find the relation between curvature and interpolation we need to recall the definition given in [5, Definition 3.5] of the space $A_p(V)$ and introduce new spaces $A_p^{(k)}(V)$. We consider the space $\mathcal{P}$ of all polynomials in 2 variables (since here $V \subseteq \mathbb{C}^2$), with the norm

$$\|P\| = \|\sum c_\alpha z^\alpha\| = \max\{|c_\alpha|\alpha!\}.$$

$\mathcal{P}^{(k)}$ is the subspace of $\mathcal{P}$ of polynomials of degree $\leq k$. If h is an entire function, then

$$|P(D)h(z)| = |\sum c_\alpha h^{(\alpha)}(z)| \leq \|P\| \sum \frac{|h^{(\alpha)}(z)|}{\alpha!}.$$

Hence, if $h \in A_p(\mathbb{C}^2)$ satisfies

$$|h(z)| \leq Ae^{B|z|}$$

we get

$$|P(D)h(z)| \leq A_1 \|P\| e^{B_1|z|},$$

where A_1, B_1 depend only on A, B. Let I be the ideal of entire functions in $A_p(\mathbb{C}^2)$ which vanish on V, denote by $\mathcal{P}_z$ the ideal of all polynomials P such that

$$(48) \qquad P(D)h(z) = 0 \qquad \text{for all } h \in I.$$

The ideal $\mathcal{P}_z$ remains the same if we replace in (48) the ideal I by the ideal of all entire functions which vanish on V, or even if I is replaced by I_z the ideal of (germs of) analytic functions defined in a neighborhood of z which vanish on the germ of variety V_z defined by V. Note that $z \in V$ if and only if $\mathcal{P}_z \neq \{0\}$. We denote $\mathcal{P}_z^{(k)} = \mathcal{P}_z \cap \mathcal{P}^{(k)}$. If $\varphi \in A(V)$ is an analytic function on V then $P(D)\varphi(z)$ is well defined for every $z \in V$ (or even $z \in \mathbb{C}^2$). The space $A_p(V)$ is then defined by (recall $p(z) = |z|$)

$$(49) \quad A_p(V) := \{\varphi \in A(V) : \text{there are constants } A,\ B \text{ such that } |P(D)\varphi(z)| \leq A\|P\|e^{B|z|} \text{ for all } P \in \mathcal{P}_z, \text{ all } z \in V\}.$$

We can also define

$$(50) \quad A_p^{(k)}(V) := \{\varphi \in A(V) : \text{there are constants } A,\ B \text{ such that } |P(D)\varphi(z)| \leq A\|P\|e^{B|z|} \text{ for all } P \in \mathcal{P}_z^{(k)}, \text{ all } z \in V\}.$$

Further we define $A_p^{(\infty)}(V) = \bigcap_{k\geq 0} A_p^{(k)}(V)$. We clearly have

$$(51) \qquad A_p(V) \subseteq A_p^{(\infty)}(V) \subseteq \cdots \subseteq A_p^{(0)}(V).$$

Note that the space defined by condition (1) is precisely $A_p^{(0)}(V)$. In this notation, V is an interpolating variety precisely when the restriction map

$$\rho : A_p(\mathbb{C}^2) \longrightarrow A_p^{(0)}(V)$$

is onto. The remarks above (48) show that

$$(52) \qquad \rho(A_p(\mathbb{C}^2)) \subseteq A_p(V).$$

Hence, the following is self-evident.

PROPOSITION 3. A necessary condition for V to be an interpolating variety is that $A_p(V) = A_p^{(0)}(V)$. In particular, it is necessary that $A_p^{(k)}(V) = A_p^{(0)}(V)$ for $k = 1,2,\ldots,\infty$.

As a consequence of Proposition 2 one can easily prove the following.

PROPOSITION 4. If $V = \{f = 0\}$ is a Riemann surface defined as above and satisfies (47), then $A_p^{(\infty)}(V) = A_p^{(0)}(V)$.

Problem 4. Is the converse of Proposition 4 true?

Problem 5. If (47) holds and V is an interpolating manifold does (44) follow?

Problem 6. If (47) holds and V is an interpolating manifold does (45) follow?

Problem 7. Assume (47) holds, given $\varepsilon, c > 0$ does there exist $A,B > 0$ such that for every pair $z_0, z_1 \in V$,

$$d(z_0,z_1) \geq \varepsilon e^{-c|z_0|}$$

implies there exist $\varphi \in A(V)$, $\varphi(z_0) = 0$, $\varphi(z_1) = 1$ and

$$|\varphi(z)| \leq Ae^{B|z|}, \qquad z \in V.$$

An affirmative answer to Problem 7 would help to answer Problem 6 affirmatively.

Finally, there is one more consequence of the inequality (44) that we want to point out, and it is the solvability of the $\bar\partial$-equation on V with exponential type bounds. In fact, as a consequence of the Kodaira identity [12, 24] and the method developed in [14], one has the following proposition.

PROPOSITION 5. Let $d\Omega$ denote the element of volume in V, V defined as above by $\{f = 0\}$, $f \in A_p(\mathbb{C}^2)$, $p(z) = |z|$. If w is (0,1) form in V satisfying for some $c \geq 0$,

$$\int_V \frac{|w|^2}{|\nabla f|^2} e^{-c|z|} d\Omega(z) < \infty$$

then there is a solution u of the equation $\bar\partial u = w$ such that

$$\int_V \frac{|w|^2}{|\nabla f|^2} \frac{e^{-c|z|}}{(1+|z|^2)^2} d\Omega(z) \leq \frac{1}{2} \int_V \frac{|w|^2}{|\nabla f|^2} e^{-c|z|} d\Omega(z).$$

In particular, if (44) and $\int_V |w|^2 e^{-c|z|} d\Omega(z) < \infty$ for some $c \geq 0$, there are constants $A,B > 0$ and solution u to the above equation such that

$$\int_V |u|^2 e^{-A|z|} d\Omega(z) \leq B \int_V |w|^2 e^{-c|z|} d\Omega(z).$$

(Here $|w|$ denotes the length of w in the induced metric.)

One can also prove the Kodaira identity in this case directly by elementary considerations similar to [3, Lemma 2]. We omit the details.

Problem 8. Suppose the last assertion of Proposition 5 holds, i.e. solvability of the $\bar{\partial}$-equation with exponential bounds. Does it follow that ∇f satisfies the inequality (44)? What if we assume further that the curvature satisfies (47)? This question is related to the following.

Problem 9. Let p, q be two weights in $\mathbb{C}$ (or $\mathbb{C}^n$) such that for every ($\bar{\partial}$-closed) (0,1) form w satisfying

$$\int |w|^2 e^{-p} d\lambda < \infty$$

we have a solution u to the problem $\bar{\partial}u = w$ with the bound

$$\int |u|^2 e^{-q} d\lambda \leq \int |w|^2 e^{-p} d\lambda,$$

what is the relation between the weights p and q? (Here $d\lambda$ denotes Lebesgue measure.)

The result [14, Theorem 4.4.2] is the statement that $q \geq p + 2\log(1+|z|^2)$ suffices. This has been improved by Skoda [25] to $q \geq p + (1+\varepsilon)\log(1+|z|^2) + C(\varepsilon)$. The above question is really the converse of these two results, that is, can we conclude for instant that $q \geq p$?

5. The authors gratefully acknowledge the support received from the National Science Foundation in the preparation of this paper. M. Berenstein voudrait aussi remercier l'Université de Paris VI et spécialement le Professeur P. Lelong pour leur hospitalité.

REFERENCES

1. L. Ahlfors, Conformal invariants, Mc Graw Hill, 1973.
2. C. A. Berenstein and M. A. Dostal, Analytically uniform spaces and their applications to convolution equations, Springer Verlag, 1972.
3. C. A. Berenstein and B. A. Taylor, A new look at interpolation theory for entire functions of one variable, Advances in Math. 33 (1979), 109-143.
4. _____, Mean-periodic functions, to appear in the Intern. J. Math. and Math. Sciences 3 (1980), 199-236.
5. _____, Interpolation problems in $\mathbb{C}^n$ with applications to harmonic analysis, to appear in J. Analyse Math. 38 (1980), 188-254.
6. J. E. Björk, Rings of differential operators, North-Holland, 1979.
7. E. Calabi, Metric Riemann surfaces, in Contributions to the theory of Riemann Surfaces (ed. by L. Ahlfors et al), 77-86, Princeton University Press, 1953.
8. J. P. Demailly, Fonctions holomorphes à croissance polynomiale sur la surface d'equation $e^x + e^y = 1$, Bull. Sci. Math. 103 (1979), 179-191.
9. _____, Scindage holomorphe d'un morphisme de fibrés vectoriels semi-positifis avec estimations L^2, manuscript.
10. L. Ehrenpreis, Fourier analysis in several complex variables, Interscience-Wiley, 1970.
11. R. Greene and H. Wu, Function theory on manifolds which posses a pole, Springer-Verlag, 1979.
12. P. Griffiths and J. Harris, Principles of algebraic geometry, Wiley Interscience, 1978.
13. H. Holmann, Local properties of holomorphic mappings, in Proceedings of the conference in complex analysis (ed. by A. Aeppli et al) 94-109, Springer-Verlag, 1965.
14. L. Hörmander, An introduction to complex analysis in several variables, van Nostrand, 1966.
15. _____, Generators for some rings of analytic functions, Bull. Amer. Math. Soc. 73 (1967), 943,949.
16. J. Horváth, Topological vector spaces, Addison-Wesley, 1963.
17. B. Jeannane, Extension d'une fonction définie sur une sous-variété avec contrôle de la croissance, Séminaire Pierre Lelong-Henri Skoda (Analyse) Annèe 1976/77, 126-133.
18. P. Lelong, Fonction entières (n variables) et fonctions plurisous-harmoniques d'ordre fini dans $\mathbb{C}^n$, J. Analyse Math. 12 (1964), 365-407.
19. B. Ja. Levin, Distribution of zeroes of entire functions, Amer. Math. Soc., 1964.
20. O. Liess, On the Fundamental Principle of Ehrenpreis-Palamodov, preprint.
21. V. P. Palamodov, Linear differential operators with constant coefficients, Springer-Verlag, 1970.
22. Y. T. Siu and S. T. Yau, Complete Kähler manifolds with nonpositive curvature of faster than quadratic decay, Ann. of Math. 105 (1977), 225-264.

23. H. Skoda, Sous-ensembles analytiques d'ordre fini on infini dans $\mathbb{C}^n$, Bull. Soc. Math. France 100 (1972), 353-408.

24. _____, Morphismes surjectifs de fibrés vectoriels semi-positifs, Ann. Scient. École Norm. Sup. 11 (1978), 577-611.

25. _____, Croissance des fonctions entières s'annulant sur une hypersurface donnée de $\mathbb{C}^n$, Séminaire Pierre Lelong (Analyse), 1970/71, 82-105.

26. W. A. Squires, Ph.D. thesis, University of Michigan, 1974.

27. W. Stoll, The growth of the area of a transcendental analytic set, Math. Annalen 156 (1964), 47-48 and 144-170.

28. A. Vitter, On the curvature of complex hypersurfaces, Indiana Univ. Math. J. 23 (1974), 813-826.

29. B. L. van der Waerden, Modern algebra, Frederick Ungar Publishing Co., 1970.

30. M. Waldschmidt, Nombres transcendants, Springer-Verlag, 1974.

31. R. O. Wells, Differential analysis on complex manifolds, Prentice-Hall, 1973.

32. P. Yang, Curvature of complex submanifolds of $\mathbb{C}^n$, J. Diff. Geom. 12 (1977), 499-511.

33. C. A. Berenstein and M. A. Dostal, A lower estimate for exponential sums, Bull, Amer. Math. Soc. 80 (1974), 687-691.

Added in proof : The answer to problem 3 (and hence problem 2) is negative. Take $f(z,w) = w^2 - \sin z \sin \lambda z/z$, where λ is a Liouville number. It satisfies (32) - (34) but doesn't satisfy (44).

C. A. Berenstein
Department of Mathematics
University of Maryland
College Park, MD 20742

B. A. Taylor
Department of Mathematics
University of Michigan
Ann Arbor, MI 48109

FONCTIONS PLURISOUSHARMONIQUES ET IDÉAL DÉFINISSANT UN ENSEMBLE ANALYTIQUE

par Mongi B L E L

INTRODUCTION .

On se donne un courant θ positif fermé de dimension pure p $(0 \leq p \leq n-1)$. On lui associe un potentiel canonique U obtenu en recolant par une partition de l'unité $[(\rho_j)_{j\in J}$ famille localement finie $]$ les potentiels canoniques locaux ,

$$U_j(z) = \frac{-1}{\omega_p} \int_{\mathbb{C}^n} \frac{\eta_j(x)d\sigma(x)}{|z-x|^{2p}}$$

avec :

. ω_p la mesure de la sphère unité dans $\mathbb{C}^p$.

. σ la mesure trace associée au courant θ définie par

$\sigma = \theta \wedge \frac{\beta^p}{p!}$ et $\beta = \frac{i}{2} d'd''|z|^2$.

. η_j une fonction indéfiniment différentielle à support compact, positive, le support de ρ_j est inclu dans le support de η_j et η_j est identiquement égale à 1 dans le support de ρ_j .

H.SKODA [8] montre que U est presque psh dans le sens qu'il existe une fonction psh continue φ_o telle que $U + \varphi_o = W$ soit une fonction plurisousharmonique. Dans le paragraphe 7 de [8] , il étudie la liaison entre le fait que e^{-tU} soit sommable ou non pour $t > 0$ au point z_o et la densité (ou nombre de Lelong) $\nu(z_o)$ du courant θ.

Dans le cas $p = n-1$ et θ est le courant d'intégration sur un ensemble analytique, P.LELONG [4] montre que le potentiel canonique vérifie l'équation $id'd''U = \theta$.

Soit X un ensemble analytique de dimension pure p $(p + k = n)$ et soit J le faisceau d'idéaux des fonctions holomorphes nulles sur X . Un théorème profond d'Oka dit que J est cohérent, de sorte que les théorèmes A et B de H.Cartan s'appliquent à J .

En vue de construire des sections globales de croissance donnée; H.Skoda redémontre en fait le théorème A pour J sans utiliser ni le théorème d'Oka ni le théorème de H.Cartan, mais en utilisant le potentiel canonique et les estimations L^2 .

Il serait intéressant de suivre la même procédure pour le théorème B pour J de manière par exemple à obtenir des versions à croissance. Ce théorème est utile pour l'étude des équations de convolution (problème de la restriction avec croissance). On est alors amené à poser le problème suivant :

Si on se donne une fonction f analytique nulle sur X ; alors la fonction $|f|^2 e^{-2kU}$ est-elle localement sommable ?

La réponse est positive sur le complémentaire de $\sigma(X)$ où $\sigma(X)$ désigne l'ensemble des points singuliers de X.

On désigne par X^* l'ensemble des points réguliers de X. Dans la première partie de ce travail on va utiliser le potentiel $V = 2kU$.
Ce potentiel a été normalisé de sorte que e^{-V} soit non localement sommable sur X mais possède la propriété limite (i.e. : $e^{-\alpha V}$ est localement sommable sur X^* pour tout α avec $0 < \alpha < 1$).

Le problème est de savoir si f s'annule suffisamment sur X pour "tuer" la singularité de e^{-V}.

Il est aisé de voir que la réponse est positive si X est une hypersurface ou si X est sans singularités.

Dans la première partie nous montrons que la réponse au problème posé est négative en codimension quelconque lorsqu'il y a des singularités. Plus précisément on démontre le théorème suivant et on donne un exemple qui vérifie les conditions du théorème.

THÉORÈME 1.2.

Soit X un ensemble analytique de $\mathbb{C}^n$ de dimension pure p, $(p < n-1)$. On suppose que $\nu_X(0) = c$, $c > 0$. Et soit f une fonction analytique nulle en 0 telle que $\nu_f(0) = m$. Alors si $m \leq kc - n$ ($k = n - p$ la codimension de X) $|f|^2 e^{-V}$ n'est pas localement sommable au voisinage de 0.

Puis on donne une condition suffisante pour que la réponse au problème posé soit positive.

Le problème B pour J semblerait donc échapper à une méthode semblable à celle de H.Skoda.

Dans la deuxième partie de ce travail on étudie la liaison entre la sommabilité de $e^{-\varphi}$ (si φ est une fonction plurisousharmonique) et le comportement de ν_φ sur les branches irréductibles de $X_\varphi = \{x / \nu_\varphi(x) > 0\}$ (ensemble des points de densité de φ) et leur codimension. Dans le cas où X_φ est un ensemble analytique de dimension pure p, on met en évidence la différence entre le comportement de U et φ, si U désigne le potentiel canonique associé à $[X]$. Pour étudier ce problème on suggère deux questions :

1°/ Si X_φ est un ensemble analytique, irréductible de dimension pure p $(0 \leqslant p \leqslant n-1)$ et si $c = \inf_{x \in X_\varphi} \nu_\varphi(x)$, $c > 0$, alors $\varphi - cU$ est-elle une fonction presque plurisousharmonique "presque psh" dans le sens qu'il existe une fonction φ_o psh telle que $\nu_{\varphi_o} \equiv 0$ et $\varphi - cU + \varphi_o$ soit psh.

2°/ Sous les mêmes hypothèses que la première question,on se demande par analogie avec le potentiel canonique de X :

. Si $\nu_\varphi(x) < 2k$ alors $e^{-\varphi}$ est-elle localement sommable, $(p + k = n)$?

Alors on démontre que le meilleur résultat possible est celui de H.Skoda à savoir que pour tout φ psh :

. Si $\nu_\varphi(x) < 2$; $e^{-\varphi}$ est localement sommable

. Si $\nu_\varphi(x) \geqslant 2n$; $e^{-\varphi}$ n'est pas localement sommable.

Nous renvoyons le lecteur à P.Lelong [5] pour les définitions et propriétés générales des courants positifs et des fonctions plurisousharmoniques, "psh".

Je tiens à exprimer toute ma reconnaissance à M. H.SKODA pour ses conseils avisés et l'aide constante qu'il m'a apportée.

Ière partie

L'objet de ce paragraphe est de motiver le problème posé en montrant que si X possède la propriété suivante :

"Pour toute fonction f analytique nulle sur X , $|f|^2 e^{-2kU}$ localement sommable sur X" . Alors on a une démonstration de $H^1(\Omega, J) = 0$ n'utilisant que le potentiel canonique et les estimations L^2 .

Position du problème.

Soient Ω un ouvert de Stein $(\Omega \subset \mathbb{C}^n)$ et $(\Omega_j)_{j \in J}$ un recouvrement assez fin, localement fini, de Ω par des ouverts de Stein.

Soient $(\varphi_j)_{j\in J}$ une partition de l'unité subordonnée au recouvrement $(\Omega_j)_{j\in J}$, et soit une donnée de Cousin additive $(f_{ij})_{j,i\in J\times J}$ avec f_{ij} nulle sur X et $f_{ij}\in H(\Omega_i\cap\Omega_j)$. $f_{ij}+f_{jk}+f_{ki}\equiv 0$. On notera $\Omega_{jk}=\Omega_j\cap\Omega_k$ $\forall_{j,k}$.

On cherche une solution $(f_j)\in H(\Omega_j)$ pour tout j telle que $f_j - f_k = f_{jk}$ et f_j nulle sur X pour tout j.

Dans la démonstration on utilisera le théorème suivant dû à Hörmander.

THÉORÈME 1.1.

Soit f une forme $\bar\partial$ fermée dans Ω (f de type $(0,1)$) vérifiant :

$$\int_\Omega |f|^2 e^{-\varphi}\, d\lambda < +\infty$$

où φ est une fonction plurisousharmonique dans Ω. Il existe u telle que $\bar\partial u = f$ et

$$2\int_\Omega |u|^2 \frac{e^{-\varphi}}{(1+|z|^2)^2}\, d\lambda \leqslant \int_\Omega |f|^2 e^{-\varphi}\, d\lambda .$$

LEMME 1.1.

Soit Ω un ouvert pseudo-convexe de $\mathbb{C}^n$, $(\varphi_j)_{j\in J}$ une partition de l'unité subordonnée au recouvrement $(\Omega_j)_{j\in J}$ de Ω. Alors il existe une fonction plurisousharmonique ψ de classe C^∞ dans Ω telle que $\sum_{j\in J} |\bar\partial \varphi_j|^2 \leqslant e^\psi$ dans Ω. La somme étant localement finie.

Démonstration du lemme 1.1.

Soit ρ une fonction psh dans Ω exhaustive de classe C^∞ sur Ω. On prend ψ de la forme $\psi = \chi\circ\rho$ où χ est une application de $\mathbb{R}\to\mathbb{R}$, convexe, croissante de classe C^∞ restreinte à la condition

$$\chi(t) \geqslant \sup_{\rho(z)<t} \log\Big(\sum_{j\in J} |\bar\partial\varphi_j|^2\Big) .$$

La borne supérieure étant finie puisque $\{z\in\Omega/\ \rho(z)\leqslant t\}$ est un compact.

Ce qui termine la démonstration du lemme.

On revient au problème posé : une solution C^∞ du problème de Cousin g_j est donnée par $g_j = \sum_\ell \varphi_\ell\, f_{\ell j}$.

Soit f la forme $\bar\partial$ fermée sur Ω définie par $f=\bar\partial g_j$ dans Ω_j (car $\bar\partial g_j - \bar\partial g_\ell = \bar\partial f_{j\ell} = 0$) de sorte que $f = \sum_\ell (\bar\partial\varphi_\ell) f_{\ell j}$ sur Ω_j.

On a encore $f = \sum_j \varphi_j f = \sum_{\ell,j} \varphi_j (\bar\partial\varphi_\ell)\tilde f_{\ell j}$, avec $\tilde f_{\ell j} = \begin{cases} f_{\ell j} \text{ sur } \Omega_{j\ell} \\ 0 \text{ ailleurs.}\end{cases}$

Par l'inégalité de Cauchy

$$|f|^2 \leqslant \left(\sum_{\ell,j} \varphi_j^2 \quad |\bar\partial \varphi_\ell|^2\right) \cdot \left(\sum_{\ell,j} \chi_\ell \cdot \chi_j \cdot |\tilde f_{\ell j}|^2\right)$$

où χ_j (respectivement χ_ℓ) désigne la fonction caractéristique du support de φ_j (resp. φ_ℓ). $\varphi_j^2 \leqslant \varphi_j$ donc, d'après le lemme 1.1.

$$|f|^2 \leqslant e^{\psi} \left(\sum_{\ell,j} \chi_j \quad \chi_\ell |\tilde f_{\ell,j}|^2\right) .$$

Donc

$$(1) \quad \boxed{\int_\Omega |f|^2 e^{-2\psi} e^{-V} d\lambda \leqslant \sum_{\ell,j} \int_{\Omega_{j\ell}} \chi_\ell \chi_j |f_{\ell j}|^2 e^{-V} e^{-\psi} d\lambda} .$$

On a supposé que pour toute fonction h analytique nulle sur X, $|h|^2 e^{-V}$ est localement sommable sur X .

Le support de $\chi_\ell \chi_j$ est compact dans $\Omega_{\ell j}$, $f_{\ell j}$ est une fonction holomorphe nulle sur X .

Donc $\displaystyle\int_{\Omega_{j\ell}} \chi_\ell \chi_j |f_{\ell j}|^2 e^{-V} e^{-\psi} d\lambda < +\infty$.

On choisit alors ψ de sorte que en plus du lemme 1.1., on ait

$$\sum_{\ell,j} \int_{\Omega_{j\ell}} \chi_\ell \chi_j |f_{\ell j}|^2 e^{-V} e^{-\psi} d\lambda < +\infty .$$

i.e. on choisit ψ à croissance assez rapide pour faire converger la série

$$\sum_{\ell,j} \int_{\Omega_{j\ell}} \chi_\ell \chi_j |f_{j\ell}|^2 e^{-V} e^{-\psi} d\lambda .$$

Donc d'après le théorème de Hörmander, il existe une fonction g qui vérifie

(2) $\quad \bar\partial g = f$

(3) et $\quad \displaystyle\int_\Omega |g|^2 \frac{e^{-\psi} e^{-V}}{(1+|z|^2)^2} d\lambda < +\infty$.

(2) $\iff$ $\bar\partial g = \bar\partial g_j$ sur Ω_j pour tout j . Donc $\bar\partial(g - g_j) = 0$ sur Ω_j pour tout j .

Soit

(4) $\quad \boxed{f_j = g_j - g}$; $\bar\partial f_j = 0$ et $f_j - f_k = g_j - g_k = f_{jk}$

reste à montrer que f_j est nulle sur X .

D'après (4) il suffit de montrer que g_j et g sont nulles sur X .

$g_j = \sum_\ell \varphi_\ell f_{\ell j}$ est nulle sur X car $f_{\ell j}$ est nulle pour tout j et ℓ de J .

Reste à montrer que g est nulle sur X . D'après (3) $|g|^2 e^{-V} \in L^1_{loc}(\Omega)$ et e^{-V} est non sommable en tout point z de X . Donc $g(z) = 0$ pour tout z de X .

Alors f est nulle sur X .

Ainsi on donne une démonstration de $H^1(\Omega, J) = 0$ en utilisant le potentiel canonique et les estimations L^2 à condition que la propriété suivante soit vérifiée.

. Pour tout ensemble analytique X de dimension pure p et pour toute fonction analytique nulle sur X ; $|f|^2 e^{-V}$ est localement sommable sur X .

DÉFINITION.

On dit qu'une fonction f sur Ω est faiblement analytique sur X si pour tout point régulier de X , il existe un voisinage de ce point sur lequel f est analytique. Aux points singuliers la fonction est supposée seulement continue.

LEMME 1.2.

Soit X l'ensemble analytique de dimension pure p définie par $X = \{z \in \mathbb{C}^n \ / \ z_{p+1} = \dots = z_n = 0\}$. Alors au voisinage de tout point de X, $e^{-V(z)}$ est équivalente à $\dfrac{K}{d(z,X)^{2k}}$ avec K une constante strictement postive et k la codimension de l'ensemble analytique X .

Démonstration.

Pour r assez petit, on prend le potentiel U au voisinage du point 0

$$U(Z) = \frac{-2k}{\omega_p} \int_{X \cap B(0,r)} \frac{\beta^p(x)/p!}{(|z_1-x_1|^2 + \dots + |z_p-x_p|^2 + d^2)^p} .$$

$x = (x_1,x_2,\dots,x_p,0,\dots,0)$

avec $d^2 = |z_1|^2 + |z_2|^2 + \dots + |z_n|^2 = d(0,X)^2$.

$d(z,X)$ désigne la distance du point z à l'ensemble analytique X .

On pose $z' = (z_1,\dots,z_p,0,\dots,0) \in \mathbb{C}^n$. On suppose $|z| < r_o$ et $r > r_o$.

Donc

$B(z',r - r_o) \subset B(0,r) \subset B(z',r + r_o)$, on pose $r_1 = r - r_o$, $r_2 = r + r_o$.

Donc

$$U(z) \leq \frac{-2k}{\omega_p} \int_{B(0,r_1) \cap X^*} \frac{\beta^p(x)/p!}{(|x_1|^2 + \dots + |x_p|^2 + d^2)^p} .$$

et

$$U(z) \geq \frac{-2k}{\omega_p} \int_{B(0,r_2) \cap X^*} \frac{\beta^p(x)/p!}{(|x_1|^2 + \dots + |x_p|^2 + d^2)^p} .$$

Soit l'intégrale $J_r = -\frac{2k}{\omega_p}\int_{B(0,r)\cap X^*} \frac{\beta^p(x)/p!}{(|n|^2+d^2)^p}$.

Alors

$$J_r = -2k \int_0^r \frac{t^{2p-1}\,dt}{(t^2+d^2)^p} = -k \int_0^r \frac{2t \,.\, t^{2p-2}\,dt}{(t^2+d^2)^p} .$$

On fait des intégrations par parties successives , on trouve :

(1.2.1.) $J_r = I_p + I_{p-1} + \dots + I_1$ avec

(1.2.2.) $I_j = \frac{k}{j-1}\left(\frac{r^{2j-2}}{(r^2+d^2)^{j-1}}\right)$ pour $p > j \geqslant 2$

et $I_1 = -k \,.\, \int \frac{2t\,dt}{(t^2+d^2)} = -k \log \frac{r^2+d^2}{d^2}$.

(1.2.3.) $I_1 = 2\,k \log d - k \log(r^2+d^2)$.

Donc au voisinage de 0

$$e^{-V(z)} \simeq \frac{K}{d^{2k}} = \frac{K}{d(z/X)^{2k}} \quad \text{avec} \quad K > 0 .$$

LEMME 1.2.1.

Soit X un ensemble analytique de dimension pure p dans $\mathbb{C}^n.(p+k=n)$. Alors au voisinage de tout point régulier x_o de X, on a :

$\forall\, \varepsilon > 0$, il existe $(r>0,\ K_1>0,\ K_2>0)$ telles que :

$$\frac{K_2}{[d(z,X)]^{2k(1+\varepsilon)}} \geqslant e^{-V(z)} \geqslant \frac{K_1}{[d(z,X)]^{2k/(1+2\varepsilon^2)^p}} \qquad (1.3.0.)$$

avec $V(z) = -\frac{2k}{\omega_p}\int_{B(x,r)\cap X^*} \frac{\beta^p(x)/p!}{|z-x|^{2p}}$.

Démonstration.

Soit $x_o \in X^*$; $x_o = 0$.

Pour r assez petit; $X \cap B(0,2r)$ est le graphe d'une application h :

$h : B(0,2r)\cap \mathbb{C}^p\times\{0\}\subset\mathbb{C}^n \longrightarrow \mathbb{C}^k$ telle que $h(0) = 0$ et $h'(0) = 0$.

On suppose de plus que $h(x) \not\equiv 0$ au voisinage de 0.

$$T_o(X) = \{z \in \mathbb{C}^n / z_{p+1} = \dots = z_n = 0\}$$

désigne l'espace tangent à X en 0.

Soit π la projection canonique

$$\pi : \mathbb{C}^n = \mathbb{C}^p \times \mathbb{C}^k \longrightarrow \mathbb{C}^p$$
$$(x,y) \longmapsto x$$

et soit φ l'application de $\mathbb{C}^p \times \{0\} \subset \mathbb{C}^n$ dans $\mathbb{C}^n$ telle que $\varphi(x,0) = (x,h(x))$.

Alors $\varphi(B(0,2r) \cap \mathbb{C}^p \times \{0\}) \subset X \cap B(0,2r+\delta)$ et $\varphi(B(0,2r) \cap \mathbb{C}^p \times \{0\}) \supset X \cap B(0,2r)$.

On peut supposer que $\delta = r$, en effet pour $x \leqslant r$ (r assez petit) $h(x) \leqslant Kr^2$.

Soit alors pour $|x| < r < \frac{1}{K}$, on a $|h(x)| < r$.

D'autre part, on a :

$$(1.3.1.) \quad U(z) \simeq \frac{-2k}{\omega_p} \int_{B(0,r)\cap T_o(X)} \frac{A(x)\, \beta^p(x)}{p!\, |z - (x,h(x)|^{2p}} = V(z),$$

avec $A(x) = (\sum_I{}' J_I^2(x))^{1/2}$; $I = (i_1,\ldots,i_p)$
$1 < i_1 < i_2 \ldots < i_p < n$.

J_I est un déterminant extrait d'ordre p de la matrice (n,p)

$$\begin{bmatrix} I_p \\ \cdots\cdots\cdots \\ \dfrac{\partial h_i}{\partial x_j} \end{bmatrix} \qquad \begin{array}{l} J_I(0) = 1 \text{ si } I = (1,\ldots,p) \\ J_I(0) = 0 \text{ pour } I \neq (1,2,\ldots,p) . \end{array}$$

Alors pour tout $\varepsilon > 0$; il existe r assez petit telle que

$$1 \leqslant A(x) \leqslant 1 + \varepsilon \qquad (1.3.2.)$$

1°/ On démontre l'inégalité de droite de (1.3.0.) d'après (1.3.1.) et (1.3.2.), on a

$$(1.3.3.) \quad V(z) \leqslant -\frac{2k}{\omega_p} \int_{B(0,r)\cap T_o(X)} \frac{\beta^p(x)/p\,!}{(|z_1-x|^2 + |z_2-h(x)|^2)^p} .$$

avec $z = (z_1,z_2) \in \mathbb{C}^p \times \mathbb{C}^k$

et r le rayon telle que (1.3.2.) soit vérifiée.

On prend z telle que $|z| < r_o$ et $r > r_o$.

Soit $(x_o,h(x_o)$ le point de $X \cap B(0,r)$ telle que $|z - (x_o,h(x_o))| = d(z,X) = d$.

On fait le changement de variables $y = x - z_1$ dans (1.3.3.) on aura :

$$(1.3.4.) \quad V(z) \leqslant -\frac{2k}{\omega_p} \int_{B(0,r-r_o)\cap T_o(X)} \frac{\beta^p(y)/p!}{(|y|^2 + |z_2 - h(y+z_1)|^2)^p}$$

car $|z| < r_o$.

On applique tout simplement l'inégalité du triangle on aura :

$$|z_2 - h(y + z_1)| \leqslant |z_2 - h(x_o)| + |h(x_o) - h(z_1)| + |h(z_1) - h(y + z_1)|.$$

Mais . $|z_2 - h(x_o)| \leqslant d$

. $|h(x_o) - h(z_1)| \leqslant \varepsilon\,|x_o - z_1| \leqslant \varepsilon d$.

. $|h(z_1) - h(y\ z_1)| \leqslant \varepsilon|y|$.

Donc $|z_2 - h(y+z_1)|^2 \leqslant 2(1 + \varepsilon)^2 d^2 + 2\varepsilon^2 |y|^2$ (1.3.5.)

Alors d'après (1.3.4.) et (1.3.5.)

$$V(z) \leqslant - \frac{2k}{\omega_p} \int_{B(0,r-r_o)\cap T_o(X)} \frac{\beta^p(y)/p!}{((1+2\varepsilon^2)|y|^2 + 2(1+\varepsilon)^2 d^2)^p} .$$

Donc d'après le lemme (1.2.) , il existe une constante $K_1' > 0$ telle que

$$e^{-V(z)} \geqslant \frac{K_1'}{d(z,X)^{2k/(1+2\varepsilon^2)^p}} .$$

Donc d'après (1.3.1.), il existe une $K > 0$ telle que

$$e^{-U(z)} \geqslant \frac{K}{d(z,X)^{2k/(1+2\varepsilon^2)^p}}$$

C.Q.F.D.

2°/ On démontre l'inégalité gauche de (1.3.0.), d'après (1.3.2.) et (1.3.1.)

$$- V(z) \leqslant \frac{2k(1+\varepsilon)}{\omega_p} \int_{B(0,r)\cap T_o(X)} \frac{\beta^p(x)/p!}{(|z_1-x|^2 + |z_2 - h(x)|^2)^p} .$$

On fait le même changement de variables que dans la première partie. On aura

$$(1.3.5.) \quad - V(z) \leqslant \frac{2k(1+\varepsilon)}{\omega_p} \int_{B(0,r+r_o)\cap T_o(X)} \frac{\beta^p(x)/p!}{(|x|^2 + |z_2-h(x+z_1)|^2)^p}$$

car $|z| < r_o$.

LEMME 1.2.2.

Pour r *et* $\varepsilon > 0$ *les deux constantes données ci-dessus* (*on suppose* $\varepsilon < 1/2$), *on a* : $\forall x \in B(0,d)$

$$(1.3.6.) \qquad |z_2 - h(x + z_1)| \geqslant d/2 .$$

Démonstration du lemme (1.3.1.)

On a $|z_2 - h(z_1)| \geqslant d$, car $(z_1, h(z_1))$ est un point de X et

$|h(z_1) - h(x,z_1)| \leqslant \varepsilon|x| \leqslant d/2$.

Donc $|z_2 - h(x + z_1)| \geqslant |z_2 - h(z_1)| - |h(z_1) - h(x + z_1)| \geqslant d - \frac{d}{2} = \frac{d}{2}$. On revient à la démonstration du 2°/ .

C.Q.F.D.

(1.3.7.) $$-V(z) \leqslant \frac{2k(1+\varepsilon)}{\omega_p} \int_{B(0,d)\cap T_o(X)} \frac{\beta^p(x)/p!}{(|x|^2 + (d/2)^2)^p} + \frac{2k(1+\varepsilon)}{\omega_p} \int_{d<|x|<r+r_o} \frac{\beta^p(x)/p!}{(|x|^2 + |z_2 - h(x+z_1)|^2)^p} .$$

Soit $$V_1 = \frac{2k(1+\varepsilon)}{\omega_p} \int_{B(0,d)\cap T_o(X)} \frac{\beta^p(x)/p!}{(x^2 + (d/2)^2)^p}$$

d'après le changement de variable $x = dy$.

(1.3.8.) $V_1(z) = C_1 > 0$ avec C_1 indépendante de d .

Soit $$V_2(z) = \frac{2k(1 + \varepsilon)}{\omega_p} \int_{d<|x|<r+r_o} \frac{\beta^p(x)/p!}{(|x|^2 + |z_2 - h(x+z_1)|^2)^p}$$

puisque $|z_2 - h(x + z_1)| \geqslant 0$.

On a : $$V_2(z) \leqslant \frac{2k(1 + \varepsilon)}{\omega_p} \int_{d<|x|r+r_o} \frac{\beta^p(x)/p!}{|x|^{2p}} .$$

$$V_2(z) \leqslant 2k(1 + \varepsilon) \int_d^{r+r_o} \frac{t^{2p-1}dt}{t^{2p}}$$

$$= -2k(1 + \varepsilon) \log d + 2k(1 + \varepsilon) \log(r + r_o) .$$

Donc il existe $K_2' > 0$ telle que

$$e^{+V_2(z)} \leqslant \frac{K_2'}{d(z,X)^{2k(1+\varepsilon)}} \qquad (1.3.9.)$$

Donc d'après (1.3.7.) ; (1.3.8.) ; (1.3.9.) , il existe une constante K_3 0 telle que

$$e^{-V(z)} \leqslant \frac{K_3}{d(z,X)^{2k(1+\varepsilon)}} .$$

Donc il existe $K_2 > 0$ telle que $e^{-U(z)} \leqslant \dfrac{K_2}{d(z,X)^{2k(1+\varepsilon)}}$ C.Q.F.D.

COROLLAIRE 1.1.

Si f est une fonction analytique nulle sur X , alors $|f|^2 e^{-U}$ est localement sommable sur X^* .

Démonstration.

Puisque f est nulle sur X ; il existe une constante $C > 0$ telle que $f(z) \leq C\, d(z,X)$

$$|f|^2 e^{-U(z)} \leq \frac{CK_2}{d(z,X)^{2(k-1)+2k\varepsilon}},$$

il suffit de prendre ε assez petit telle que $\varepsilon < \frac{1}{2k}$, ainsi $\dfrac{CK_2}{d(z,X)^{2(k-1)+2k\varepsilon}}$ sera localement sommable sur X^*.

Autre formulation du lemme 1.2.1.

Soit X un ensemble anlytique de dimension pure p dans $\mathbb{C}^n$. $(p+k=n)$. Alors au voisinage de tout point régulier x_o, on a :

$\forall \varepsilon > 0$; il existe $(r>0,\ K_1>0,\ K_2>0)$ telles que :

$$\frac{K_2}{|d(z,X)|^{2k+\varepsilon}} \geq e^{-U(z)} \geq \frac{K_1}{|d(z,X)|^{2k-\varepsilon}} \quad \text{avec } U(z) = \frac{-2k}{\omega_p}\int_{B(x_o,r)\cap X^*} \frac{\beta^p(x)/p!}{|z-x|^{2p}}.$$

et z assez voisin de x_o.

On s'intéresse maintenant aux points singuliers de X et à voir le comportement de $|f|^2 e^{-V}$ au voisinage de ces points.

Notons que la propriété plus forte : $d(z,X)^2 e^{-V}$ localement sommable est fausse déjà pour une hypersurface.

Exemple :

Dans $\mathbb{C}^2$, on considère la fonction $f(z_1,z_2)$ définie par $f(z_1,z_2) = z_1(z_2^2-z_1^3)$

0 est un point singulier de $X = f^{-1}(0)$.

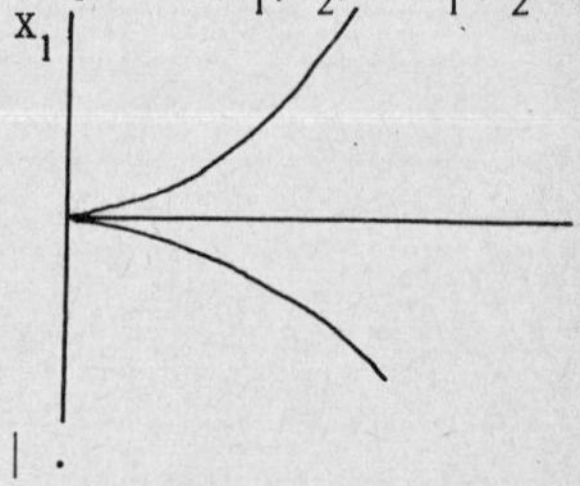

$X = X_1 \cup X_2$ avec

$X_1 = \{(z_1,z_2)\in\mathbb{C}^2 / z_1 = 0\}$

$X_2 = \{(z_1,z_2)\in\mathbb{C}^2 / z_2^2 - z_1^3 = 0\}$.

Et on a $d(z,X) = \min\,(d(z,X_1)\ ,\ d(z,X_2))$ et $d(z,X_1) = |z_1|$.

On va montrer qu'il existe un ouvert dans la boule $B(0,r)$, r positif assez petit telle que $d(z,X) = d(z,X_1)$ pour tout z dans cet ouvert.

Supposons que $d(z,X) < |z_1|$ avec $|z_1| < 1/2$ (1)

Il existe $x \in \mathbb{C}^2$ telle que $x_1^3 = x_2^2$ et $|z_1 - x_1| < |z_1|$ (2)

$|z_2 - x_2| < |z_1|$ (3)

(1) et (2) nous donnent que $|x_1| < 2|z_1| < 1$.

Comme $|x_1|^3 = |x_2|^2$, on en déduit que $|x_2| = |x_1|^{3/2} < (2|z_1|)^{3/2} < 2|z_1|$.

(car $2|z_1| < 1$ et $3/2 > 1$) d'où d'après (3) $|z_2| < 3|z_1|$.

Par suite si $|z_2| > 3|z_1|$ et $|z_1| < 1/2$, $d(z,X) > |z_1|$.

Soit $V = \{(z_1,z_2)$ tq $|z_2| > 3|z_1|\} \cap B(0,r)$.

Alors $I = \int_{B(0,r)} d(z,X)^2 e^{-V(z)} d\lambda(z) \geqslant \int_V d(z,X)^2 e^{-V(z)} d\lambda(z)$.

Puisque X est une hypersurface , alors

$V(z) = \log |f|^2 + \omega$ où ω est harmonique, donc $e^{-V(z)} \geqslant \frac{K}{|f|^2}$ où K est une constante positive.

$V \subset B(0,r)$. Donc $I \geqslant K \int_V \frac{|z_1|^2 \, d\lambda(z)}{|z_1|^2 \, |z_2^2 - z_1^3|^2}$

Comme $|a+b| \leqslant |a| + |b|$

$|z_2| > 3|z_1|$.

Alors $I \geqslant K \int_V \frac{d\lambda(z)}{|z_2^2 - z_1^3|^2} \geqslant K \int_V \frac{d\lambda(z)}{|z_2|^4 (1+|z_2|/3^3)^2}$.

Dans V ; $1 + |z_2|/3^3 < C$ avec $C > 0$.

Donc $I \geqslant C\,K. \int_V \frac{d\lambda(z)}{|z_2|^4}$.

On intègre par rapport à z_1

$2\pi \int_0^{|z_2|/3} t\,dt = \frac{\pi}{9} |z_2|^2$ et $\int_{B(0,r)^2} \frac{d\lambda(z_2)}{|z_2|^2} = +\infty$ Donc $I = +\infty$.

Pour résoudre le problème, on va distinguer 3 cas, suivant que $p = 0$, $p = n-1$ ou $0 \leqslant p \leqslant n-2$. Dans les 2 premiers cas la réponse est positive tandis que dans le 3e cas, la réponse est en général négative.

<u>Ier cas</u> : <u>$p = 0$</u>.

Alors X est la réunion de points isolés . Donc il n'y a pas de points singuliers et la réponse est positive.

<u>2e cas</u> : <u>$p = n-1$</u>.

Rappelons que si X est une hypersurface $I(\underline{X}_o)$ est principal avec $I(\underline{X}_o)$ et l'ensemble des fonctions holomorphes en 0 et nulle sur $\underline{X}_o$.

On en déduit que si f est une fonction holomorphe nulle sur X, $|f|^2 e^{-V}$ est localement sommable sur X. En effet :

D'après H.Skoda [7] ; $V = \log |g|^2 + \omega$ avec : ω est une fonction harmonique

. $X = g^{-1}(0)$ et g engendre l'idéal $I(\underline{X}_o)$.

Donc si $f \in I(\underline{X}_o)$, il existe une fonction h holomorphe telle que $f = g.h$

$|f|^2 e^{-V} = |h|^2 . e^{-\omega}$ qui est localement sommable sur X.

<u>Conséquences.</u>

<u>Si</u> X <u>est une hypersurface</u> $H^q(\Omega, J) = 0$ <u>pour tout</u> $q \geqslant 1$. <u>Et toute fonction holomorphe sur</u> X <u>est la restriction d'une fonction holomorphe sur</u> Ω.

<u>3e cas</u> : <u>$0 < p \leqslant n-2$</u>.

Nous faisons un rappel de la démonstration pour le lecteur des lemmes 1.3., 1.4. qui se trouve bien détaillé dans H.Skoda [8], Proposition 7.1.et lemme 7.1.

<u>LEMME 1.3.</u>

Soit Θ un courant positif fermé de dimension pure p. On suppose $\nu_\Theta(0) > c$ avec $c > 0$. Alors l'intégrale suivante

$$\frac{k}{P} \cdot \frac{P!}{\pi^P} \int_0^R \frac{d\sigma(t)}{(|z|+t)^{2p}} \geqslant 2k\,c \log\left(1 + \frac{R}{|z|}\right) - 4p.k\,\frac{p!}{\pi^p}\,\frac{\sigma(R)}{R^{2p}} . \tag{2.1.}$$

<u>Démonstration.</u>

On intègre par partie l'intégrale de 2-1. On trouve :

$$\int_0^R \frac{d\sigma(t)}{(|z|+t)^{2p}} = \frac{\sigma(R)}{(|z|+R)^{2p}} + 2p \int_0^R \frac{\sigma(t)dt}{(|z|+t)^{2p+1}} . \tag{2.2.}$$

On en déduit

$$\int_0^R \frac{d\sigma(t)}{(|z|+t)^{2p}} \geqslant 2p \int_0^R \frac{\sigma(t)dt}{t^{2p}(t+|z|)} - 2p \int_0^R \left| \frac{1}{t^{2p}} - \frac{1}{(t+|z|)^{2p}} \right| \frac{\sigma(t)}{t+|z|} dt. \tag{2.3.}$$

En remarquant que $(t + |z|)^{2p} - t^{2p} \leqslant 2p\, |z|(t + |z|)^{2p-1}$.

On obtient :

$$(2.4.) \quad \int_0^R \frac{d\sigma(t)}{(t + |z|)^{2p}} \geqslant 2p \int_0^R \frac{\sigma(t)dt}{t^{2p}(t + |z|)} - 4p^2 |z| \int_0^R \frac{\sigma(t)dt}{t^{2p}(t + |z|)^2} .$$

Comme $\sigma(t)/t^{2p}$ est une fonction croissante de t , on a

$$(2.5.) \quad \frac{p!}{\pi^p} \frac{\sigma(R)}{R^{2p}} \geqslant \frac{p!}{\pi^p} \frac{\sigma(t)}{t^{2p}} \geqslant c .$$

Alors on a :

$$\frac{k}{p} \frac{p!}{\pi^p} \int_0^R \frac{d\sigma(t)}{(t + |z|)^{2p}} \geqslant 2k\, c \int_0^R \frac{dt}{t + |z|} - 4pk \frac{p!}{\pi^p} . \frac{\sigma(R)}{R^{2p}} . |z| . \int_0^R \frac{dt}{(t + |z|)^2}$$

On minore en sommant de 0 à $+\infty$ dans la dernière intégrale , on aura :

$$(2.6.) \quad \frac{k}{p} \frac{p!}{\pi^p} \int_0^R \frac{d\sigma(t)}{(t + |z|)^{2p}} \geqslant 2k\, c \log\left(1 + \frac{R}{|z|}\right) - 4pk \frac{p!}{\pi^p} \frac{\sigma(R)}{R^{2p}} . \qquad \text{C.Q.F.D.}$$

LEMME 1.4.

Soit X un ensemble analytique de dimension p et si σ est la mesure trace associée au courant $\Theta = [X]$. Alors

$$V(z) = \frac{-2k}{\omega_p} \int_{|x|<R} \frac{d\sigma(x)}{|z - x|^{2p}}$$

$$(2.7.) \quad V(z) = \frac{k}{\pi^p} \int_{|x|<R} \log |z - x|^2 \alpha^p(z - x) \wedge \Theta(x) + W(z) .$$

avec

- $W(z)$ une fonction continue
- $\alpha(z) = \frac{i}{2} d'd'' \log |z|^2$.

Soient $\mu(z) = \int_{|x|<R} \alpha^p(z - x) \wedge \Theta(x)$ et $\chi(z) = \frac{k\mu(z)}{\pi^p}$.

Alors $\forall \varepsilon > 0\ \exists\ r_o > 0$ telle que pour tout $r < r_o$ et presque tout z tel que $|z| < r$, on ait :

$$\left(\frac{R - 2r}{R - r}\right)^{2p} \nu(R - 2r) \leqslant \frac{\mu(z)}{\pi^p} < c(1 + \varepsilon) ,$$

la deuxième inégalité étant vraie pour tout z tel que $|z| < r$, ainsi que

$\chi(z) \leqslant kc\,(1 + \varepsilon)$.

Démonstration.

$$\mu(z) = \int_{|x| < R} \alpha^p(z - x) \wedge \Theta(x)$$
$$\leqslant \int_{|z-x|<R+r} \alpha^p(z - x) \wedge \Theta(x) .$$

$$(2.8.) \leq \frac{p!}{(R+r)^{2p}} \int_{|z-x|<R+r} d\sigma(x) \leq \frac{p!}{(R+r)^{2p}} \cdot \sigma(R+2r) .$$

Pour la majoration on utilise la formule classique suivante :

$$(2.9.)\quad \nu(z) + \frac{1}{\pi^p} \int_{0<|z-x|<r} \alpha^p(z-x) \wedge \Theta(x) = \frac{p!\ \sigma(z,r)}{\pi^p . r^{2p}} .$$

Soit alors $\mu(z) \leq (\frac{2r+R}{R+r})^{2p} \nu(R+2r)\ \pi^p$.

Et comme $\nu(r) \xrightarrow[r \to o]{} \nu(0) = c < c(1+\varepsilon)$ pour tout $\varepsilon > 0$.

Donc on peut choisir R et r de façon que , pour $r < r_o$, $(\frac{R+2r}{R+r})^{2p} \nu(R+2r) < c(1+\varepsilon)$

Donc

$$2.10.\quad \frac{\mu(z)}{\pi^p} < c(1+\varepsilon)$$

$$\mu(z) \geq \int_{|z-x|<R-r} \alpha^p(z-x) \wedge \Theta(x)$$

si $\nu(z) = 0$ on a d'après (2.8.)

$$\int_{|z-x|<R-r} \alpha^p(z-x) \wedge \Theta(x) = \frac{p!}{(R-r)^{2p}} \int_{|Z-x|<R-r} d\sigma(x) = \pi^p \nu(z, R-r) .$$

Donc il résulte que

$$\mu(z) \geq p!(R-r)^{-2p} \int_{|x|<R-2r} d\sigma(x)$$

$$2.11.\quad \mu(z) \geq \pi^p (\frac{R-2r}{R-r})^{2p} \nu(R-2r) ,$$ ceci pour $\nu(z) = 0$.

Et puisque l'ensemble des z telle que $\nu(z) = 0$ est de mesure de Lebesgue nulle alors

$$\mu(z) \geq \pi^p (\frac{R-2r}{R-r})^{2p} \nu(R-2r) \text{ pp} .$$

D'après (2.9.) :

$$\chi(z) = \frac{k\mu(z)}{p} < k\ c\ (1+\varepsilon), \text{ pour } |z| < r_o .$$

THÉORÈME 1.2.

Soit X un ensemble analytique de $\mathbb{C}^n$ de dimension pure p ; on suppose que $\nu_X(0) = c$, $c > 0$, et soit f une fonction analytique nulle en 0 telle que $\nu_f(0) = m$. Alors si $m \leq k\ c - n$, $k = n - p)$, $|f|^2 e^{-V}$ n'est pas localement sommable au voisinage de 0 .

Démonstration.

Comme f est nulle en 0 de multiplicité m on a un développement de $f(z)$ de la forme

$f(z) = P_m(z) + P_{m+1}(z) + \ldots$, avec $P_j(z)$ un polynôme homogène de degré j. $P_m \neq 0$.

On va distinguer deux cas :

Ier cas :

On suppose dans ce cas que P_m ne dépend que de z_1 . Alors $|P_m(z)| = \lambda |z_1|^m$ avec $\lambda > 0$.

On pose $h(z) = \sum_{j=m+1}^{\infty} P_j(z)$. h est une fonction analytique nulle en 0 d'ordre supérieur à $m+1$. Donc d'après le lemme de Schwartz, il existe $\lambda_1 > 0$ telle que $|h(z)| \leqslant \lambda_1 |z|^{m+1}$ pour $|z|$ assez petit.

On considère le domaine D défini par $D = \{z \in \mathbb{C}^n \,/\, \tau > |z_1| > |z'|$ avec $z = (z_1, z')$ $z_1 \in \mathbb{C}$ et $\tau > 0\}$

Sur D $|h(z)| \leqslant \lambda_1 |z_1|^{m+1} \cdot 2^{m+1/2}$ et $|f(z)| > (\lambda |z_1|^m - \lambda_1 |z_1|^m \cdot 2^{m+1/2} \cdot |z_1|)$

$(\lambda - \lambda_1 |z_1| 2^{m+1/2}) |z_1|^m$.

En choisissant bien $\tau > 0$, on aura $|f(z)| \geqslant \lambda_2 |z_1|^m$ sur D pour un $\lambda_2 > 0$.

On va montrer maintenant que $|f|^2 e^{-V}$ n'est pas sommable au voisinage de 0

$V(z) = 2k\, U(z) + H(z)$ avec H une fonction plurisousharmonique continue. Donc sur un voisinage borné de 0, $e^{-H(z)}$ est bornée. Au voisinage de 0 , $U(z)$ est équivalente à

$$U_1(z) = -\frac{1}{\omega_p} \int_{B(0,R)} \frac{d\sigma(x)}{|z-x|^{2p}} .$$

Donc au voisinage de 0 , $e^{-V(z)}$ a le même comportement que $e^{-2\,k\,U_1(z)}$

$$-U_1(z) = \frac{1}{\omega_p} \int_{B(0,R)} \frac{d\sigma(x)}{|z-x|^{2p}} \leqslant \frac{1}{\omega_p} \int_0^R \frac{d\sigma(t)}{(|z| + t)^{2p}} .$$

Donc d'après (2.1.), il existe $\lambda_3 > 0$ telle que :

$$e^{-V(z)} \geqslant \frac{\lambda_3}{|z|^{2kc}}$$

Alors $I = \int_{|z|<r} |f(z)|^2 \cdot e^{-V(z)}\, d\lambda(z) \geqslant \int_D |f(z)|^2\, e^{-V(z)}\, d\lambda(z)$.

$$I \geqslant \int_D \frac{\lambda_2\, \lambda_3 |z_1|^{2m}}{|z|^{2kc}}\, d\lambda(z) \geqslant \frac{\lambda_2\, \lambda_3}{2^{kc}} \int_D \frac{d\lambda(z)}{|z_1|^{2kc-2m}} .$$

$$\int_D \frac{d\lambda(z)}{|z_1|^{2kc-2m}} = \int_{|z_1|<\tau_2} \frac{d\lambda(z_1)}{|z_1|^{2kc-2m}} \left|\int_{|z'|<|z_1|} d\lambda(z')\right|$$

2.12. $$\int_{|z_1|<\tau} \frac{d\lambda(z_1)\ |z_1|^{2(n-1)}}{|z_1|^{2kc-2m}} \cdot \frac{(n-1)!}{\pi^{n-1}} .$$

Si <u>$m \leqslant kc - n$</u> l'intégrale 2.12. est divergente.

<u>2e cas.</u>

On écrit $P_m(z)$ sous forme de série entière au voisinage de 0.

$$P_m(z) = \sum_{\substack{\nu = (\nu_1,\ldots,\nu_n) \\ |\nu| = m}} a_\nu z^\nu .$$

Soit $\tau_1 = \sup|a_\nu|$ et $\tau = \max(\tau_1, 1)$.

On peut toutefois supposer $P_m(z) = z_1^m + \sum_{\substack{|\nu|=m \\ \nu_1 < m}} a_\nu z^\nu$.

On considère le domaine D_1 défini par $D_1 = \{z \in \mathbb{C}^n / \ |z_1| > 2N\ |z'|$ avec N le nombre de termes a_ν non nuls dans la somme $\sum_{\substack{|\nu|=m \\ \nu_1<m}} a_\nu z^\nu$.

On a dans D_1 : $|a_\nu z^\nu| = |a_\nu z'^{\nu'} z_1^{\nu_1}| < \tau.(\frac{|z_1|}{2N}).\ |z_1|^{|\nu_1|} < \frac{|z_1|^m}{2N}$.

Donc $\sum_{\substack{|\nu|=m \\ \nu_1<m}} |a_\nu z^\nu| \leqslant \sum_{\substack{|\nu|=m \\ \nu_1<m}} \frac{|z_1|^m}{2N} = \frac{|z_1|^m}{2}$.

Donc $|P_m(z)| \geqslant \frac{|z_1|^m}{2}$ sur D_1 .

Le même calcul que celui fait dans le premier cas nous donne le résultat .

<u>Exemple 1.</u>

On construit un ensemble analytique vérifiant les propriétés du théorème (2.1.).

On considère l'ensemble analytique X défini par $X = \{(t^p, t^q, t^s) \in \mathbb{C}^3$ telle que $p+q = s$ et $t \in \mathbb{C}\}$

On suppose $2 \leqslant p < q$ et (p,q,s) trois entiers premiers entre eux deux à deux.

Il est aisé de montrer que X est un ensemble analytique irréductible admettant l'origine comme seul point singulier.

La dimension de X est 1 dans $\mathbb{C}^3$. On pose $z = (x,y,z)$, alors $f(z) = xy - z$ est analytique dans $\mathbb{C}^3$ et nulle sur X .

LEMME 1.5.

Si $\nu_X(0)$ désigne le nombre de Lelong de X au point 0 alors $\nu_X(0) = p > 2$.

Rappelons la définition de $\nu_X(z)$. Si Y est un ensemble analytique de dimension pure p ; alors

$$\nu_Y(z) = \lim_{r\to o} \frac{\sigma(z,r)}{\tau_p\, r^{2p}} = \lim_{r\to o} \frac{\mathrm{vol}(Y\cap B(z,r))}{\tau_p\, r^{2p}}, \qquad \tau_p = \frac{\pi^p}{p!}$$

le volume de la boule unité dans $\mathbb{C}^p$.

Démonstration du lemme 1.5.

$$\nu_X(0) = \lim_{r\to o} \frac{1}{\pi r^2} \int_{X\cap B(0,r)} \beta(z)$$

En utilisant le paramétrage

$$I = \int_{X\cap B(0,r)} \beta = \int_{B(0,r')} \left[p^2|t|^{2(p-1)} + q^2|t|^{2(q-1)} + s^2|t|^{2(p-1)}\right] \frac{i}{2}\, dt \wedge d\bar{t}$$

avec $[r^2 = r'^{2p} + r'^{2q} + r'^{2s}]$.(1)

$$I = 2\pi \int_o^{r'} (p^2 x^{2p-1} + q^2 x^{2q-1} + s^2 x^{2s-1})\, dx \quad , \; x \text{ réelle}$$

$$= 2\pi \left(p^2 \frac{r'^{2p}}{2p} + q^2 \frac{r'^{2q}}{2q} + s^2 \frac{r'^{2s}}{2s}\right) .$$

$$\nu_X(0) = \lim_{r\to o} 2\, \frac{\pi}{\pi} \left(\frac{pr'^{2p} + qr'^{2q} + sr'^{2s}}{2 \;\; r'^{2p} + r'^{2q} + r'^{2s}} \right) = p \quad \text{car } p<q<s .$$

Donc $\nu_X(0) = p$, ce qui termine la démonstration du lemme 1.5.

Pour l'ensemble analytique X ci-dessus et f la fonction analytique $f(z) = xy-z$.

On a

$$\int_{B(0,r)} |f|^2\, e^{-V(z)}\, d\lambda(z) = +\infty \quad \forall r > 0 .$$

En effet $\nu_X(0) = p$ et $\nu_f(0) = 1$ et $kc - n = 2\times 2 - 3 = 1$.

Donc d'après le théorème 2.1. l'intégrale ci-dessus diverge.

La question se pose maintenant pour les ensembles dits "intersections complètes au sens des idéaux".

DÉFINITION.

Un ensemble analytique de dimension pure p est une intersection complète au sens des idéaux s'il existe une fonction f de $\mathbb{C}^n$ dans $\mathbb{C}^{n-p}$ telle que $f^{-1}(0) = X$ et $I(\underline{X}_o) = (f_1,\ldots,f_k) = f$. $k = n - p$.

Exemple 2.

On construit un ensemble analytique ; intersection complète au sens des idéaux telle qu'il existe une fonction f nulle sur X et $|f|^2 e^{-V}$ non localement sommable en application du théorème 2.1.

L'exemple 1 répond à la question mais le calcul fait pour faire la démonstration est long. On donne un autre exemple plus simple.

Soit $X = \{(t^p, t^{2p}, t^q)\ ,\ t \in \mathbb{C}\ ,\ q \equiv 1[p]\ ,\ 2 \leqslant p < q\}$.

LEMME 1.6.

On pose $f_1(x,y,z) = x^2 - y$

$f_2(x,y,z) = x^q - z^p$.

f_1 et f_2 s'annulent sur X et elles engendrent $I(X)$.

Pour démontrer le lemme on a besoin du théorème suivant qui est dû à H.Cartan.

THÉORÈME 1.3.

Tout idéal de l'anneau des fonctions analytiques sur un ouvert d'holomorphie (ou un ouvert de Stein) est fermé.

Démonstration du lemme 1.6.

Soit f une fonction analytique s'annulant sur X . On écrit f sous la forme d'une série entière

$$f(x,y,z) = \sum_{|\nu|=0}^{\infty} a_\nu Z^\nu . \qquad \nu = (\nu_1, \nu_2, \nu_3)$$

$$|\nu| = \nu_1 + \nu_2 + \nu_3 .$$

$$z^\nu = x^{\nu_1} . y^{\nu_2} . Z^{\nu_3} .$$

Alors si $\nu_3 = s + tp$ avec $(0 < s < p)$

$$(1)\quad x^{\nu_1}.y^{\nu_2}.z^{\nu_3} = x^{\nu_1+2\nu_2}.z^{\nu_3} \ (\text{mod. } f_1)$$

$$= x^{\nu_1+2\nu_2+tq}.z\ (\text{mod.}(f_1, f_2)).$$

On note $Q_m(Z) = \sum_{|\nu|=o}^{m} a_\nu Z^\nu$; $\lim_{m\to+\infty} Q_m(Z) = f(Z)$.

D'après (1) on déduit

(2) $Q_m(Z) = P_{o,m}(x) + P_{1,m}(x)z + \ldots + P_{p-1,m}(x)\, z^{p-1} + A_{1,m}f_1 + A_{2,m}f_2$

avec $P_{j,m}(x) = \sum_{|\nu|=j,\nu_3=j+tp}^{m} a_\nu\, x^{\nu_1+2\nu_2+tq}$.

On note $r = E(\frac{q}{p}) + 1$ où $E(\frac{q}{p})$ est la partie entière de $\frac{q}{p}$.

On va montrer que la limite uniforme sur tout compact de $P_{j,m}(x)$ quand m tend vers l'infini existe. Il suffit de montrer ceci pour $|x| > 1$.

Pour $|x| > 1$: $I = \sum_{|\nu|=j,\nu_3=tp+j} |a_\nu|\,|x|^{\nu_1+2\nu_2+tq} \leqslant \sum_{|\nu|=o}^{\infty} |a_\nu|\,|x|^{\nu_1+2\nu_2+r\nu_3}$

Comme $r \geqslant 2$ $(q = 1 \bmod p)$.

Donc $I < \sum_{|\nu|=o} |\alpha_\nu|\;|x|^{r|\nu|} < +\infty$ car la somme $\sum_{|\nu|=o}^{\infty} |a_\nu|\;|y|^{|\nu|} < +\infty$ pour $|y| > 1$. En particulier pour $y = (x^r,x^r,x^r)$. Donc $\lim_{m\to+\infty} P_{j,m}(x) = P_j(x)$ est une fonction entière dans $\mathbb{C}$. Et, d'après (2) $Q_m(Z) = \sum_{j=o}^{p-1} P_{j,m}(x).z^j = H_m(Z)$ appartient à l'idéal engendré par f_1 et f_2 et $H(Z) = \lim H_m(Z) = f(Z) - \sum_{j=o}^{p-1} P_j(x).z^j$ est une fonction entière qui,d'après le théorème de H.Cartan appartient à l'idéal engendré par f_1 et f_2 .

Remarque.

Un calcul direct mais long montre que la limite des $A_{j,m}(z) = A_j(z)$ existe uniformément sur tout compact et elle est analytique.

On revient au problème posé : on veut montrer que si f est une fonction analytique nulle sur X , alors elle appartient à l'idéal (f_1,f_2) .

f nulle sur X si et seulement si pour tout $t \in \mathbb{C}$, $f(t^p,t^{2p},t^q) \equiv 0$ donc

$$\sum_{j=o}^{p-1} P_j(t^p.t^{qj}) = 0 .$$

Les degrés des monomes apparaissant dans $P_o(t^p)$, ..., $P_j(t^p)t^{qj}$, ..., $P_{p-1}(t^p)t^{q(p-1)}$ sont respectivement congrus modulo p à $0,1,2$, $p-1$, car $q \equiv 1[p]$.

Donc $P_j \equiv 0$ pour $0 \leqslant j \leqslant p-1$. Et ainsi on montre que f appartient à l'idéal (f_1,f_2) .

Donc $I(\underline{X}) = (f_1,f_2)$. C.Q.F.D.

On déduit de ce lemme que X est une intersection complète au sens des idéaux et on a encore $|f_1|^2 e^{-V}$ n'est pas localement sommable sur X (avec V le potentiel associé à X indiqué au début de ce travail). En effet :

$$\nu_X(0) = P \quad , \quad \nu_f(0) = 1 \quad \text{et} \quad 1 \leqslant 2p - 3$$

et d'après le théorème 2.1 $|f_1|^2 e^{-V}$ n'est pas localement sommable.

REMARQUE 1.1.

La démonstration du théorème 2.1. utilise la nullité de f en 0 et non pas la nullité de f sur X tout entier bien que le résultat ne soit significatif que pour des fonctions nulles sur X.

PROPOSITION 1.1.

Soit X un ensemble analytique de dimension pure p et f une fonction analytique s'annulant sur X de multiplicité au moins m sur $X \cap B(0,r)$ pour r assez petit. Si $\nu_X(0) = c$, $c \neq 1$. Alors si $m \geqslant k(c-1) + 1$

$$\int_{B(0,r)} |f|^2 e^{-V} d\lambda(z) < +\infty .$$

Si $c = 1$; 0 est un point régulier et il suffit de prendre $m = 1$ i.e. pour toute fonction analytique nulle sur X

$$\int_{B(0,r)} |f|^2 e^{-V} d\lambda(z) < +\infty .$$

COROLLAIRE 1.2.

Pour toute fonction analytique nulle sur X, $|f|^{2k(c-1)+2} . e^{-V}$ est localement sommable au voisinage d'un point x telle que $\nu_X(x) = c$.

Cod $X = k$, dim $X = p$.

Démonstration de la proposition 1.1.

D'après 2.7., on a :

$-V(z) = -U_1(z) + W(z)$ avec $W(z)$ une fonction continue et

$$-U_1(z) = \int_{|x|<R} \log|z-x|^{-2\chi(z)} \frac{\alpha^p(z-x)}{\mu(z)} \wedge \Theta(x).$$

D'après l'inégalité de convexité de l'exponentielle, $\forall z \notin X$;

$$e^{-U_1(z)} \leqslant \int_{|x|<R} |z-x|^{-2\chi(z)} \frac{\alpha^p(z-x)}{\mu(z)} \wedge \Theta(x)$$

D'après 2.6., 2.10., 2.11., il existe une constante $\lambda > 0$ et $\varepsilon > 0$ telle que :

$$e^{-U_1(z)} \leqslant \lambda \int_{|x|<R} \frac{d\sigma(x)}{|z-x|^{2p+2kc+2\varepsilon}} .$$

Si $m \geqslant k(c-1) + 1$

$$|f(z)| \leqslant \lambda_1 \; |z - x|^{2k(c-1)+2}$$ pour $\lambda_1 > 0$, ceci d'après la formule de Taylor.

Alors : $$\int_{|z|<r} |f(z)|^2 \; e^{-V(z)} \; d\lambda(z) \leqslant \lambda_2 \int_{|z|<r} \int_{|x|<R} \frac{d\sigma(x).\; d\lambda(z)}{|z-x|^{2n-2+2\varepsilon}} < +\infty \text{ pour } \lambda_2 > 0.$$

Ce qui termine la démonstration de la proposition 2.1.

On pose la question suivante :

Problème.

Pour tout ensemble analytique X de dimension pure p ; $(0 < p < n-1)$ admettant des points singuliers ($\sigma(X) \neq \emptyset$) ; existe-t-il une fonction analytique nulle sur X telle que $|f|^2 \; e^{-V}$ n'est pas localement sommable sur X , c'est-à-dire est-ce que les exemples précédents ne sont que des cas particuliers d'un phénomène général ?

2ème partie

On rappelle que l'objet de cette partie est l'étude de la liaison entre la sommabilité de $e^{-\varphi}$ (si φ est une fonction plurisousharmonique) et le comportement de ν_φ sur les branches irréductibles de $X_\varphi = \{x/\nu_\varphi(x) > 0\}$ (ensemble des points de densité de φ) et leur codimension .

On rappelle qu'il convient de dire qu'une fonction ψ de $\mathbb{C}^n \to \mathbb{R}$ est dite "presque plurisousharmonique" s'il existe une fonction φ_o plurisousharmonique telle que

$\nu_{\varphi_o} \equiv 0$ et $\psi + \varphi_o$ est plurisousharmonique.

THÉORÈME 2.1. , H.SKODA [8]

Si Θ est un courant positif fermé de dimension pure p dans $\mathbb{C}^n$ ($p+k=n$) . Si U désigne le potentiel canonique associé à Θ , on a :

1°/ e^{-U} est non sommable au voisinage de x_o si $\nu_\Theta(x_o) \geqslant 2n$.

2°/ e^{-U} est sommable au voisinage de x_o si $\nu_\Theta(x_o) < 2k$.

COROLLAIRE 2.1.

Pour toute fonction φ plurisousharmonique dans $\mathbb{C}^n$ on a :

1°/ Si $\nu_\varphi(x_o) < 2$, $e^{-\varphi}$ est sommable au voisinage de x_o.

2°/ Si $\nu_\varphi(x_o) \geqslant 2n$, $e^{-\varphi}$ n'est pas sommable au voisinage de x_o.

Pour la démonstration du Corollaire, il suffit d'appliquer le théorème précédent au courant $u = \frac{i}{\pi} d'd''\varphi.(p = n - 1)$.

THÉORÈME 2.2., Y.T.SIU [9].

Soit Θ un courant positif fermé de type (k,k) sur un ouvert Ω de $\mathbb{C}^n$ et soit Y une variété complexe sans singularités de codimension k. On suppose que $Y = \{z \in \mathbb{C}^n / z_1 = \ldots = z_k = 0\}$. On note $c = \inf_{x \in Y} \nu_\Theta(x)$. On suppose $c > 0$, alors $\Theta - c[Y]$ est un courant positif fermé et $\nu_\Theta(x) = c$ presque partout sur Y (par rapport à la mesure σ associée au courant $[Y]$).

LEMME 2.1.

Le théorème précédent s'étend pour Y un ensemble analytique de dimension pure p.

LEMME 2.2.

Si Γ est un courant fermé d'ordre nul sur Ω et positif sur $\Omega \setminus Z$ où Z est fermé et $H_{2p}(Z) = 0$. Alors T est positif sur Ω.

Démonstration du lemme 2.2.

$\chi_{\Omega\setminus Z}.T$ est un courant localement plat car il est limite pour la masse des courants normaux, tronqués de T à support dans $\Omega \setminus Z$ ($\chi_{\Omega\setminus Z}$ désigne la fonction caractéristique de $\Omega \setminus Z$). Par suite $\chi_Z.T = T - \chi_{\Omega\setminus Z}.T$ est localement plat, à support dans Z. Le théorème de support nous donne que $\chi_Z.T = 0$ i.e. $T = \chi_{\Omega\setminus Z}.T$ qui est positif donc T est positif.

THÉORÈME (2.3.) de support.

Si T est un courant localement plat dans un ouvert Ω de $\mathbb{C}^n$ tel que la bidimension de $T = (m,m)$ et $H_{2m}(\text{supp } T) = 0$, alors $T = 0$.

Démonstration du Lemme 2.1.

On pose $T = \Theta - c[Y]$; T est un courant d'ordre nul fermé sur Ω positif sur $\Omega \setminus \sigma(Y)$ où $\sigma(Y)$ désigne le lieu des points singuliers de Y. Donc positif sur Ω d'après le Lemme 2.2.

PROPOSITION 2.1.

Soit φ une fonction plurisousharmonique dans $\mathbb{C}^n$. On suppose que $\varphi = \varphi_1 + \varphi_2$ avec φ_1 et φ_2 deux fonctions plurisousharmoniques vérifiants :

a/ $\nu_{\varphi_1}(0) = 0$

b/ $\{t \in \mathbb{R} \mid e^{-t\varphi_2}$ est sommable au voisinage de 0 est un ouvert$\}$.

Alors on a l'équivalence des deux propositions suivantes :

1°/ $e^{-\varphi}$ sommable au voisinage de 0 .

2°/ $e^{-\varphi_2}$ sommable au voisinage de 0 .

Démonstration.

1 ⟹ 2 comme φ_1 est localement majorée, il existe une constante $c > 0$ telle que $e^{-\varphi_1} > c$.

Pour r assez petit on a $+\infty \int_{B(0,r)} e^{-\varphi(z)}\, d\lambda(z) > c \int_{B(0,r)} e^{-\varphi_2(z)}\, d\lambda(z)$.

Donc $e^{-\varphi_2}$ est sommable au voisinage de 0 .

2 ⟹ 1.

On va utiliser l'inégalité de Hölder et le Corollaire 2.1 . L'hypothèse b/ affirme qu'il existe $\varepsilon > 0$ telle que $e^{-(1+\varepsilon)\varphi_2}$ est sommable au voisinage de 0 .

Soit $p = 1 + \frac{1}{\varepsilon}$; $q = 1 + \varepsilon$.

$$\frac{1}{p} + \frac{1}{q} = 1 .$$

D'après l'inégalité de Hölder on a :

$$\int_{B(0,r)} e^{-\varphi(z)}\, d\lambda(z) \leq \left[\int_{B(0,r)} e^{-p\varphi_1(z)}\, d\lambda(z)\right]^{1/p} \bullet \left[\int_{B(0,r)} e^{-q\varphi_2(z)}\, d\lambda(z)\right]^{1/q}$$

$p\varphi_1$ est une fonction plurisousharmonique et $\nu_{p\varphi_1}(0) = 0$. D'après le Corollaire 2.1. $e^{-p\varphi_1}$ est sommable au voisinage de 0 . Donc $e^{-\varphi}$ est sommable au voisinage de 0 .

C.Q.F.D.

PROPOSITION 2.2. (P.LELONG [3]).

G un domaine pseudo-convexe. Soit $G_\rho \subset\subset G$ et $M(G_\rho) = \sup V(z); z \in G_\rho$. Alors pour $\|y\| < \rho$, $x \in G_{2\rho}$, $r < \rho$ posons $\text{Reg sup}_x f(x) = \lim \sup f(x')$, pour $x' \to x$

On a : $\nu(x) = \lim\limits_{r\to o} \frac{1}{\log r}[\lambda(V,x,r) - M(G_\rho)]$. $r < d(x,G_\rho)$, bG_ρ désigne le bord de G_ρ .

$$\nu(\delta,\xi) = \lim_{r\to o} \frac{1}{\log r}\left[\frac{1}{2\pi}\int_o^{2\pi} V(\delta + r\xi e^{i\varphi})d\varphi - M(G_\rho)\right].$$

a/ $\nu(\delta, \xi)$ est définie pour $\delta \in G_{2\rho}$; $\|\xi\| < \rho$, on a $-\nu(\delta, \xi) = -\infty$ si et seulement si $\nu(\delta + u\xi) = -\infty$.

b/ $-\nu(\delta,\xi)$ est limite croissante de fonctions plurisousharmoniques de (δ, ξ) négatives.

c/ On a $\nu(\delta, \xi) \geqslant \nu(\delta)$ et $\text{Reg Sup}_\xi[-\nu(\delta,\xi)] = -\nu(\delta)$ et pour chaque $x \in G$, l'ensemble des ξ où $\nu(\delta, \xi)$ est supérieur à $\nu(\delta)$ est un cône polaire dans $\mathbb{C}^n$.

COROLLAIRE 2.2.

Soit $\varphi(\delta, \xi) = \log(|\delta|^2 + |\xi|^4)$ pour $(\delta, \xi) \in \mathbb{C}\times\mathbb{C}$. Alors $\nu_\varphi(0) = 2$.

Démonstration.

$$\frac{1}{2\pi}\int_o^{2\pi} \varphi(r(\delta,\xi)\,.\,e^{i\Theta})d\Theta = \log(|\delta|^2 + r^2|\xi|^4) + \log r^2 .$$

Donc pour $\delta \neq 0$; $\nu((0,0),(x,y)) = \lim\limits_{r\to 0}\left[2 \frac{2\log r - N(G)}{\log r}\right] + \lim\limits_{r\to 0}\left(\frac{|x|^2 + r^2|y|^2}{\log r}\right)$

et $\lim\limits_{r\to 0} \frac{\log|x|^2 + r^2|y|^4}{\log r} = 0$ car $x \neq 0$.

Donc sauf pour une seule direction $(x = 0)$

$\nu((0,0),(x,y)) = 2$, donc $\nu_\varphi(0) = 2$.

PROPOSITION 2.3.

Dans $\mathbb{C}^2$ la fonction plurisousharmonique φ ; $\varphi = \log(|x|^2 + |y|^4)$ vérifie :

1°/ $\varphi - 2\log|z|$ n'est pas une fonction plurisousharmonique

2°/ pour toute fonction φ_o plurisousharmonique telle que $\nu_{\varphi_o} = 0$ dans $\mathbb{C}^2$ la fonction $\psi = \varphi - 2\log|z| + \varphi_o$ n'est pas plurisousharmonique (i.e. n'est pas "presque plurisousharmonique").

Démonstration.

La démonstration repose essentiellement sur la proposition 2.1.

On suppose que la fonction ψ est plurisousharmonique (avec φ_o une fonction psh telle que $\nu_{\varphi_o} = 0$. Donc $\nu_\psi(0) = 0$ (1)

et $\frac{3}{2}\psi + 3\log|z| = \frac{3}{2}\varphi + \frac{3}{2}\varphi_o$ (2)

$3\log|z|$ vérifie l'hypothèse (b) de la proposition 2.1. car $\alpha\log|z|$ est sommable au voisinage de 0 dans $\mathbb{C}^2$ si et seulement si $\alpha < 4$.

Donc d'après (2), (1) et la proposition 2.1. $\exp(-\frac{3}{2}\psi - 3\log|z|)$ est sommable au voisinage de 0.

$\nu_{\varphi_o}(0) = 0$. Si $\varphi_o > 0$, elle est bornée et la sommabilité de $\exp(-\frac{3}{2}\varphi - \frac{3}{2}\varphi_o)$ au voisinage de 0 est équivalente à la sommabilité de $\exp(-\frac{3}{2}\varphi)$ au voisinage de 0. Si $\varphi_o \leq 0$ $\exp(-\frac{3}{2}\varphi - \frac{3}{2}\varphi_o) \geq \exp(-\frac{3}{2}\varphi)$. Mais $\exp(-\frac{3}{2}\varphi)$ n'est pas sommable au voisinage de 0, en effet : pour $z \in \mathbb{C}^2$; $z = (x,y)$.

$$I = \int_{B(0,r)} \frac{d\lambda(z)}{(|x|^2+|y|^4)^{3/2}} = \int_{B(0,r)\subset\mathbb{C}} \left[\int_{B(0,r')\subset\mathbb{C}} \frac{d\lambda(x)}{(|x|^2+|y|^4)^{3/2}}\right] d\lambda(y)$$

avec $r'^2 = r^2 - |y|^2$.

Par le changement de coordonnées polaires

$$I = \pi . \int_{B(0,r)} \left[\int_0^{r'} \frac{2t\,dt}{(t^2+y^4)^{3/2}}\right] d(y) = 2\pi . \int_{B(0,r)\subset\mathbb{C}} \left[\frac{1}{y^2} - \frac{1}{(r^2-y^2+y^4)^{1/2}}\right]$$

$$= 2\pi \int_{B(0,r)} \frac{d\lambda(y)}{|y|^2} - 2\pi \int_{B(0,r)} \frac{d\lambda(y)}{(r^2-|y|^2+|y|^4)^{1/2}} .$$

Le terme $\int_{B(0,r)} \frac{d\lambda(y)}{(r^2-|y|^2+|y|^4)^{1/2}}$ est fini car $r^2 - y^2 + y^4$ ne s'annule pas pour r assez petit. Le terme $\int_{B(0,r)\subset\mathbb{C}} \frac{d\lambda(y)}{|y|^2}$ est infini. Donc $I = +\infty$.

Donc ψ n'est pas une fonction plurisousharmonique .

Remarques.

1°/ On a le même résultat que dans la proposition 2.3. pour la fonction φ plurisousharmonique suivante dans $\mathbb{C}^n$, $\varphi(z_1,\dots,z_n) = |z_1|^2 + \dots + |z_{n-1}|^2 + |z_n|^4$,
$\varphi - 2\log|z|$ n'est pas une fonction "presque plurisousharmonique et $\nu_\varphi(0) = 2$.

2°/ L'hypothèse b de la proposition 2.1. est vérifiée pour toute fonction φ plurisousharmonique dans $\mathbb{C}$.

En effet d'après le corollaire 2.1. $n = 1$,
$e^{-\varphi}$ est sommable au voisinage de 0 est équivalent à $\nu_\varphi(0) < 2$ et l'application $(x \longmapsto \nu_\varphi(x)$ est semi-continue supérieurement.

PROPOSITION 2.4.

Soit X l'ensemble analytique dans $\mathbb{C}^n$ défini par $X = \{z \in \mathbb{C}^n / z_1 = \dots = z_k = 0\}$ $(p+k = n)$.

Soit $\varphi = \log(|z_1|^2 + \dots + |z_{k-1}|^2 + |z_k|^4)$. φ est une fonction plurisousharmonique. Si U désigne le potentiel canonique associé à X. Alors $\varphi - 2U$ n'est pas une fonction "presque plurisousharmonique".

Démonstration.

La démonstration est identique à celle de la proposition 2.3. Elle repose essentiellement sur la proposition 2.1.

Si φ_o est une fonction plurisousharmonique dans $\mathbb{C}^n$ telle que $\nu_{\varphi_o}(0) = 0$. Si $\psi = \varphi - 2U + \varphi_o$ est une fonction plurisousharmonique, alors $\nu_\psi(0) = 0$ et

$$\frac{2k-1}{2}\psi + (2k-1)U = \frac{2k-1}{2}\varphi + \frac{2k-1}{2}\varphi_o .$$

$U(z) = \frac{1}{2}\log(|z_1|^2 + \dots + |z_k|^2)$. $U(z)$ vérifie l'hypothèse (b) de la proposition 2.1. et $\exp(-(2k-1)U)$ est sommable au voisinage de 0 ; d'après le théorème 2.2. on aurait donc $\exp(-\frac{2k-1}{2}\varphi)$ localement sommable. Mais $\exp(-\frac{2k-1}{2}\varphi)$ n'est pas sommable au voisinage de 0. En effet, par récurrence :

1°/ $k = 2$

$$I = \int_{B(0,r)} \frac{d\lambda(z)}{(|z_1|^2+|z_2|^4)^{3/2}} = +\infty \text{ avec } z = (z_1, z_2, \dots, z_n).$$

(Le même calcul fait dans la démonstration de la proposition 2.3.).

2°/ On suppose le résultat vrai pour $k-1$ et on veut le démontrer pour k

$$I = \int_{B(0,r)} \frac{d\lambda(z)}{(|z_1|^2 + \dots + |z_{k-1}|^2 + |z_k|^4)^{k-1/2}}$$

$$= \int_{B(0,r)\subset\mathbb{C}^{n-1}} d\lambda(z') \left[\int_{B(0,r')} \frac{d\lambda(z_1)}{(|z_1|^2 + \dots + |z_{k-1}|^2 + |z_k|^4)^{k-1/2}} \right]$$

avec $z' = (z_2, \dots, z_n)$ et $r'^2 = r^2 - (|z_2|^2 + \dots + |z_n|^2)$.

$$\text{Donc } I = \frac{\pi}{k-3/2}\left[\int_{B(0,r)} \frac{d\lambda(z')}{(|z_2|^2+\dots+|z_k|^4)^{k-1-1/2}} - \int_{B(0,r)} \frac{d\lambda(z')}{(r^2-|z_k|^2+|z_k|^4)^{k-3/2}}\right]$$

$r^2 - |z_k|^2 + |z_k|^4$ ne s'annule pas. Donc le terme suivant $\displaystyle\int_{B(0,r)} \frac{d\lambda(z')}{(|r^2-|z_k|^2+|z_k|^4)^{k-3/2}}$

est fini et le terme $\int_{B(0,r)} \frac{d\lambda(z')}{(|z_2|^2 + \ldots + |z_k|^4)^{k-3/2}}$ est infini (par récurrence).

Donc $I = +\infty$.

On déduit de ce travail que si φ est une fonction psh, et si $X_\varphi = \{x / \nu_\varphi(x) > 0\}$ est un ensemble analytique de dimension pure p, alors $\varphi - cV$ n'est pas toujours "presque psh", avec V le potentiel associé à X. De ce résultat négatif on pose trois problèmes qui sont liés à cette question dans le but de bien savoir le comportement de φ et de ν_φ au voisinage des points singuliers de X.

Ier problème. Si φ est une fonction psh. A-t-on $\nu_\varphi \geqslant c\,\nu_V$ avec $c = \inf_{x \in X_\varphi} \varphi(x)$ et V le potentiel associé à X_φ, X de dimension pure p.

2e problème. Si φ est une fonction psh et X_φ de dimension pure p, existe-t-il un courant Θ positif fermé de bidimension (p,p) telle que $\nu_\Theta = \nu_\varphi$.

3e problème. Si Θ est un courant positif fermé de bidimension (p,p), existe-t-il un courant Θ_q positif fermé de bidimension (q,q) telle que $\nu_\Theta = \nu_{\Theta_q}$, pour tout $q \geqslant p$.

Il est facile de voir que le problème 2 entraîne le problème 1 et 3.

On va montrer que le problème 2 est faux en donnant un exemple qui ne vérifie pas le problème 1. Le problème 1 est une condition nécessaire pour le problème 2.

Le problème 3 reste alors ouvert.

Exemple. Dans $\mathbb{C}^3$ on considère la courbe gauche

$X = \{z \in \mathbb{C}^3 / z = (t^2, t^3, t^5)\ ;\ t \in \mathbb{C}\}$. X est irréductible ; $\nu_X(0) = 2$, et 0 est le seul point singulier de X.

On considère le courant $\omega = \frac{i}{2\pi} d'd'' \log(|x^3 - y^2|^2 + |xy - z|^2)$
$= 1/2 \log(|x^3 - y^2|^2 + |xy - z|^2)$.

Remarque. X est intersection complète et

$$\begin{cases} f_1 = xy - z \\ f_2 = x^3 - y^2 \end{cases}$$ engendrent l'idéal $I(X)$.

LEMME 1.

$$\nu_\varphi(x) = 1 \text{ sur } X.$$

LEMME 2.

Soient $f_1, \ldots, f_m$ des fonctions analytiques sur $\mathbb{C}^n$ et $\varphi = 1/2 \log(|f_1|^2 + \ldots + |f_m|^2)$.

Alors $\nu_\varphi(x) = \inf_j \nu_{f_j}(x)$.

La démonstration du lemme est faite dans un article nouveau de P.Lelong [10] .

Le lemme 2 entraîne le lemme 1.

Donc on a

$$1 = \nu_\varphi(0) < \nu_X(0) = 2.$$

Donc le problème 1 n'est pas vérifié pour $\varphi = 1/2 \log |xy - z|^2 + |x^3-y^2|^2$.

Donc de même pour le problème 2.

Conclusions.

1°/ Soit φ une fonction plurisousharmonique dans $\mathbb{C}^n$ telle que l'ensemble des points de densité de φ soit un ensemble analytique de dimension pure p ; (p+k = n) .

On note $X_\varphi = \{x \ / \ \nu_\varphi(x) > 0\}$.

Il résulte du théorème de Y.T.SIU et du résultat de H.SKODA que si Θ est un courant (k,k) positif fermé et W un potentiel plurisousharmonique associé à Θ, ($\nu_W = \nu_\Theta$) , si W' est un potentiel plurisousharmonique associé à X ($\nu_{W'} = \nu_X$) , si $c = \inf_{x \in X} \nu_\Theta(x)$, alors

i) $\Theta - c\,[X]$ est un courant positif fermé

ii) $W - c\,W'$ est une fonction presque plurisousharmonique.

On vient de montrer que le résultat (ii) ne peut être généralisé (au cas où $\dim X_\varphi < n-1$ et X_φ pure) de la manière suivante . $\varphi - c\,W'$ n'est pas "presque plurisousharmonique" en général avec $c = \inf_{x \in X} \nu_\varphi(x)$.

2°/ On considère une fonction φ plurisousharmonique vérifiant les hypothèses de la première conclusion . Alors les propositions 2.3. et 2.4. montrent que même si $\nu_\varphi(x_o) < 2k$ (avec codimension de X en x_o est k) $k \neq 1$; $e^{-\varphi}$ n'est pas localement sommable.

3°/ Le meilleur résultat possible entre la liaison de ν_φ en x_o et la sommabilité de $e^{-\varphi}$ est celui de H.SKODA à savoir le corollaire 2.1.

BIBLIOGRAPHIE

[1] HÖRMANDER (L.). - An introduction to complex analysis in several variables. New York, Van Nostrand Company, 1966.

[2] LELONG (P.). - Intégration sur un ensemble analytique complexe. Bull. Soc. Math. de France, 85, pp. 239-262, 1957.

[3] LELONG (P.). - Sur la structure des courants positifs fermés. Séminaire P.Lelong, 1975-76, p. 136-156.

[4] LELONG (P.). - Fonctions entières (n variables) et fonctions plurisousharmoniques d'ordre fini dans $\mathbb{C}^n$. J. Analyse Math. Jérusalem, t. 12, 1964, p. 365-407.

[5] LELONG (P.). - Fonctions plurisousharmoniques et formes différentielles positives. Paris, Londres, New York, Gordon and Breach, Dunod, 1968.

[6] SKODA (H.). - Croissance des fonctions entières s'annulant sur une hypersurface donnée de $\mathbb{C}^n$. Séminaire P.Lelong, 1970-71, p. 82-105.

[7] SKODA (H.). - Nouvelle méthode pour l'étude des potentiels associés aux ensembles analytiques. Séminaire P.Lelong, 1972-73, p. 117, 141.

[8] SKODA (H.). - Sous-ensembles analytiques d'ordre fini ou infini dans $\mathbb{C}^n$. Bull. Soc. Math. France, 100, p. 353-408, 1972.

[9] SIU (Y.T.). - Analycity of sets associated to Lelong number and the extention of closed positive currents. Inv. Math. , t. 27, p. 53-156, 1974.

[10] LELONG (P.). - Ensembles analytiques complexes définis comme ensembles de densité (à paraître).

RELATIONS ENTRE LES DIFFÉRENTES NOTIONS DE FIBRÉS ET DE COURANTS POSITIFS.

par J.-P. DEMAILLY

0. INTRODUCTION.

Nous nous proposons de généraliser les résultats de l'article [2], consacré à l'étude des relations entre les notions de positivité de P.A. Griffiths et de S. Nakano pour les fibrés vectoriels. Etant donné une forme hermitienne θ sur un produit tensoriel $T \otimes E$, il y a trois manières naturelles de définir la positivité de θ, calquées sur les définitions usuelles concernant les courants positifs. Dans le cas où θ est la forme de courbure d'un fibré vectoriel holomorphe hermitien E au-dessus d'une variété analytique X, on retrouve les notions de positivité de P.A. GRIFFITHS [4] et de S. NAKANO [6] relatives aux fibrés, ainsi qu'une troisième notion de positivité plus restrictive, appelée ici positivité forte. Notre objectif essentiel est la démonstration du résultat suivant, contenu implicitement dans [2] : si le fibré E est positif au sens de Griffiths, alors le fibré $E \otimes \det E$ est positif fortement (donc aussi au sens de Nakano). Ce type de résultat est lié étroitement aux calculs de courbure intervenant dans la théorie des morphismes surjectifs de fibrés vectoriels semi-positifs de H. SKODA [8] (cf. aussi [1]). Nous montrons dans le dernier paragraphe comment ces techniques peuvent s'appliquer aux formes et aux courants pour établir des relations entre positivité faible et forte.

1. FORMES HERMITIENNES POSITIVES SUR UN PRODUIT TENSORIEL.

Soit θ une forme hermitienne sur un produit tensoriel $T \otimes E$ d'espaces vectoriels complexes.

DÉFINITION 1. θ sera dite

(1) semi-positive au sens de Griffiths, si pour tout vecteur décomposable $x \in T \otimes E$, $x = \xi \otimes u$, avec $\xi \in T$, $u \in E$, on a

$$\theta(x,x) \geqslant 0 ;$$

(2) semi-positive au sens de Nakano, si elle est semi-positive au sens usuel sur $T \otimes E$, c'est-à-dire si

$$\theta(x,x) \geqslant 0 \quad \text{pour tout} \quad x \in T \otimes E ;$$

(3) semi-positive fortement, si on peut écrire

$$\theta(x,x) = \sum_{j=1}^{N} |x_j^\star(x)|^2$$

pour une famille finie $\{x_j^\star\}_{1 \leqslant j \leqslant N}$ de formes linéaires $x_j^\star$ décomposables sur $T \otimes E$, $x_j^\star = \xi_j^\star \otimes u_j^\star$, avec $\xi_j^\star \in T^\star$, $u_j^\star \in E^\star$.

On désignera par $\geqslant_G$, $\geqslant_N$, $\geqslant_S$ les inégalités de semi-positivité de Griffiths, de Nakano, et de semi-positivité forte. On dira que θ est (strictement) positive, et on écrira respectivement $\theta >_G 0$, $\theta >_N 0$, $\theta >_S 0$ si toute petite perturbation de θ est encore semi-positive dans le sens considéré.

Il est clair que $\theta \geqslant_S 0$ entraîne $\theta \geqslant_N 0$, et que $\theta \geqslant_N 0$ entraîne $\theta \geqslant_G 0$, mais les réciproques sont fausses en général comme on le verra au § 2. Les trois notions coïncident toutefois si l'un des espaces E ou T est de dimension 1.

On suppose maintenant que l'espace E est muni d'une forme hermitienne définie positive φ ; on désigne par n la dimension de T, par r celle de E, et on définit $\mathrm{Tr}_E\,\theta$ comme la forme hermitienne sur T telle que

$$\mathrm{Tr}_E\,\theta\,(\xi,\xi') = \sum_{j=1}^{r} \theta(\xi \otimes e_j \,,\, \xi' \otimes e_j)$$

pour toute base orthonormée $(e_j)_{1\leq j\leq r}$ de E, et tout couple $(\xi,\xi') \in T^2$; la forme $\mathrm{Tr}_E\ \theta$ est indépendante de la base orthonormée (e_j) choisie, et elle est semi-positive dès que $\theta \geqslant_G 0$. Les semi-positivités forte et de Griffiths sont reliées par le théorème suivant.

THÉORÈME 1. - Si la forme hermitienne θ sur $T \otimes E$ est semi-positive au sens de Griffiths, alors la forme

$$\theta + \mathrm{Tr}_E\ \theta \otimes \varphi$$

est semi-positive fortement (donc aussi au sens de Nakano).

La démonstration sera une conséquence aisée du lemme suivant.

LEMME 1. - Soient q un entier $\geqslant 3$, u_j et v_k, $1 \leqslant j,k \leqslant r$ des nombres complexes. σ décrivant l'ensemble $\mathfrak{F}$ des applications de $\{1,2,\ldots,r\}$ dans le groupe des racines q-ièmes de l'unité, on pose

$$u'_\sigma = \sum_{\ell=1}^{r} u_\ell\ \overline{\sigma(\ell)}\ ,\quad v'_\sigma = \sum_{m=1}^{r} v_m\ \overline{\sigma(m)}\ .$$

Alors pour tout couple (j,k), $1 \leqslant j,k \leqslant r$, on a l'identité

$$q^{-r} \sum_{\sigma\in\mathfrak{F}} u'_\sigma\ \overline{v'_\sigma}\ \sigma(j)\ \overline{\sigma(k)} = u_j\ \overline{v_k} \quad \text{si } j \neq k$$

$$= \sum_{\ell=1}^{r} u_\ell\ \overline{v_\ell} \quad \text{si } j = k\ .$$

<u>Démonstration.</u> Le coefficient de $u_\ell\ \overline{v_m}$ dans la quantité

$$q^{-r} \sum_{\sigma\in\mathfrak{F}} u'_\sigma\ \overline{v'_\sigma}\ \sigma(j)\ \overline{\sigma(k)}$$

est donné par

$$q^{-r} \sum_{\sigma\in\mathfrak{F}} \sigma(j)\ \overline{\sigma(k)}\ \overline{\sigma(\ell)}\ \sigma(m)\ .$$

Ce coefficient vaut 1 lorsque les paires $\{j,m\}$ et $\{k,\ell\}$ coïncident (puisqu'alors $\sigma(j)\ \overline{\sigma(k)}\ \overline{\sigma(\ell)}\ \sigma(m) = 1$ pour chacun des q^r éléments $\sigma \in \mathfrak{F}$).

Il s'agit de montrer que

$$\sum_{\sigma\in\mathfrak{F}} \sigma(j)\ \overline{\sigma(k)}\ \overline{\sigma(\ell)}\ \sigma(m) = 0$$

lorsque les paires $\{j,m\}$, $\{k,\ell\}$ sont distinctes.

Si $\{j,m\} \neq \{k,\ell\}$, l'un des éléments de l'une des paires n'appartient pas à l'autre paire. Comme les quatre indices j,k,ℓ,m jouent le même rôle (quitte à changer éventuellement σ en $\bar{\sigma}$) , on peut supposer par exemple que j n'appartient pas à $\{k,\ell\}$.

Effectuons sur σ la substitution $\sigma \mapsto \tau$, où τ est défini par

$$\tau(j) = e^{\frac{2i\pi}{q}} \sigma(j) \ , \ \tau(s) = \sigma(s) \quad \text{pour} \quad s \neq j \ .$$

On obtient

$$\begin{aligned}\sum_{\sigma \in \mathfrak{F}} \sigma(j)\,\overline{\sigma(k)}\,\overline{\sigma(\ell)}\,\sigma(m) &= \sum_{\tau \in \mathfrak{F}} \\ &= e^{\frac{2i\pi}{q}} \sum_{\sigma \in \mathfrak{F}} \qquad \text{si} \quad j \neq m \\ &= e^{\frac{4i\pi}{q}} \sum_{\sigma \in \mathfrak{F}} \qquad \text{si} \quad j = m \ .\end{aligned}$$

Comme $q \geqslant 3$ par hypothèse, il en résulte bien

$$\sum_{\sigma \in \mathfrak{F}} \sigma(j)\,\overline{\sigma(k)}\,\overline{\sigma(\ell)}\,\sigma(m) = 0 \ .$$

<u>Démonstration du théorème 1</u> .

Etant donné une base de T et une base orthonormée (e_j) de E , on désigne par (ξ_λ) , $1 \leqslant \lambda \leqslant n$, les coordonnées de $\xi \in T$, par (u_j) , $1 \leqslant j \leqslant r$, celles de $u \in E$, et par $(x_{\lambda j})$ celles de $x \in T \otimes E$.

Si les nombres complexes $a_{\lambda\mu jk}$ sont les coefficients de θ (avec $\bar{a}_{\lambda\mu jk} = a_{\mu\lambda kj}$) , on a les formules

$$\theta(\xi \otimes u, \xi \otimes u) = \sum_{\lambda,\mu,j,k} a_{\lambda\mu jk}\, \xi_\lambda\, \bar{\xi}_\mu\, u_j\, \bar{u}_k \ ,$$

$$\theta(x,x) = \sum_{\lambda,\mu,j,k} a_{\lambda\mu jk}\, x_{\lambda j}\, \bar{x}_{\mu k} \ ,$$

$$\mathrm{Tr}_E\, \theta \otimes \varphi\,(x,x) = \sum_{\lambda,\mu,j,k} a_{\lambda\mu jj}\, x_{\lambda k}\, \bar{x}_{\mu k} \ ,$$

avec $\quad 1 \leqslant \lambda,\mu \leqslant n \ , \quad 1 \leqslant j,k \leqslant r \ .$

Par hypothèse, $\theta(\xi \otimes u, \xi \otimes u)$ est $\geqslant 0$.

σ décrivant comme dans le lemme 1 l'ensemble $\mathcal{F}$ des applications de $\{1,\ldots,r\}$ dans le groupe des racines q-ièmes de l'unité, on pose

$$x'_{\lambda\sigma} = \sum_{\ell=1}^{r} x_{\lambda\ell}\, \overline{\sigma(\ell)} \ .$$

D'après le lemme 1 , on a

$$q^{-r} \sum_{\sigma\in\mathcal{F}} \sum_{\lambda,\mu,j,k} a_{\lambda\mu jk}\, x'_{\lambda\sigma}\, \overline{x'_{\mu\sigma}}\, \sigma(j)\, \overline{\sigma(k)}$$

$$= \sum_{\lambda,\mu,j\neq k} a_{\lambda\mu jk}\, x_{\lambda j}\, \overline{x_{\mu k}} + \sum_{\lambda,\mu,j,k} a_{\lambda\mu jj}\, x_{\lambda k}\, \overline{x_{\mu k}}$$

$$= \theta(x,x) + \mathrm{Tr}_E\, \theta \otimes \varphi(x,x) - \sum_{\lambda,\mu,j} a_{\lambda\mu jj}\, x_{\lambda j}\, \overline{x_{\mu j}} \ .$$

On obtient donc

$$\theta(x,x) + \mathrm{Tr}_E\, \theta \otimes \varphi(x,x) =$$

$$q^{-r} \sum_{\sigma\in\mathcal{F}} \sum_{\lambda,\mu,j,k} a_{\lambda\mu jk}\, x'_{\lambda\sigma}\, \overline{x'_{\mu\sigma}}\, \sigma(j)\, \overline{\sigma(k)} + \sum_{\lambda,\mu,j} a_{\lambda\mu jj}\, x_{\lambda j}\, \overline{x_{\mu j}} \geqslant 0$$

d'après l'hypothèse de positivité de Griffiths de θ . Il nous reste à vérifier que le second membre est somme de carrés de formes linéaires décomposables sur $T \otimes E$. Par hypothèse, la forme hermitienne de coefficients $\left(\sum_{j,k} a_{\lambda\mu jk}\, \sigma(j)\, \overline{\sigma(k)}\right)_{\lambda,\mu}$ est semi-positive sur T , donc somme de carrés de formes linéaires $\xi^\star_{\nu\sigma} \in T^\star$, $1 \leqslant \nu \leqslant n$. De même la forme de coefficients $(a_{\lambda\mu jj})_{\lambda,\mu}$ est somme de carrés de formes linéaires $\xi^\star_{\nu j} \in T^\star$, $1 \leqslant \nu \leqslant n$, $1 \leqslant j \leqslant r$.

Pour tout vecteur décomposable $x = \xi \otimes u \in T \otimes E$, on peut écrire, en notant $e_\sigma = \sum_{j=1}^{r} \sigma(j)\, e_j$:

$$x'_{\lambda\sigma} = \xi_\lambda\, \varphi(u,e_\sigma) \ ,$$

$$\theta(x,x) + \mathrm{Tr}_E\, \theta \otimes \varphi(x,x) =$$

$$q^{-r} \sum_{\sigma\in\mathcal{F}} \sum_{\nu=1}^{n} |\xi^\star_{\nu\sigma}(\xi)|^2\, |\varphi(u,e_\sigma)|^2$$

$$+ \sum_{j=1}^{r} \sum_{\nu=1}^{n} |\xi^\star_{\nu j}(\xi)|^2\, |\varphi(u,e_j)|^2 \ ,$$

de sorte que

$$\theta + Tr_E\, \theta \otimes \varphi = q^{-r} \sum_{\sigma \in \mathcal{F}} \sum_{\nu=1}^{n} |\xi^{\star}_{\nu\sigma} \otimes \varphi(?,e_\sigma)|^2$$

$$+ \sum_{j=1}^{r} \sum_{\nu=1}^{n} |\xi^{\star}_{\nu j} \otimes \varphi(?,e_j)|^2 \ .$$

La démonstation est achevée. Le corollaire qui suit est une généralisation du lemme fondamental (3,5) de H. SKODA [8], relatif au cas où $-\theta$ est la forme de courbure d'un sous-fibré E d'un fibré trivial.

COROLLAIRE 1. - Si la forme hermitienne θ est semi-positive au sens de Griffiths sur $T \otimes E$, où $\dim T = n$, $\dim E = r$, alors

$$\theta \leqslant_S \mathrm{Inf}(n,r)\ .\ Tr_E\, \theta \otimes \varphi\ .$$

<u>Démonstration</u>. Montrons tout d'abord le

LEMME 2. - $Tr_E\, \theta \otimes \varphi - \theta \geqslant_G 0$.

En effet, tout vecteur décomposable $x \in T \otimes E$ peut s'écrire $x = \xi \otimes u$ où $\|u\|^2 = \varphi(u,u) = 1$; si l'on choisit une base orthonormée $(e_j)_{1\leqslant j\leqslant r}$ de E telle que $e = u$, il vient

$$\theta(x,x) = \theta(\xi \otimes e_1, \xi \otimes e_1)\ ,$$

$$Tr_E \theta \otimes \varphi(x,x) = \sum_{j=1}^{r} \theta(\xi \otimes e_j, \xi \otimes e_j)\ \|u\|^2 \geqslant \theta(x,x)\ ,$$

grâce à l'hypothèse $\theta \geqslant_G 0$. Le lemme 2 est démontré. ■

D'après le théorème 1, on a donc

$$Tr_E\, \theta \otimes \varphi - \theta + Tr_E(Tr_E\, \theta \otimes \varphi - \theta) \otimes \varphi = r\, Tr_E\, \theta \otimes \varphi - \theta \geqslant_S 0\ .$$

Il nous reste à montrer qu'on a également

$$\theta \leqslant_S n\, Tr_E\, \theta \otimes \varphi\ ,$$

ce qui est plus difficile.

Munissons T de la forme hermitienne semi-positive $\omega = Tr_E\, \theta$, que nous supposons pour l'instant <u>non dégénérée</u>. Soit $\hat{\theta} = \omega \otimes \varphi - \theta \geqslant_G 0$ la forme considérée dans le lemme 2. Les coefficients de $\hat{\theta}$, relativement à un couple de bases orthonormées de T et E, sont donnés en fonction des coefficients $a_{\lambda\mu jk}$ de θ et des symboles de Kronecker $\delta_{\lambda\mu}$, δ_{jk}, par

$$\hat{a}_{\lambda\mu jk} = \delta_{\lambda\mu}\,\delta_{jk} - a_{\lambda\mu jk} .$$

On applique le procédé de sommation du lemme 1, mais cette fois par rapport à l'espace T (indices λ et μ). Si $\mathcal{F}$ est l'ensemble des applications de $\{1,\dots,n\}$ dans le groupe des racines q-ièmes de l'unité, et si $1 \leqslant \lambda,\mu \leqslant n$, $1 \leqslant j,k \leqslant r$, il vient d'après le lemme 1 :

$$q^{-n} \sum_{\sigma\in\mathcal{F}} \sum_{\lambda,\mu,j,k} \hat{a}_{\lambda\mu jk}\, x'_{\sigma j}\, \overline{x'_{\sigma k}}\, \sigma(\lambda)\, \overline{\sigma(\mu)}$$

$$= \sum_{\lambda\neq\mu,j,k} \hat{a}_{\lambda\mu jk}\, x_{\lambda j}\, \overline{x_{\mu k}} + \sum_{\lambda,\mu,j,k} \hat{a}_{\lambda\lambda jk}\, x_{\mu j}\, \overline{x_{\mu k}}$$

$$= - \sum_{\lambda\neq\mu,j,k} a_{\lambda\mu jk}\, x_{\lambda j}\, \overline{x_{\mu k}} - \sum_{\lambda,\mu,j,k} a_{\lambda\lambda jk}\, x_{\mu j}\, \overline{x_{\mu k}} + n \sum_{\mu,j} |x_{\mu j}|^2 ,$$

ce qui donne

$$n \sum_{j,\mu} |x_{\mu j}|^2 - \sum_{\lambda,\mu,j,k} a_{\lambda\mu jk}\, x_{\lambda j}\, \overline{x_{\mu k}}$$

$$= q^{-n} \sum_{\sigma\in\mathcal{F}} \sum_{\lambda,\mu,j,k} \hat{a}_{\lambda\mu jk}\, \sigma(\lambda)\, \overline{\sigma(\mu)}\, x'_{\sigma j}\, \overline{x'_{\sigma k}} + \sum_{\lambda\neq\mu,j,k} a_{\lambda\lambda jk}\, x_{\mu j}\, \overline{x_{\mu k}} .$$

Comme dans la démonstration du théorème 1, on voit donc que

$$\theta \leqslant_S n\,\omega \otimes \varphi = n\, \mathrm{Tr}_E\, \theta \otimes \varphi .$$

Lorsque ω est dégénérée, de noyau K, il est facile de voir grâce au lemme 2 que θ induit une forme hermitienne Θ sur $T/K \otimes E$. En remplaçant θ par Θ, et n par $N = \dim T/K \leqslant n$, on obtient

$$\Theta \leqslant_S N \,.\, \mathrm{Tr}_E\, \Theta \otimes \varphi ,$$

ce qui entraîne

$$\theta \leqslant_S N \,.\, \mathrm{Tr}_E\, \theta \otimes \varphi \leqslant_S n \,.\, \mathrm{Tr}_E\, \theta \otimes \varphi .$$

Nous allons voir maintenant comment ces notions se traduisent dans le cadre des fibrés vectoriels hermitiens.

2. FIBRÉS POSITIFS.

Si E est un fibré vectoriel holomorphe hermitien au dessus d'une variété analytique complexe X , on peut définir une connexion canonique D sur E , hermitienne et holomorphe (cf. A. DOUADY et J.L. VERDIER [3], P.A. GRIFFITHS [4]). D envoie l'espace $C^{\infty}_{p,q}(X,E)$ des formes de type (p,q) à valeurs dans E , dans l'espace $C^{\infty}_{p+1,q}(X,E) \oplus C^{\infty}_{p,q+1}(X,E)$; la forme de courbure $c(E)$ du fibré E est alors définie par la propriété suivante

$$D^2 u = c(E).u$$

pour toute section C^{∞} u de E , de sorte que $i\,c(E)$ est une $(1,1)$ forme à valeurs dans le fibré Herm (E,E) des endomorphismes hermitiens de E . On identifiera $i\,c(E)$ à la forme hermitienne θ sur $TX \otimes E$ qui lui est canoniquement associée.

DÉFINITION 2. - Le fibré E est dit semi-positif (respectivement positif) au sens de Griffiths, au sens de Nakano, ou au sens fort, s'il en est ainsi pour la forme hermitienne θ sur chaque fibre $T_zX \otimes E_z$, $z \in X$.

La forme de courbure $c(E^{\star})$ du fibré dual $E^{\star}$ est donnée par

$$c(E^{\star}) = -\ {}^{t}c(E) ,$$

où ${}^{t}c(E) \in \mathrm{Herm}\,(E^{\star},E^{\star})$ désigne l'endomorphisme transposé de $c(E)$. Le lecteur en déduira aisément la proposition suivante.

PROPOSITION 1. Le fibré E est (semi-) positif au sens de Griffiths (resp. au sens fort) si et seulement si le fibré dual $E^{\star}$ est (semi-) négatif au sens de Griffiths (resp. au sens fort).

Le résultat analogue pour la positivité de Nakano n'est pas vrai (voir l'exemple ci-dessous).

Il est classique d'autre part (P.A. GRIFFITHS [4]) qu'un fibré quotient d'un fibré $E >_G 0$ est encore positif au sens de Griffiths. De même un sous-fibré d'un fibré $E <_G 0$ est $<_G 0$. On peut vérifier que cette deuxième propriété subsiste au sens de Nakano ; aucune par contre n'est vraie au sens fort en général. Les diverses notions sont reliées grâce au théorème 1 et au corollaire 1, qui impliquent le

<u>THÉORÈME 2</u>. - Soit E un fibré hermitien (semi-) positif au sens de Griffiths. On désigne par r le rang de E et par n la dimension de la variété X. Alors les fibrés

$$E \otimes \text{dét } E \ , \ E^{\star} \otimes (\text{dét } E)^{\text{Inf}(n,r)}$$

sont fortement (semi-) positifs. En particulier, ils sont (semi-) positifs au sens de Nakano, et les fibrés

$$E^{\star} \otimes (\det E)^{-1} \ , \ E \otimes (\text{dét } E)^{-\text{Inf}(n,r)}$$

sont (semi-) négatifs au sens de Nakano.

<u>Démonstration</u>. Il est bien connu que la courbure du fibré $\text{dét } E = \Lambda^r E$ est reliée à la courbure de E par la formule

$$c(\text{dét } E) = \text{Tr}_E\, c(E) = - \text{Tr}_{E^\star} c(E^{\star}) \ ,$$

et que pour deux fibrés vectoriels hermitiens E_1 et E_2, on a

$$c(E_1 \otimes E_2) = c(E_1) \otimes \text{Id}_{E_2} + \text{Id}_{E_1} \otimes c(E_2).$$

Le théorème 2 se déduit alors du théorème 1 et de son corollaire en prenant successivement $\theta = i\, c(E)$, $\theta = -\, i\, c(E^{\star}) = i\,{}^t c(E)$.

<u>Exemple</u>. Soient V un espace vectoriel hermitien de dimension $n+1$, $\mathbb{P}_n = \mathbb{P}(V)$ l'espace projectif associé, $O(-1)$ le sous-fibré linéaire canonique du fibré trivial V sur $\mathbb{P}_n$, $Q = V/O(-1)$ le fibré quotient de rang n. On munit $O(-1)$ et Q de leurs métriques naturelles, induites par celle de V, et $\mathbb{P}_n$ de sa métrique kählérienne usuelle. On a classiquement les isomorphismes métriques

$$\text{dét } Q \simeq O(-1)^{\star} = O(1)$$

$$T\,\mathbb{P}_n \simeq Q \otimes O(1) \simeq Q \otimes \text{dét } Q \ .$$

Q est semi-positif au sens de Griffiths, comme quotient du fibré trivial V. On retrouve donc d'après le théorème 2 que le fibré tangent $T\,\mathbb{P}_n$ est semi-positif au sens de Nakano (cf. M. SCHNEIDER [7]), et même au sens fort.

Etant donné une base orthonormée $(e_o,e_1,\ldots,e_n)$ de V, $(e_1,\ldots,e_n)$ définit une base orthonormée de la fibre Q_z au dessus du point $z = [e_o]$ de $\mathbb{P}_n$. Si l'on munit $T_z\,\mathbb{P}_n$ de la base orthonormée $(\eta_1,\ldots,\eta_n)$ correspondante, déduite de l'isomorphisme canonique $T\,\mathbb{P}_n \simeq Q \otimes \text{dét}\, Q$, la forme de courbure $\theta = i\, c(Q)$ s'explicite en coordonnées par les formules

$$\theta(\xi \otimes u, \xi \otimes u) = |\sum_{j=1}^{n} \xi_j\, \bar{u}_j|^2 ,$$

$$\theta(x,x) = \sum_{1\leqslant j,k\leqslant n} x_{jk}\, \overline{x_{kj}} ,$$

$$\mathrm{Tr}_Q\theta(\xi,\xi) = |\xi|^2 = \sum_{j=1}^{n} |\xi_j|^2 ,$$

pour $\xi = \sum_{j=1}^{n} \xi_j\, \eta_j \in T_z\,\mathbb{P}_n$, $u = \sum_{j=1}^{n} u_j\, e_j \in Q_z$,

$$x = \sum_{1\leqslant j,k\leqslant n} x_{jk}\, \eta_j \otimes e_k \in T_z\,\mathbb{P}_n \otimes Q_z .$$

La forme de courbure de $T\,\mathbb{P}_n$ s'écrit donc

$$<(\theta + \mathrm{Tr}_Q\, \theta \otimes \mathrm{Id}_Q)(x),x> = \sum_{j,k} x_{jk}\, \overline{x_{kj}} + x_{jk}\, \overline{x_{jk}}$$

$$= 2 \sum_j |x_{jj}|^2 + \sum_{j<k} |x_{jk} + x_{kj}|^2 .$$

La forme $\mathrm{Tr}_Q\, \theta$ est définie positive, mais la forme $\theta + \mathrm{Tr}_Q\, \theta \otimes \mathrm{Id}_Q$ est seulement <u>semi</u>-positive au sens de Nakano si $n \geqslant 2$. Il n'existe donc pas de constante $\gamma < 1$ telle que $\theta + \gamma\, \mathrm{Tr}_Q \otimes \mathrm{Id}_Q \geqslant_N 0$, et en ce sens le résultat du théorème 1 est le meilleur possible. En ce qui concerne le corollaire 1, le lecteur pourra vérifier que l'inégalité $Q^\star \otimes (\text{dét}\, Q)^n \geqslant_N 0$ est optimale, mais que dans cet exemple $Q \otimes (\text{dét}\, Q)^{-1} \leqslant_N 0$. Lorsque $n \geqslant 2$, il apparaît qu'on a $Q \geqslant_G 0$ sans avoir $Q \geqslant_N 0$, et qu'on a $Q^\star \leqslant_N 0$ (puisque $Q^\star$ est un sous-fibré du fibré trivial $V^\star$) sans avoir $Q^\star \leqslant_S 0$ (ce qui entraînerait $Q \geqslant_S 0$!).

3. COURANTS POSITIFS.

Seules les propriétés ponctuelles des courants seront étudiées dans ce paragraphe, de sorte qu'on se limitera à la considération des formes différentielles. La première définition des formes et des courants positifs a été donnée par P. LELONG [5].

Soient T un espace vectoriel complexe de dimension n, $F = \mathrm{Hom}_{\mathbb{R}}(T,\mathbb{C})$ le complexifié du $\mathbb{R}$-dual de T, $\Lambda^{p,q} F$ l'espace des formes de type (p,q) sur T. Pour tout entier p, on pose

$$\varepsilon_p = \left(\frac{i}{2}\right)^p (-1)^{\frac{p(p-1)}{2}} = 2^{-p}\, i^{p^2} .$$

DÉFINITION 3. - Une forme $\alpha \in \Lambda^{p,q} F$ est dite

(1) positive, si elle peut s'écrire

$$\alpha = \sum_{j=1}^{N} \varepsilon_p\, \alpha_j \wedge \overline{\alpha_j}$$

avec des éléments $\alpha_j \in \Lambda^{p,o} F$;

(2) fortement positive, si on a une écriture analogue avec des formes α_j décomposables ;

(3) faiblement positive, si pour toute forme $\beta \in \Lambda^{k,k} F$ fortement positive, où $p+k = n$, la (n,n)-forme $\alpha \wedge \beta$ est positive.

Nous noterons P^p (resp. SP^p, WP^p) le cône des (p,p)-formes positives (resp. fortement, faiblement positives), et nous désignerons par $\geqslant$ (resp. $\geqslant_S$, $\geqslant_W$) l'inégalité de positivité (resp. de positivité forte, faible).

On vérifie aisément à partir de cette définition qu'on a les inclusions

$$SP^p \subset P^p \subset WP^p ,$$

et que les éléments de WP^p sont réels.

Si l'on a choisi une (n,n)-forme τ positive, non nulle, il y a une forme bilinéaire naturelle

$$\Lambda^{p,p} F \times \Lambda^{k,k} F \to \mathbb{C}$$

(avec $p+k = n$) qui à $\alpha \in \Lambda^{p,p} F$, $\beta \in \Lambda^{k,k} F$ associe l'unique nombre complexe γ

tel que

$$\alpha \wedge \beta = \gamma \, . \, \tau \, .$$

WP^p s'identifie alors par définition au cône dual $(SP^k)^o$ de SP^k, et on peut montrer d'autre part que

$$SP^p = (WP^k)^o \quad , \quad P^p = (P^k)^o \, .$$

Enfin, si $p = 0,1$, $n-1$, ou n, toutes les formes de $\Lambda^{p,o} F$ sont décomposables, donc

$$SP^p = P^p$$
$$WP^p = (SP^k)^o = (P^k)^o = P^p \, .$$

On suppose désormais que l'espace T est muni d'une métrique hermitienne, représentée par la $(1,1)$-forme positive

$$\omega = \frac{i}{2} \sum_{j=1}^{n} dz_j \wedge d\bar{z}_j$$

dans une base orthonormée $(dz_j)_{1 \leqslant j \leqslant n}$ du dual $E^\star$ de E. L'espace $F = \mathrm{Hom}_{\mathbb{R}}(T,\mathbb{C})$ et l'agèbre extérieure ΛF sont munis des métriques usuelles correspondantes. On désigne classiquement par L l'opérateur de multiplication extérieure par ω dans l'espace hilbertien ΛF, et par Λ son adjoint ; on a donc :

$$L\alpha = \omega \wedge \alpha \ , \quad (\Lambda\alpha|\beta) = (\alpha, L\beta) = (\alpha|\omega \wedge \beta)$$

pour toutes formes $\alpha, \beta \in \Lambda F$.

On peut écrire en coordonnées, pour tout $\alpha \in \Lambda^{p,p} F$,

$$\alpha = \varepsilon_p \sum_{J,K}{}' \alpha_{J,K} \, dz_J \wedge d\bar{z}_K \ ,$$

$$\frac{1}{r!} L^r\alpha = \varepsilon_{p+r} \sum_{J,K,M}{}' \alpha_{J,K} \, dz_{JM} \wedge d\bar{z}_{KM} \ ,$$

$$\frac{1}{r!} \Lambda^r\alpha = \varepsilon_{p-r} \sum_{M,N,P}{}' \alpha_{NM,PM} \, dz_N \wedge d\bar{z}_P \ ,$$

où la notation $\sum'$ signifie que les sommes sont étendues à tous les multi-indices <u>croissants</u> J,K,M,N,P, avec ici $|J| = |K| = P$, $|M| = r$, $|N| = |P| = p-r$. On convient que les symboles $\alpha_{J,K}$, dz_J, $d\bar{z}_K$ sont définis pour des multi-indices non nécessairement croissants J,K, de sorte que leurs signes soient alternés en J,K.

On rappelle enfin que l'opérateur $\star$ de Hodge-de Rham-Poincaré est défini sur ΛF par la relation

$$\alpha \wedge \star \beta = (\alpha|\bar{\beta}) \frac{\omega^n}{n!} .$$

Les propriétés des formes positives sont intimement liées aux opérateurs L,Λ et $\star$; la proposition 2 ci-dessous est classique, et de démonstration aisée.

PROPOSITION 2. - Soit $\alpha \in \Lambda^{p,p} F$ une forme positive (resp. fortement, faiblement positive). Alors les formes

$$L^r\alpha \in \Lambda^{p+r,p+r} F , \Lambda^r\alpha \in \Lambda^{p-r,p-r} F , \star \alpha \in \Lambda^{n-p,n-p} F$$

sont positives (resp. fortement, faiblement positives).

Les méthodes du § 1 conduisent d'autre part au résultat suivant.

PROPOSITION 3. - Soit α une (p,p)-forme faiblement positive sur l'espace hilbertien (T,ω). Alors la (p,p)-forme

$$\sum_{r=o}^{p-1} \frac{p-r}{p} \frac{(p-r)!}{p!r!} L^r \Lambda^r\alpha$$

est fortement positive.

Nous aurons besoin des notations suivantes : σ décrivant l'ensemble $\mathcal{F}$ des applications de $\{1,2,\dots,n\}$ dans le groupe des racines q-ièmes de l'unité $(q \geqslant 3)$, on pose

$$w_\sigma = \sum_{\ell=1}^{n} \overline{\sigma(\ell)}\, dz_\ell ,$$

et pour tout $\sigma = (\sigma_1,\dots,\sigma_p) \in \mathcal{F}^p$, on pose

$$W_\sigma = w_{\sigma_1} \wedge \dots \wedge w_{\sigma_q} = \sum_L \overline{\sigma(L)}\, dz_L$$

où la somme $\sum_L$ est étendue à tous les multi-indices $L = (\ell_1,\dots,\ell_p)$, non nécessairement croissants, et où

$$\sigma(L) = \sigma_1(\ell_1) \dots \sigma_p(\ell_p) .$$

LEMME 3. - Pour tout couple de multi-indices (J,K) tels que $|J| = |K| = p$, on a l'égalité

$$q^{-np} \sum_{\sigma \in \mathcal{F}^p} \sigma(J)\, \overline{\sigma(K)}\, W_\sigma \wedge \bar{W}_\sigma = \sum_{L,M} dz_L \wedge d\bar{z}_M$$

où la somme $\sum_{L,M}$ est étendue à tous les L,M tels que

$$\{j_s, m_s\} = \{k_s, \ell_s\} \text{ pour tout } s,\ 1 \leqslant s \leqslant p .$$

Pour $p = 1$, le lemme 3 se réduit au lemme 1, et on obtient aussitôt le cas général en observant que

$$\sum_{\sigma \in \mathcal{F}^p} = \sum_{\sigma_1 \in \mathcal{F}} \cdots \sum_{\sigma_p \in \mathcal{F}} .$$

<u>Démonstration de la proposition 3</u>. Ecrivons

$$\alpha = \varepsilon_p \sum_{|J|=|K|=p}' \alpha_{J,K}\, dz_J \wedge d\bar{z}_K$$

$$= \frac{\varepsilon_p}{p!^2} \sum_{|J|=|K|=p} \alpha_{J,K}\, dz_J \wedge d\bar{z}_K$$

Dire que $\alpha \in WP^p$ équivaut par définition à dire que pour toute famille (x_J) décomposable, $x_J = x^1_{j_1} x^2_{j_2} \dots x^p_{j_p}$, on a

$$\sum_{J,K} \alpha_{J,K}\, x_J\, \overline{x_K} \geqslant 0 .$$

Pour chaque élément $\sigma \in \mathcal{F}^{p-1}$, la $(1,1)$-forme

$$\frac{i}{2} \sum_{j,k,|J|=|K|=p-1} \alpha_{jJ,kK}\, \sigma(J)\, \overline{\sigma(K)}\, dz_j \wedge d\bar{z}_k$$

est donc positive, ce qui entraîne que la forme $\beta(\alpha)$ suivante appartient à SP^p :

$$\beta(\alpha) = \frac{\varepsilon_p}{p!^2 q^{n(p-1)}} \sum_{\sigma \in \mathcal{F}^{p-1}} \sum_{j,k,|J|=|K|=p-1} \alpha_{jJ,kK}\, \sigma(J)\, \overline{\sigma(K)} dz_j \wedge W_\sigma \wedge d\bar{z}_k \wedge \bar{W}_\sigma .$$

D'après le lemme 3, il vient

$$\beta(\alpha) = \frac{\varepsilon_p}{p!^2} \sum_{j,k,|J|=|K|=|L|=|M|=p-1} \alpha_{jJ,kK}\, dz_{jL}\, d\bar{z}_{kM} ,$$

la somme étant prise sur les J,K,L,M tels que

la somme étant prise sur les J,K,L,M tels que

$$\{j_s,m_s\} = \{k_s,\ell_s\} \quad \text{pour tout } s \,, \quad 1 \leqslant s \leqslant p-1 \,.$$

Si $\{j_s,m_s\} = \{k_s,\ell_s\}$, on a ou bien $(j_s,k_s) = (\ell_s,m_s)$, ou bien $j_s = k_s \neq \ell_s = m_s$, les deux cas s'excluant mutuellement. On obtient donc

$$\beta(\alpha) = \frac{\varepsilon_p}{p!^2} \sum_{r=o}^{p-1} \binom{p-1}{r} \sum_{\substack{|J|=|K|=|p-r| \\ |N|=|P|=r}} \alpha_{JN,KN}\, dz_{JP} \wedge d\bar{z}_{KP} \,,$$

où la somme est restreinte aux multi-indices N,P tels que $n_s \neq p_s$ pour tout $s \in \{1,\dots,r\}$. Le coefficient binomial $\binom{p-1}{r}$ apparaît parce qu'il faut choisir r indices $s \in \{1,\dots,p-1\}$ pour lesquels on aura $j_s = k_s \neq \ell_s = m_s$. Pour tout multi-indice M de longueur $|M| = m$, définissons la "contraction" $\alpha\,[M] \in WP^{p-m}$ de α par :

$$\alpha\,[M] = \frac{\varepsilon_{p-m}}{p!^2} \sum_{|J|=|K|=p-m} \alpha_{JM,KM}\, dz_J \wedge d\bar{z}_K \,,$$

et considérons la forme

$$\gamma(\alpha) = \frac{\varepsilon_p}{p!^2} \sum_{r=o}^{p-1} \binom{p-1}{r} \sum_{\substack{|J|=|K|=p-r \\ |N|=|P|=r}} \alpha_{JN,KN}\, dz_{JP} \wedge d\bar{z}_{KP} \,,$$

dans laquelle la sommation est prise sans restriction sur les multi-indices N et P . Pour chaque couple (N,P) , on peut écrire $N = N'M$, $P = P'M$ où les multi-indices N' , P' ont la propriété que $n'_s \neq p'_s$ pour tout s; on observera que dans cette notation, M n'est pas nécessairement constitué des $m = |M|$ derniers indices de N et P , mais de m quelconques des r indices possibles. Si on réordonne en écrivant M à la fin, chaque couple $(N'M,P'M)$ proviendra d'exactement $\binom{r}{m}$ couples (N,P) , obtenus après avoir "mélangé" M à N' , M à P' .

De l'égalité $\binom{p-1}{r}\binom{r}{m} = \binom{p-1}{m}\binom{p-m-1}{r-m}$ résulte alors

$$\gamma(\alpha) = \frac{\varepsilon_p}{p!^2} \sum_{\substack{0\leq r\leq p-1\\ 0\leq m\leq r}} \binom{p-1}{r}\binom{r}{m} \sum_{\substack{|J|=|K|=p-r\\ |N'|=|P'|=r-m\\ |M|=m}} \alpha_{JN'M,KP'M}\, dz_{JN'M} \wedge d\bar{z}_{KP'M}$$

$$= \frac{1}{p!^2} \sum_{m=0}^{p-1} \binom{p-1}{m} \sum_{0\leq r-m\leq p-m-1} \binom{p-m-1}{r-m}$$

$$\times \sum_{\substack{|J|=|K|=p-m-(r-m)\\ |N'|=|P'|=r-m\\ |M|=m}} \varepsilon_{p-m}\, \alpha_{JN'M,KP'M}\, dz_{JN'} \wedge d\bar{z}_{KP'} \wedge \varepsilon_m\, dz_M \wedge d\bar{z}_M$$

$$= \sum_{m=0}^{p-1} \binom{p-1}{m} \sum_{|M|=m} \beta(\alpha[M]) \wedge \varepsilon_m\, dz_M \wedge d\bar{z}_M \ ,$$

de sorte que $\gamma(\alpha) \in SP^p$. D'autre part, on a

$$\frac{1}{r!^2} L^r \Lambda^r \alpha = \varepsilon_p \sum_{\substack{|J|=|K|=p-r\\ |N|=|P|=r}}' \alpha_{JN,KN}\, dz_{JP} \wedge d\bar{z}_{KP}$$

$$= \frac{\varepsilon_p}{(p-r)!^2 r!^2} \sum_{\substack{|J|=|K|=p-r\\ |N|=|P|=r}} \alpha_{JN,KN}\, dz_{JP} \wedge d\bar{z}_{KP} \ ;$$

on obtient donc l'égalité suivante, qui prouve la proposition 3 :

$$\gamma(\alpha) = \sum_{r=0}^{p-1} \binom{p-1}{r} \frac{(p-r)!^2}{p!^2} L^r \Lambda^r \alpha \ .$$

On peut démontrer de même un résultat analogue au corollaire 1.

<u>PROPOSITION 4</u>. - Si α est une (p,p)-forme faiblement positive $(p \geq 2)$, on a l'inégalité forte

$$\alpha \leq_S \sum_{r=1}^{p-1} \left[n\binom{p-2}{r-1} - \binom{p-2}{r}\right] \frac{(p-r)!^2}{p!^2} L^r \Lambda^r \alpha$$

<u>Démonstration</u>. Avec les notations de la proposition 3 , on pose

$$\tilde{\alpha}_{J,K} = \sum_{j=1}^{n} \alpha_{J'j,K'j} - \alpha_{J,K} \quad \text{si} \quad J = J'j_p,\ K = K'k_p \quad \text{et} \quad j_p = k_p \ ,$$

$$= -\alpha_{J,K} \quad \text{si} \quad j_p \neq k_p \ .$$

On prendra garde au fait que $\tilde{\alpha}_{J,K}$ <u>n'est pas</u> alterné en J et K (mais seulement par rapport aux p-1 premiers indices de J et K). Il est clair que l'expression

$$\sum_{J,K} \tilde{\alpha}_{J,K}\, x_J\, \bar{x}_K$$

est ≥ 0 pour toute famille (x_J) décomposable (cf. lemme 2) ; il en résulte comme précédemment que SP^p contient la forme

$$\hat{\beta}(\alpha) = \frac{\varepsilon_p}{p!^2 q^{n(p-1)}} \sum_{\sigma \in \mathfrak{F}^{p-1}} \sum_{j,k,|J|=|K|=p-1} \tilde{\alpha}_{jJ,kK}\, \sigma(J)\, \overline{\sigma(K)}\, dz_j \wedge W_\sigma \wedge d\bar{z}_k \wedge \bar{W}_\sigma$$

$$= \frac{\varepsilon_p}{p!^2} \sum \tilde{\alpha}_{jJ,kK}\, dz_{jL} \wedge d\bar{z}_{kM}$$

les sommations sur J,K,L,M étant prises pour

$|J| = |K| = |L| = |M| = p-1$, $\{j_s, m_s\} = \{k_s, \ell_s\}$ si $s \in \{1,\dots,p-1\}$;

par conséquent

$$\hat{\beta}(\alpha) = -\frac{\varepsilon_p}{p!^2} \sum_{j,k,\ell\neq m} \alpha_{jJ\ell,kKm}\, dz_{jL\ell} \wedge d\bar{z}_{kMm}$$

$$+ (n-1)\frac{\varepsilon_p}{p!^2} \sum_{j,k,\ell,m} \alpha_{jJ\ell,kK\ell}\, dz_{jLm} \wedge d\bar{z}_{kMm} ,$$

avec $|J| = |K| = |L| = |M| = p-2$, $\{j_s, m_s\} = \{k_s, \ell_s\}$ si $s \in \{1,\dots,p-2\}$.

Définissons la forme $\tilde{\beta}(\alpha) \in SP^p$ par

$$\tilde{\beta}(\alpha) = \hat{\beta}(\alpha) + \sum_{\ell\neq m} \beta(\alpha[\ell]) \wedge \frac{i}{2}\, dz_m \wedge d\bar{z}_m$$

$$= \hat{\beta}(\alpha) + \frac{\varepsilon_p}{p!^2} \sum_{j,k,\ell\neq m} \alpha_{jJ\ell,kK\ell}\, dz_{jLm} \wedge d\bar{z}_{kMm}$$

$$= -\frac{\varepsilon_p}{p!^2} \sum_{j,k,\ell,m} \alpha_{jJ\ell,kKm}\, dz_{jL\ell} \wedge d\bar{z}_{kMm}$$

$$+ n\frac{\varepsilon_p}{p!^2} \sum_{j,k,\ell,m} \alpha_{jJ\ell,kK\ell}\, dz_{jLm} \wedge d\bar{z}_{kMm} .$$

On vérifie comme dans la démonstration de la proposition 3 qu'on a l'égalité

$$\sum_{m=0}^{p-2} \binom{p-2}{m} \sum_{|M|=m} \tilde{\beta}(\alpha[M]) \wedge \varepsilon_m\, dz_M \wedge d\bar{z}_M$$

$$= -\sum_{r=0}^{p-2} \binom{p-2}{r} \frac{(p-r)!^2}{p!^2}\, L^r \Lambda^r \alpha$$

$$+ n \sum_{r=0}^{p-2} \binom{p-2}{r} \frac{(p-r-1)!^2}{p!^2}\, L^{r+1} \Lambda^{r+1} \alpha .$$

Comme le premier membre est une (p, p) - forme fortement positive, la proposition 4 est démontrée.

Nous pouvons maintenant énoncer le résultat essentiel de ce paragraphe.

THÉORÈME 3. - Soit α une (p,p)-forme faiblement positive $(1 \leqslant p \leqslant n)$. On a les inégalités fortes

$$- C'(n,p) \frac{1}{p!^2} L^{p-1} \Lambda^{p-1} \alpha \leqslant_S \alpha \leqslant_S C(n,p) \frac{1}{p!^2} L^{p-1} \Lambda^{p-1} \alpha$$

où les constantes positives $C(n,p)$, $C'(n,p)$ sont définies par

$$C(n,1) = 1 \quad , \quad C'(n,1) = 0 \ ,$$

$$C(n,2) = n \quad , \quad C'(n,2) = 1 \ ,$$

$$C(n,p) + C'(n,p) = \frac{(n+1)!}{(n-p+2)!} \ ,$$

et la relation de récurrence

$$C'(n,p) = \sum_{r-1}^{p-1} \binom{p-1}{r} C(n-r,p-r) \ .$$

Démonstration. On raisonne par récurrence sur p , en utilisant les propositions 3 et 4. Lorsque $p = 1$, on peut choisir

$$C(n,p) = 1 \quad , \ C'(n,p) = 0 \ .$$

On observe que, pour tout entier $r \geqslant 1$,

$$\Lambda^r \alpha = \frac{p!^2}{(p-r)!^2} \sum_{|M|=r} \alpha[M] \ ,$$

et que $\alpha[M]$ ne fait intervenir que les variables dont les indices appartiennent au complémentaire $\complement M$ de M . Si $L_{\complement M}$ et $\Lambda_{\complement M}$ désignent les opérateurs L et Λ relatifs à ces variables, on obtient par hypothèse de récurrence

$$\alpha[M] \leqslant_S C(n-r,p-r) \frac{1}{(p-r)!^2} L_{\complement M}^{p-r-1} \Lambda_{\complement M}^{p-r-1} \alpha[M] \ .$$

Comme $\Lambda_{\complement M}^{p-r-1} \alpha[M] = \Lambda^{p-r-1} \alpha[M]$, il vient

$$\alpha[M] \leqslant_S C(n-r,p-r) \frac{1}{(p-r)!^2} L_{\complement M}^{p-r-1} \Lambda^{p-r-1} \alpha[M]$$

$$\leqslant_S C(n-r,p-r) \frac{1}{(p-r)!^2} L^{p-r-1} \Lambda^{p-r-1} \alpha[M] \ ,$$

car la $(1,1)$-forme $\Lambda^{p-r-1} \alpha[M]$ est dans $WP^1 = SP^1$; en appliquant l'opérateur L^r qui conserve les inégalités fortes (proposition 2), on obtient après sommation sur M :

$$L^r \Lambda^r \alpha \leqslant_S \frac{C(n-r,p-r)}{(p-r)!^2} L^{p-1} \Lambda^{p-1} \alpha \quad .$$

Les propositions 3 et 4 montrent qu'on peut prendre

$$C(n,p) = \sum_{r=1}^{p-1} [n\binom{p-2}{r-1} - \binom{p-2}{r}] \; C(n-r,p,r)$$

(car le terme entre crochets est toujours $\geqslant 0$) et

$$C'(n,p) = \sum_{r=1}^{p-1} \binom{p-1}{r} \; C(n-r,p-r) \; .$$

Grâce à la relation $\binom{p-1}{r} = \binom{p-2}{r} + \binom{p-2}{r-1}$, on voit que

$$\begin{aligned} C(n,p) &= \sum_{r=1}^{p-1} [(n+1)\binom{p-2}{r-1} - \binom{p-1}{r}] C(n-r,p-r) \\ &= (n+1)\;(C'(n-1,p-1) + C(n-1,p-1)) - C'(n,p) \; . \end{aligned}$$

On en déduit aussitôt

$$C(n,p) + C'(n,p) = (n+1)\ldots(n-p+3) = \frac{(n+1)!}{(n-p+2)!} \; . \; \blacksquare$$

Nous avons maintenant besoin de la majoration suivante, dont la démonstration est immédiate (se ramener au cas d'une forme $\beta = \varepsilon_p \, dz_J \wedge d\bar{z}_J$, $|J| = p$).

LEMME 4. - Pour toute forme $\beta \in SP^p$ et tout entier $k \leqslant p$, on a

$$\beta \leqslant_S \frac{(p-k)!}{p!k!} L^k \Lambda^k \beta \quad .$$

En appliquant cette inégalité pour $k = 1$ à la forme $\beta = \Lambda^{p-1}\alpha \in WP^1 = SP^1$, on obtient (la notation $\mathrm{Tr}\,\alpha$ désignant le scalaire $\frac{1}{p!} \Lambda^p \alpha$) :

COROLLAIRE 2. - Si $\alpha \in WP^p$, on a les inégalités fortes

$$- C'(n,p) \,\mathrm{Tr}\,\alpha \,.\, \frac{\omega^p}{p!} \leqslant_S \alpha \leqslant_S C(n,p) \,\mathrm{Tr}\,\alpha \,.\, \frac{\omega^p}{p!}$$

avec les constantes du théorème 3 , et

$$C(n,0) = 1 \;,\; C'(n,0) = 0 \; .$$

Comme l'opérateur $\star$ conserve la trace $\operatorname{Tr}\alpha$ et la positivité forte, on peut d'ailleurs remplacer dans le corollaire 2

$$C(n,p) \quad \text{par} \quad \operatorname{Inf}(C(n,p), C(n,n-p)) ,$$

$$C'(n,p) \quad \text{par} \quad \operatorname{Inf}(C'(n,p), C'(n,n-p)) .$$

Ces dernières constantes ne sont malheureusement pas optimales. Ainsi pour $p = 2$:

PROPOSITION 5. - Si $\alpha \in WP^2$, alors $\alpha \leqslant_S \frac{n}{2} \cdot \operatorname{Tr}\alpha \cdot \frac{\omega^2}{2}$.

<u>Démonstration</u>. Si $\alpha = \frac{\varepsilon_2}{4} \sum_{j,k,\ell,m} \alpha_{j\ell,km}\, dz_j \wedge dz_j \wedge d\bar{z}_k \wedge d\bar{z}_m$, nous savons (voir la démonstration de la proposition 3) que la forme

$$\beta(\alpha) = \alpha + \frac{\varepsilon_2}{4} \sum_{j,k,\ell\neq m} \alpha_{j\ell,k\ell}\, dz_j \wedge dz_m \wedge d\bar{z}_k \wedge d\bar{z}_m$$

est fortement positive. En utilisant le lemme 4 avec $p = k = 2$, on trouve

$$\alpha \leqslant_S \beta(\alpha) \leqslant_S \operatorname{Tr}(\beta(\alpha)) \cdot \frac{\omega^2}{2} ,$$

avec $$\operatorname{Tr}(\beta(\alpha)) = \operatorname{Tr}\alpha + \frac{1}{4} \sum_{j\neq\ell\neq m\neq j} \alpha_{j\ell,j\ell} = \frac{n}{2} \cdot \operatorname{Tr}\alpha .$$

=-=-=-=-=-=-=-=-=-=-=-=-=-=-=

BIBLIOGRAPHIE

[1] DEMAILLY (J.P.).
Scindage holomorphe d'un morphisme de fibrés vectoriels semi-positifs avec estimations L^2, à paraître.

[2] DEMAILLY (J.P.) et SKODA (H.).
Relations entre les notions de positivités de P.A. GRIFFITHS et de S. NAKANO pour les fibrés vectoriels, Séminaire P. LELONG-H. SKODA. (Analyse), 19ème année, 1978-79, Lecture Notes n° 822,1980,Springer-Verlag,Berlin,Heidelberg,New York.

[3] DOUADY (A.) et VERDIER (J.L.).
Séminaire de Géométrie analytique, E.N.S., 1972-73, Différents aspects de la positivité, Astérisque 17, 1974, Société Mathématique de France.

[4] GRIFFITHS (P.A.).
Hermitian differential geometry, Chern classes and positive vector bundles, Global Analysis, Princeton University Press, p. 185-251, 1969.

[5] LELONG (P.).
Integration of a differential form on an analytic complex subvariety, Proc. Nat. Acad. Sc. of U.S.A., t. 43, p. 246-248, 1957.

[6] NAKANO (S.).
Vanishing theorems for weakly 1-complete manifolds II, Publ. RIMS, Kyoto University, p. 133-139, 1974.

[7] SCHNEIDER (M.).
Lefschetz theorems and a vanishing theorem of Grauert-Riemenschneider, Proceedings of Symposia in Pure Mathematics, vol. 30, Several Complex Variables, Amer, Math. Society Providence, t. 2, p. 35-39, 1977.

[8] SKODA (H.).
Morphismes surjectifs de fibrés vectoriels semi-positifs, Annales scient. de l'Ecole Normale Supérieure, 4e série, t.11,p.577-611, 1978.

SCINDAGE HOLOMORPHE D'UN MORPHISME DE FIBRÉS VECTORIELS SEMI-POSITIFS AVEC ESTIMATIONS L^2

par J.-P.DEMAILLY

Table des matières.

0. Introduction et notations.

1. Rappel sur les différentes notions de positivité.

2. Estimations a priori et inégalités L^2.

3. Calcul de courbure.

4. Construction de rétractions holomorphes.

5. Extension des fonctions holomorphes avec contrôle de la croissance.

0. Introduction et notations.

Le présent travail réétudie dans un cas particulier les techniques développées par H.SKODA [11] , [12] , [13] , [14] , [15] pour l'étude des morphismes surjectifs de fibrés vectoriels holomorphes semi-positifs. On considère une suite exacte

$$(1) \qquad 0 \to S \to E \xrightarrow{g} Q \to 0$$

de fibrés vectoriels holomorphes, de rangs respectifs s,p,q (avec $s = p-q$) , au-dessus d'une variété analytique complexe X de dimension n . On dit qu'un morphisme de Q dans E réalise un scindage holomorphe de la suite exacte (1) si

$$g \circ h = \mathrm{Id}_Q ,$$

de sorte qu'on a alors la décomposition

$$E = S \oplus h(Q) .$$

Plus généralement, étant donné un fibré linéaire M sur X , et une section f du fibré $\mathrm{Hom}(Q, Q \otimes M)$, on recherche s'il existe une section h du fibré $\mathrm{Hom}(Q, E \otimes M)$ telle que

$$g \circ h = f .$$

Pour obtenir de tels résultats, nous serons amenés à faire des hypothèses de convexité sur la variété X , et de positivité sur les fibrés E et M . Nous supposerons, comme H.SKODA [13] , [14] , [15] , que X est une <u>variété kählérienne</u>, munie d'une métrique de Kähler ω non nécessairement complète, et que X est <u>faiblement pseudoconvexe</u>, c'est-à-dire qu'il existe sur X une fonction de classe C^2, plurisousharmonique et exhaustive. Les variétés compactes, les variétés de Stein, l'espace total d'un fibré holomorphe semi-négatif an sens de Griffiths au-dessus d'une variété compacte, sont faiblement pseudoconvexes. On suppose de plus que les fibrés E, M sont munis de métriques hermitiennes, et que les fibrés S, Q, $\mathrm{Hom}(Q, Q\otimes M)$, $\mathrm{Hom}(Q, E\otimes M)$ sont munis des métriques naturelles déduites de celles de E et M.

Les hypothèses de positivité sont les suivantes (voir le paragraphe 1 pour les définitions concernant la courbure et la positivité) : le fibré E est <u>semi-positif au sens de Griffiths</u>, et il existe un réel $k > \mathrm{Inf}(n,q) + \inf(n,s)$ tel que l'une des conditions (2) ou (3) soit réalisée :

(2) $ic(M) - ik\, c(\det Q) - ic(\det E) + i\,\mathrm{Ricci}(\omega) \geqslant 0$;

(3) le rang s de S est égal à 1, ou bien E est semi-positif au sens de Nakano, et on a

$$ic(M) - ikc(\det Q) + i\,\mathrm{Ricci}(\omega) \geqslant 0 .$$

On a alors un théorème d'existence avec estimations L^2 précises (on a noté $dV = \frac{\omega^n}{n!}$ l'élément de volume canonique sur X).

THÉORÈME 1. - <u>Pour toute section globale</u> f <u>de</u> $\mathrm{Hom}(Q, Q\otimes M)$ <u>telle que le second membre de</u> (5) <u>soit fini, il existe une section globale</u> h <u>de</u> $\mathrm{Hom}(Q, E\otimes M)$ <u>telle que</u>

(4) $g \circ h = f$,

(5) $\displaystyle\int_X |h|^2\, dV \leqslant C \int_X |f|^2\, dV$,

<u>avec</u> $C = \dfrac{k - \mathrm{Inf}(n,q)}{k-\mathrm{Inf}(n,s)-\mathrm{Inf}(n,q)}$.

En pratique, le théorème 1 s'appliquera surtout au cas où la variété X est de Stein, car les conditions de positivité et la convergence globale des intégrales semblent supposer l'existence de fonctions d'exhaustion strictement plurisousharmoniques (du moins lorsqu'on cherche à construire des scindages holomorphes).

La démonstration repose sur l'inégalité de Kodaïra-Nakano, et sur le lien qui existe entre les formes de courbure des fibrés Q, S et l'obstruction au scindage holomorphe de la suite exacte (1). Les calculs sont directement inspirés de [14] , mais utilisent de plus une relation entre les notions de positivité de P.A.GRIFFITHS et de S.NAKANO, récemment publiée par l'auteur dans [3] en collaboration avec H.SKODA , et dans [2].

Etant donné une sous-variété X de dimension n de C^p , le théorème 1 s'applique en particulier à la suite exacte

$$0 \to TX \to T\mathbb{C}^p|_X \to NX \to 0$$

qui définit le fibré normal NX de X . Cette idée , déjà utilisée par C.A.BERENSTEIN et B.A.TAYLOR [1] , nous permet de montrer l'existence d'un voisinage tubulaire U se rétractant holomorphiquement sur X , et d'estimer la taille de U*. En appliquant des techniques analogues à celles de B.JENNANE [9] , nous obtenons au dernier paragraphe un théorème d'extension des fonctions holomorphes avec contrôle de la croissance** .

Je tiens à remercier Monsieur Henri SKODA pour ses nombreuses suggestions, qui ont permis d'améliorer la rédaction initiale.

1. Rappel sur les différentes notions de positivité.

Si E est un fibré hermitien, on peut définir une connexion canonique D sur E , hermitienne et holomorphe (cf. A.DOUADY et J.-L.VERDIER [4] , P.A.GRIFFITHS [5]) , qui envoie l'espace $C^\infty_{a,b}(X,E)$ des formes de type (a,b) à valeurs dans E , dans $C^\infty_{a+1,b}(X,E) \oplus C^\infty_{a,b+1}(X,E)$.
La forme de courbure $c(E)$ du fibré E est alors définie par la propriété suivante :

$$D^2 u = c(E).u$$

pour toute section C^∞ u de E , de sorte que $ic(E)$ est une $(1,1)$-forme à valeurs dans le fibré $Herm(E,E)$ des endomorphismes hermitiens de E ; on identifiera toujours $ic(E)$ à la forme hermitienne sur $TX \otimes E$ qui lui est associée canoniquement.

DÉFINITION 1. - Soit Θ une forme hermitienne sur un produit tensoriel $T \otimes E$ d'espaces vectoriels complexes T et E de dimensions respectives n et p . On dira que Θ est semi-positive

*(cf. § 4, théorèmes 4 et 5).
**(cf. corollaires 2,3,et 4 du théorème 6).

(6) au sens de GRIFFITHS, si pour tout vecteur décomposable $x = t \otimes e \in T \otimes E$, on a $\Theta(x,x) \geqslant 0$;

(7) au sens de NAKANO, si pour tout $x \in T \otimes E$, on a $\Theta(x,x) \geqslant 0$.

Le fibré vectoriel hermitien E sur la variété X est dit semi-positif au sens de GRIFFITHS (resp. au sens de NAKANO) si pour tout point $z \in X$, la forme hermitienne ic(E) sur la fibre $T_z X \otimes E_z$ est semi-positive dans ce sens.

Nous noterons $\geqslant_G$ la semi-positivité de GRIFFITHS, $\geqslant_N$ celle de NAKANO. Il est clair que la semi-positivité de NAKANO entraîne celle de GRIFFITHS. De plus, les deux notions coïncident si $n = 1$ ou si $p = 1$, et sont reliées en général par le théorème suivant.

THÉORÈME 2 - Soit $\Theta \geqslant_G 0$ une forme hermitienne sur $T \otimes E$. Si $(e_j)_{1 \leqslant j \leqslant p}$ est une base orthonormée quelconque de E (supposé hermitien) on définit $Tr_E\, \Theta$ par

$$Tr_E\, \Theta(t,t) = \sum_{j=1}^{p} \Theta(t \otimes e_j, t \otimes e_j)$$

pour $t \in T$. Alors on a

(8) $\Theta + Tr_E\, \Theta \otimes Id_E \geqslant_N 0$,

(9) $\Theta \leqslant_N Inf(n,p)\, Tr_E \Theta \otimes Id_E$.

Soit E un fibré vectoriel hermitien sur X , semi-positif (resp. positif) au sens de GRIFFITHS. Alors

(10) $E \otimes \text{dét } E \geqslant_N 0$ (resp. $E \otimes \text{dét } E >_N 0$) ,

$E^{\star} \otimes \text{dét } E^{\star} \leqslant_N 0$ (resp. $E \otimes \text{dét } E <_N 0$) ,

(11) $E \otimes (\text{dét } E)^{-Inf(n,p)} \leqslant_N 0$ (resp. $<_N 0$) ,

$E^{\star} \otimes (\text{dét } E)^{Inf(n,p)} \geqslant_N 0$ (resp. $>_N 0$) .

REMARQUE 1. Les deux assertions de (10) (ou de (11)) ne sont pas équivalentes car la positivité de E équivaut à la négativité de $E^{\star}$ seulement au sens de GRIFFITHS (mais pas au sens de NAKANO en général).

Les points (9) et (11) sont en fait une généralisation du lemme fondamental (3,5) de [14] . Lorsque E est quotient d'un fibré trivial, la seconde assertion de (11) est d'ailleurs conséquence de ce lemme de [14] .

Démonstration. Nous renvoyons le lecteur à [2] , [3] pour une preuve de (8).

Preuve de (9) : montrons tout d'abord que

$$\mathrm{Tr}_E\, \Theta \otimes \mathrm{Id}_E - \Theta \geqslant_G 0 \,.$$

En effet, tout vecteur décomposable $x \in T \otimes E$ peut s'écrire $x = t \otimes e$ où $\|e\| = 1$; si l'on choisit une base orthonormée $(e_j)_{1 \leqslant j \leqslant p}$ de E telle que $e_1 = e$,il vient

$$\Theta(x,x) = \Theta(t \otimes e_1, t \otimes e_1) \quad , \text{ et}$$

$$\mathrm{Tr}_E\, \Theta \otimes \mathrm{Id}_E(x,x) = \sum_{j=1}^{p} \Theta(t \otimes e_j, t \otimes e_j)\, \|e\|^2 \geqslant \Theta(x,x) \,,$$

grâce à l'hypothèse $\Theta \geqslant_G 0$. D'après (8) on a donc

$$\mathrm{Tr}_E\, \Theta \otimes \mathrm{Id}_E - \Theta + \mathrm{Tr}_E(\mathrm{Tr}_E\, \Theta \otimes \mathrm{Id}_E - \Theta) \otimes \mathrm{Id}_E = p\, \mathrm{Tr}_E\, \Theta \otimes \mathrm{Id}_E - \Theta \geqslant_N 0 \,,$$

ce qui démontre (9) lorsque $p \leqslant n$. Lorsque $n \leqslant p$, il est aisé de voir que

$$T \otimes E = \bigcup_{F \subset E,\ \dim F = n} T \otimes F \,.$$

Appliquons le résultat déjà trouvé à la restriction Θ_F de Θ à $T \otimes F$, pour tout sous-espace F de dimension n de E . (9) se déduit des inégalités suivantes

$$\Theta_F \leqslant_N n\, \mathrm{Tr}_F\, \Theta_F \otimes \mathrm{Id}_F \leqslant_N n\, \mathrm{Tr}_E \Theta \otimes \mathrm{Id}_F \,,$$

soit $\Theta \leqslant_N n\, \mathrm{Tr}_E \Theta \otimes \mathrm{Id}_E$ sur $T \otimes F$.

Preuve de (10) et (11).

Il est bien connu que la courbure du fibré $\text{dét}\, E = \Lambda^p E$ est reliée à celle de E par la formule

(12) $$c(\text{dét}\, E) = \mathrm{Tr}_E\, c(E) = - \mathrm{Tr}_E\, c(E^*) \,,$$

et que pour deux fibrés vectoriels hermitiens E_1 et E_2 , on a

(13) $$c(E_1 \otimes E_2) = c(E_1) \otimes \mathrm{Id}_{E_2} + \mathrm{Id}_{E_1} \otimes c(E_2) \,.$$

(10) et (11) se déduisent alors de (8) et (9) en prenant $\Theta = ic(E)$ ou $\Theta = -ic(E^*)$ selon le cas.

2. Estimations a priori et inégalités L^2 .

On considère , avec les notations et hypothèses de l'introduction, la suite exacte (1) :

$$0 \to S \to E \xrightarrow{g} Q \to 0 .$$

La connexion canonique D sur E se décompose suivant le scindage orthogonal C^∞ , $E = S \otimes Q$, de la manière suivante

$$(14) \qquad D = \begin{pmatrix} D_S & -\beta^* \\ \beta & D_Q \end{pmatrix} ,$$

où D_S et P_Q sont respectivement les connexions canoniques sur S et Q , et où $\beta \in C^\infty_{1,0}(X, \mathrm{Hom}(S,Q))$.

Nous renvoyons à P.A.GRIFFITHS [5] pour la démonstration de (14), et à H.SKODA [14] pour le détail des calculs qui vont suivre. On a classiquement

$$c(E) = D^2 = \begin{pmatrix} D_S^2 - \beta^* \wedge \beta & -D\beta^* \\ D\beta & D_Q^2 - \beta \wedge \beta^* \end{pmatrix}$$

d'où par définition

$$(15) \left\{ \begin{array}{l} c(S) = D_S^2 = c(E)|_S + \beta^* \wedge \beta \\ c(Q) = D_Q^2 = c(E)|_Q + \beta \wedge \beta^* . \end{array} \right.$$

Soit $K = \Lambda^n T^* X$ le fibré canonique de la variété X et M un fibré en droites sur X . Par application à (1) du foncteur $\mathrm{Hom}(Q, ? \otimes K \otimes M)$, on obtient la suite exacte

$$(16) \qquad 0 \to \mathrm{Hom}(Q, S \otimes K \otimes M) \to \mathrm{Hom}(Q, E \otimes K \otimes M) \to \mathrm{Hom}(Q, Q \otimes K \otimes M) \to 0 ,$$

et la connexion $D_{\mathrm{Hom}(Q,\, E \otimes K \otimes M)}$ se décompose suivant le scindage orthogonal de (16) , en

$$D_{\mathrm{Hom}(Q,\, E\otimes K\otimes M)} = \begin{pmatrix} D_{\mathrm{Hom}(Q,\, S\otimes K\otimes M)} & -\beta^* \\ \beta & D_{\mathrm{Hom}(Q,\, Q\otimes K\otimes M)} \end{pmatrix}.$$

Posons $R = \mathrm{Hom}(Q, S \otimes M)$ pour simplifier les notations.

PROBLÈME. - <u>Etant donné une section holomorphe</u> f <u>de</u> $\mathrm{Hom}(Q, Q\otimes K\otimes M)$, <u>chercher un</u>

<u>relèvement</u> h <u>de</u> f <u>dans</u> $\mathrm{Hom}(Q, E \otimes K \otimes M)$ <u>sous la forme</u>

$$h = f + u ,$$

<u>avec</u> $u \in C^{\infty}(X, R \otimes K) = C^{\infty}_{n,0}(X, R)$.

On aura bien par construction $g \circ h = f$, et h sera holomorphe si et seulement si

$$D''_R u = - D''_{\mathrm{Hom}(Q,\, E \otimes K \otimes M)} f = \beta^* f . \tag{17}$$

On résout cette équation par la méthode d'HÖRMANDER [6] et [7] , ce qui, modulo l'inégalité de KODAIRA-NAKANO (cf. [4], exposé III, th. 3) , nécessite l'obtention d'une estimation a priori du type suivant (avec une constante $A \geq 0$)

$$|(\beta^* f | v)|^2 \leqslant A(ic(R)\Lambda\, v \,|\, v) \tag{18}$$

pour toute $(n,1)$ forme v à valeurs dans R , de classe C^{∞} et à support compact.

Le produit scalaire $(\;|\;)$ est défini à partir du produit scalaire ponctuel $< , >$ des formes par la formule

$$(v \,|\, w) = \int_X \langle v, w \rangle \, dV ,$$

pour deux formes v et w à valeurs dans R , de classe C^{∞} , et à support compact.

Λ désigne d'autre part l'opérateur de type $(-1,-1)$, adjoint de l'opérateur L de multiplication extérieure par ω , pour le produit scalaire ponctuel $< , >$.

Nous nous servirons de la proposition suivante (cf. H.SKODA [14] , lemme (3,1) , proposition 3.1. , et conclusion (4,17)).

PROPOSITION. - <u>Sous l'hypothèse</u> (18), <u>il existe une forme</u> $u \in C^{\infty}(X, R \otimes K)$ <u>vérifiant</u> (17) : $D''_R u = \beta^* f$, et :

$$\int_X |u|^2 \, dV \leqslant A . \tag{19}$$

Le relèvement $h = f + u$ de f est donc tel que

$$\int_X |h|^2 \, dV \leqslant A + \int_X |f|^2 \, dV , \tag{20}$$

car f et u sont orthogonaux.

Le théorème 1 sera démontré si nous établissons l'inégalité (18) et déterminons la constante A .

3. Calcul de courbure.

Nous aurons besoin des notations et résultats suivants .

Le produit intérieur $a \lrcorner b$ de deux formes à valeurs scalaires est défini en tout point z de X par dualité :

$$< a \lrcorner b \, , \, c > \; = \; < b \, , \, \bar{a} \wedge c > \quad ;$$

on étend le produit intérieur au cas où a, b sont des formes à valeurs dans des fibrés vectoriels E, F par bilinéarité (le résultat étant à valeurs dans le fibré G si on a un morphisme bilinéaire $E \times F \longrightarrow G$).

LEMME 1. - (H.SKODA [14] , lemmes (3,3) et (3,4)) .

Pour toute forme $v \in C^{\infty}_{n,1}(X, R) = C^{\infty}_{0,1}(X, \mathrm{Hom}(Q, S \otimes K \otimes M))$ à support compact, toute (1,1)-forme réelle $\Theta \geqslant_N 0$ à valeurs dans $\mathrm{Herm}(R, R)$, et toute forme $\beta \in C^{\infty}_{1,0}(X, \mathrm{Hom}(S, Q))$, on a

(21) $$(\Theta \;\Lambda\; v \,|\, v) \geqslant 0$$

(22) $$(-i\beta^{*} \wedge \beta \;\Lambda\; v \,|\, v) = \|\beta \lrcorner v\|^{2}.$$

Démonstration de (21). Soit

$$\Theta = i \sum_{\lambda,\mu,j,k} a_{\lambda\mu jk}\, dz_{\lambda} \wedge d\bar{z}_{\mu} \otimes e^{*}_{j} \otimes e_{k} \; ,$$

$$v = \frac{i}{2} \sum_{\lambda,j} v_{\lambda j}\, dz_{1} \wedge \ldots \wedge dz_{n} \wedge d\bar{z}_{\lambda} \otimes e_{j} \; ,$$

l'écriture canonique de Θ et de v relativement à une base orthonormée $(dz_{\lambda})_{1 \leqslant \lambda \leqslant n}$ de $T^{*}X$, et à une base orthonormée $\{e_{j}\}$ de la fibre de R . On a en tout point

$$\Lambda v = \sum_{\lambda,j} (-1)^{n-\lambda}\, v_{\lambda j}\, dz_{1} \wedge \ldots \wedge \widehat{dz_{\lambda}} \wedge \ldots \wedge dz_{n} \otimes e_{j} \; ,$$

$$\Theta \wedge v = i \sum_{\lambda,\mu,j,k} a_{\lambda\mu jk}\, v_{\lambda j}\, dz_{1} \wedge \ldots \wedge dz_{n} \wedge d\bar{z}_{\mu} \otimes e_{k} \; ,$$

$$< \Theta \wedge v, v > = 2^{n} \sum_{\lambda,\mu,j,k} a_{\lambda\mu jk}\, v_{\lambda j}\, \bar{v}_{\mu k} \geqslant 0 \; ,$$

d'après l'hypothèse de semi-positivité de NAKANO de Θ .

Démonstration de (22). Ecrivons en tout point de X

$$\beta = \sum_{\lambda,j,\ell} \beta_{\lambda\ell j}\, dz_{\lambda} \otimes \varepsilon^{*}_{j} \otimes \eta_{\ell} \; ,$$

$$v = \frac{i}{2} \sum_{\lambda,j,m} v_{\lambda j}\, dz_{1} \wedge \ldots \wedge dz_{n} \wedge d\bar{z}_{\lambda} \otimes \varepsilon_{j} \; ,$$

avec $v_{\lambda j} \in Q^* \otimes M$, $1 \leqslant \lambda \leqslant n$, $1 \leqslant j \leqslant s$, $1 \leqslant \ell \leqslant q$, dans des bases orthonormées $(\varepsilon_j)_{1 \leqslant j \leqslant s}$, $(\eta_\ell)_{1 \leqslant \ell \leqslant q}$ des fibres de S et Q . On vérifie que

$$-i\beta^* \wedge \beta = i \sum_{\lambda,\mu,j,k,\ell} \beta_{\lambda\ell j} \overline{\beta_{\mu\ell k}} \, dz_\lambda \wedge d\bar{z}_\mu \otimes \varepsilon_j^* \otimes \varepsilon_k ,$$

de sorte que les coefficients de $\Theta = -i\beta^* \wedge \beta \in \mathrm{Herm}(S,S)$ sont donnés par

$$a_{\lambda\mu jk} = \sum_\ell \beta_{\lambda\ell j} \overline{\beta_{\mu\ell k}} .$$

D'après la première partie de la démonstration, on a

$$\begin{aligned} < -i\beta^* \wedge \beta \, \Lambda \, v,v > &= < \Theta \otimes \mathrm{Id}_{Q^* \otimes M} \Lambda \, v,v > \\ &= 2^n \sum_{\lambda,\mu,j,k,\ell} \beta_{\lambda\ell j} \overline{\beta_{\mu\ell k}} < v_{\lambda j} , v_{\mu k} > \\ &= 2^n \sum_\ell \Big| \sum_{\lambda,j} \beta_{\lambda\ell j} v_{\lambda j} \Big|^2 . \end{aligned}$$

(22) résulte donc de l'inégalité

$$\beta \lrcorner v = (-1)^n i \sum_{\lambda,j,\ell} \beta_{\lambda\ell j} v_{\lambda j} \, dz_1 \wedge \ldots \wedge dz_n \otimes \eta_\ell . \quad ■$$

Nous disposons maintenant des moyens techniques nécessaires pour effectuer le calcul de courbure.

Puisque $R = Q^* \otimes S \otimes M$, on a d'après la formule (13)

(23) $c(R) = c(Q^*) \otimes \mathrm{Id}_{S \otimes M} + c(S) \otimes \mathrm{Id}_{Q^* \otimes M} + c(M) \otimes \mathrm{Id}_{Q^* \otimes S}$.

Supposons le fibré E semi-positif au sens de GRIFFITHS.

Alors il en est de même pour le fibré Q . D'après le théorème 2 (11) on a

(24) $ic(Q^*) + \mathrm{Inf}(n,q) \, ic(\det Q) \otimes \mathrm{Id}_{Q^*} \geqslant_N 0$.

Puisque $-i\beta^* \wedge \beta \geqslant_G 0$ (en fait on a même $-i\beta^* \wedge \beta \geqslant_N 0$) , (9) et (15) entraînent successivement

(25) $$\begin{aligned} i(c(E)|_S - c(S)) = -i\beta^* \wedge \beta &\leqslant_N \mathrm{Inf}(n,s) i \, \mathrm{Tr}(-\beta^* \wedge \beta) \otimes \mathrm{Id}_S \\ &= \mathrm{Inf}(n,s) i \, \mathrm{Tr} \, \beta \wedge \beta^* \otimes \mathrm{Id}_S , \end{aligned}$$

(26) $i(c(E)|_S - c(S)) \leqslant_N \mathrm{Inf}(n,s) \, i(c(\det Q) - \mathrm{Tr}\, c(E)|_Q) \otimes \mathrm{Id}_S$,

d'où , après substitution de (24) et (26) dans (23) :

$$ic(R) \geqslant_N i[c(M) - (\mathrm{Inf}(n,q) + \mathrm{Inf}(n,s)) c(\det Q) + \mathrm{Inf}(n,s).\mathrm{Tr}\, c(E)|_Q] \otimes \mathrm{Id}_R + ic(E)|_S \otimes \mathrm{Id}_{Q^* \otimes M} .$$

Si (hypothèse (3)) S est de rang 1 , ou si $E \geqslant_N 0$, on a $ic(E)|_S \geqslant_N 0$; de façon générale, (8) implique

$$ic(E)|_S + i\,\mathrm{Tr}\,c(E)|_S \otimes \mathrm{Id}_S \geqslant_N 0 .$$

En substituant à nouveau dans (27), il vient

$$ic(R) \geqslant_N i[c(M) - (\mathrm{Inf}(n,q) + \mathrm{Inf}(n,s)c(\text{dét } Q) + \mathrm{Inf}(n,s)\mathrm{Tr}\,c(E)|_Q - \mathrm{Tr}\,c(E)|_S] \otimes \mathrm{Id}_R .$$

Posons $\varepsilon = k - \mathrm{Inf}(n,q) - \mathrm{Inf}(n,s) > 0$; comme $\mathrm{Tr}\,\beta\wedge\beta^* = c(\text{dét } Q) - \mathrm{Tr}\,c(E)|_Q$ d'après (15), on obtient

(28) $$ic(R) - i\varepsilon\,\mathrm{Tr}\,\beta\wedge\beta^* \otimes \mathrm{Id}_R \geqslant_N$$

$$i(c(M) - k\,c(\text{dét } Q) + (k - \mathrm{Inf}(n,q))\mathrm{Tr}\,c(E)|_Q - \mathrm{Tr}\,c(E)|_S) \otimes \mathrm{Id}_R ;$$

le terme $\mathrm{Tr}\,c(E)|_S$ peut même être omis sous l'hypothèse (3).

Remplaçons M par $M \otimes K^* = M \otimes \text{dét } TX$; compte tenu de ce que $\mathrm{Ricci}(\omega) = c(\text{dét } TX)$ et $E \geqslant_G 0$, le premier membre de (28) est semi-positif sous les hypothèses (2) ou (3). La propriété (21) entraîne donc l'inégalité suivante :

(29) $$(ic(R)\,\Lambda v|v) \geqslant \varepsilon(i\,\mathrm{Tr}\,\beta\wedge\beta^*\,\Lambda v|\,v) .$$

Pour obtenir l'estimation a priori (18), il nous reste à majorer $|(\beta^* f|v)|^2$ en fonction de $(i\,\mathrm{Tr}\,\beta\wedge\beta^*\Lambda v|v)$.

Pour obtenir (18), il nous reste à majorer $|(\beta^* f\,|\,v)|^2$ en fonction de $(i\,\mathrm{Tr}\,\beta\wedge\beta^*\,\Lambda\,v|v)$. Par définition du produit intérieur, et d'après (22), on a

(30) $$|(\beta^* f|v)|^2 = |(f|\beta \lrcorner v)|^2 \leqslant \|f\|^2\,\|\beta\lrcorner v\|^2 = \|f\|^2\,(-i\beta^*\wedge\beta\,\Lambda v|v) .$$

Une nouvelle application de (21) fournit à partir de (25) :

(31) $$(-i\beta^*\wedge\beta\,\Lambda\,v|v) \leqslant \mathrm{Inf}(n,s)\,(i\,\mathrm{Tr}\,\beta\wedge\beta^*\,\Lambda\,v|v) ,$$

d'où en combinant avec (29):

LEMME 2. - <u>L'estimation</u> (18) $|(\beta^* f|v)|^2 \leqslant A(i\,c(R)\,\Lambda v|v)$ <u>est satisfaite , et on peut choisir la constante</u> A <u>égale à</u>

$$\frac{\mathrm{Inf}(n,s)}{k-\mathrm{Inf}(n,q) - \mathrm{Inf}(n,s)} \int_X |f|^2\,dV .$$

Le théorème 1 résulte maintenant de la proposition (cf. (20)) et du lemme 2.

<u>REMARQUE 2</u>. Les calculs précédents montrent en fait que le théorème est vrai si l'on suppose seulement

(32) $$i(c(M)-kc(\text{dét } Q)+(k-\mathrm{Inf}(n,q))\mathrm{Tr}\,c(E)|_Q-\mathrm{Tr}\,c(E)|_S+\mathrm{Ricci}\,\omega) \geqslant 0 ,$$

le terme $\mathrm{Tr}\,c(E)|_S$ étant superflu si $s = 1$, ou si $E \geqslant_N 0$.

Mais nous avons préféré énoncer le théorème 1 avec des hypothèses géométriques qui ne supposent pas une connaissance approfondie de $c(E)$.

REMARQUE 3. Lorsque la section f du fibré $\mathrm{Hom}(Q, Q \otimes M)$ est de la forme $f = \mathrm{Id}_Q \otimes u$ pour une section u de M, on va montrer que

$$|(\beta^* f|v)|^2 \leqslant \mathrm{Inf}(\tfrac{n}{q}, s)\ \|f\|^2\ (i\ \mathrm{Tr}\ \beta\wedge\beta^*\ \Lambda\ v|v)\ ,$$

de sorte qu'on peut dans le lemme 2 prendre

$$A = \frac{\mathrm{Inf}(\tfrac{n}{q}, s)}{k - \mathrm{Inf}(n,q) - \mathrm{Inf}(n,s)} \int_X |f|^2\, dV\ ,$$

et remplacer la constante C du théorème 1 par

$$C' = 1 + \frac{\mathrm{Inf}(\tfrac{n}{q}, s)}{k - \mathrm{Inf}(n,q) - \mathrm{Inf}(n,s)}\ .$$

Ecrivons en effet en chaque point $z \in X$ la section

$v \in C^\infty_{n,1}(X,R) = C^\infty(X, \mathrm{Hom}(Q,S \otimes K \otimes M))$ sous la forme $v = w \otimes e$, où e est un vecteur unitaire de la fibre $K_z \otimes M_z$, et où w est une $(0,1)$-forme à valeurs dans $\mathrm{Hom}_z(Q,S)$. On a

$$(33)\qquad \langle\beta^* f, v\rangle = \langle \beta^* \circ \mathrm{Id}_Q \otimes u, w \otimes e\rangle = \langle\beta^*, w\rangle\ \langle u, e\rangle .$$

Dans une base orthonormée (dz_j) de $T_z^* X$, les formes β et w s'écrivent

$$\beta = \sum_{j=1}^n \beta_j\, dz_j\ ,\quad \beta_j \in \mathrm{Hom}_z(S,Q)\ ,$$

$$w = \frac{1}{2} \sum_{j=1}^n w_j\, d\bar z_j\ ,\quad w_j \in \mathrm{Hom}_z(Q,S)\ .$$

Il vient donc

$$\langle\beta^*, w\rangle = \sum_{j=1}^n \langle\beta_j^*, w_j\rangle\ ,$$

$$(34)\quad |\langle\beta^*, w\rangle|^2 \leqslant n \sum_{j=1}^n |\beta_j|^2\ |v_j|^2\ ,\ \text{avec}\ v_j = w_j \otimes e\ .$$

Si d'autre part, on a choisi la base (dz_j) de sorte que les éléments β_j soient orthogonaux (ce qui est toujours possible, car toute matrice $r \times n$, B peut s'écrire $B = U D V$, où D matrice "diagonale" $r \times n$, U et V matrices unitaires $r \times r$ et $n \times n$) on obtient successivement :

$$\beta \wedge \beta^* = \sum_{1 \leqslant j,k \leqslant n} \beta_j\, \beta_k^*\, dz_j \wedge d\bar z_k\ ,$$

$$i\ \mathrm{Tr}\ \beta \wedge \beta^* = i \sum_{j=1}^n \mathrm{Tr}(\beta_j\, \beta_j^*)\, dz_j \wedge d\bar z_j = i \sum_{j=1}^n |\beta_j|^2\, dz_j \wedge d\bar z_j\ ,$$

$$(35)\qquad <i\,\mathrm{Tr}\,\beta\wedge\beta^{*}\Lambda v, v> = \sum_{j=1}^{n} |\beta_j|^2\,|v_j|^2 \;;$$

pour établir l'égalité (35), on se reportera à la démonstration de (21). En combinant (33), (34) et (35), on voit que

$$|<\beta^{*} f, v>|^2 \leqslant n <i\,\mathrm{Tr}\,\beta\wedge\beta^{*}\,\Lambda v, v> \;\; |u|^2$$
$$= \frac{n}{q}\,|f|^2 <i\,\mathrm{Tr}\,\beta\wedge\beta^{*}\,\Lambda v, v> .$$

Après intégration sur X , l'inégalité de Cauchy-Schwarz implique

$$|\,(\beta^{*} f|v)|^2 \leqslant \frac{n}{q}\Big(\int_X |f|^2\, dV\Big)\,.\,(i\,\mathrm{Tr}\,\beta\wedge\beta^{*}\,\Lambda\, v|v)\;,$$

et (30) , (31) montrent qu'on peut remplacer $\frac{n}{q}$ par $\mathrm{Inf}(\frac{n}{q}, s)$. ∎

On va maintenant énoncer les résultats du théorème 1 en fonction d'une métrique donnée a priori sur Q , de manière à pouvoir traiter comme H.SKODA [14] le cas où le morphisme g dégénère.

Soit g^{*} le morphisme C^{∞} , transposé de g , pour les métriques données sur E et Q :

$$0 \longrightarrow Q \xrightarrow{g^{*}} E\;.$$

Le morphisme $g^{*}(gg^{*})^{-1} : Q \rightarrow E$, est le scindage C^{∞} de la suite exacte (1) qui envoie Q dans $(\mathrm{Ker}\; g)^{\perp} = S^{\perp}$; la métrique quotient $|\;\;|'$ sur Q est donc donnée en fonction de la métrique initiale $|\;\;|$ par

$$(36)\qquad |u|'^2 = |g^{*}(gg^{*})^{-1}\, u|^2 = <(gg^{*})^{-1}\, u,u>\;,\; u\in Q\;.$$

Désignons par $c'(\text{dét}\; Q)$ la forme de courbure de $\text{dét}\; Q$ relative à la métrique quotient sur Q . Il résulte aussitôt de (36) que pour tout $v \in \text{dét}\; Q$:

$$|v|'^2 = \text{dét}(gg^{*})^{-1}\;\; |v|^2\;.$$

On a donc

$$c'(\text{dét}\; Q) = c(\text{dét}\; Q) + d'd''\,\mathrm{Log}\;\text{dét}(gg^{*})\;.$$

Si l'on veut conserver telles quelles les hypothèses de positivité (2) et (3), on est amené à multiplier la métrique de M par le poids $[\text{dét}(gg^{*})]^{k}$, de sorte que l'estimation (5) du théorème 1 devient

$$(37)\qquad \int_X |h|'^2\,(\text{dét}\; gg^{*})^{-k}\, dV \leqslant C \int_X |f|'^2\,(\text{dét}\; gg^{*})^{-k}\, dV\;.$$

Pour tout élément $h\in \mathrm{Hom}(Q, E)$, la norme $|h|'$ est donnée en fonction de la norme naturelle $|h|$ par

$$|h|'^2 = |h \circ g|^2 = \langle h \circ gg^*, h \rangle ,$$

et d'après (36), on a pour tout $f \in \mathrm{Hom}(Q, Q)$:

$$|f|'^2 = \langle (gg^*)^{-1} \circ f \circ gg^* , f \rangle$$
$$= (\det gg^*)^{-1} \langle \widetilde{gg^*} \circ f \circ gg^*, f \rangle ,$$

où $\widetilde{gg^*}$ désigne l'endomorphisme cotransposé de gg^*.

L'estimation (37) s'écrit donc

$$\int_X \langle h \circ gg^*, h \rangle (\det gg^*)^{-k} \, dV \leqslant C \int_X \langle \widetilde{gg^*} \circ f \circ gg^* , f \rangle (\det gg^*)^{-k-1} \, dV .$$

Si maintenant g n'est surjectif qu'au dessus de X privé d'un ensemble analytique Z, on suppose que Z est X-négligeable au sens suivant.

DÉFINITION 2. - Nous dirons qu'un ensemble $Z \subset X$ est X-négligeable, s'il existe un sous-ensemble fermé Y de X, contenant Z, de mesure nulle, tel que l'ouvert $X \setminus Y$ soit faiblement pseudoconvexe, et tel que toute fonction de carré sommable sur un ouvert U de X, holomorphe dans $U \setminus Y$, se prolonge en une fonction holomorphe sur U.

Lorsque la variété X est de Stein ou projective, Z est toujours X-négligeable : il suffit de prendre pour Y une hypersurface quelconque de X contenant Z.

Si Z est X-négligeable, le théorème 1 s'applique à $X \setminus Y$. Comme $k \geqslant 1$, la finitude de l'intégrale

$$\int_{X \setminus Y} \langle h \circ gg^*, h \rangle (\det gg^*)^{-k} \, dV$$

entraîne que h est localement L^2, donc que h se prolonge à X. D'où le

THÉORÈME 2. - Etant donné un morphisme $g : E \to Q$, on suppose que l'ensemble analytique $Z = \{z \in X \mid g(E_z) \neq Q_z\}$ est X-négligeable[*], et que le fibré linéaire M vérifie l'une des conditions de positivité (2), (3) ou (32).

Alors, pour toute section f de $\mathrm{Hom}(Q, Q \otimes M)$ telle que le second membre de (38) soit fini, il existe une section h de $\mathrm{Hom}(Q, E \otimes M)$ telle que $g \circ h = f$ et

$$(38) \quad \int_{X \setminus Z} \langle h \circ gg^*, h \rangle (\det gg^*)^{-k} \, dV \leqslant C \int_{X \setminus Z} \langle \widetilde{gg^*} \, f \circ gg^*, f \rangle (\det gg^*)^{-k-1} \, dV ,$$

avec $C = \dfrac{k - \mathrm{Inf}(n,q)}{k - \mathrm{Inf}(n,q) - \mathrm{Inf}(n,s)}$.

---

[*] On peut démontrer que cette hypothèse est superflue.

4. Construction de rétractions holomorphes.

Soit X une sous-variété fermée de dimension n d'un ouvert pseudoconvexe Ω de $\mathbb{C}^p$. On s'intéresse à la suite exacte

$$0 \longrightarrow TX \longrightarrow T\Omega|_X \xrightarrow{g} NX \longrightarrow 0$$

où NX est le fibré normal à X. Tous ces objets sont munis des métriques induites par la métrique de $\mathbb{C}^p$. Avec les notations de l'introduction, on a

$$E = T\Omega|_X \ , \ S = TX \ , \ Q = NX \ , \ s = n \ , \ q = p - n \ .$$

De plus, dét Q = dét $NX \simeq$ (dét $TX)^{-1}$ métriquement, donc

$$c(\text{dét}\, Q) = - \operatorname{Ricci}(X) \ .$$

Choisissons pour M un fibré trivial, dont la métrique est donnée par le poids $e^{-\varphi}$ de telle sorte que $c(M) = d'd''\varphi$.

Comme le fibré $E = T\Omega|_X$ est plat, les conditions (2) et (3) s'écrivent :

$$id'd''\varphi + i(k+1)\operatorname{Ricci}(X) \geqslant 0 \ ;$$

en appliquant le théorème 1 à $f = Id_Q$, dont la norme en tant que section de $\operatorname{Hom}(Q, Q \otimes M)$ vaut $qe^{-\varphi}$, et compte-tenu de la remarque 3, on obtient le

THÉORÈME 3. - Pour toute fonction φ plurisousharmonique sur X, et tout réel $k > \operatorname{Inf}(2n,p)$ tels que

$$id'd''\varphi + i(k+1)\operatorname{Ricci}(X) \geqslant 0 \ ,$$

il existe une section $h : NX \longrightarrow T\Omega|_X$ telle que

$$g \circ h = Id_{NX} \ ,$$

$$(39) \quad \int_X |h|^2 e^{-\varphi}\, dV \leqslant \left(p-n + \frac{n}{k - \operatorname{Inf}(2n,p)}\right) \int_X e^{-\varphi}\, dV \ .$$

REMARQUE 4. Le résultat a été démontré seulement lorsque φ est de classe C^∞, mais il est immédiat de se débarrasser de cette hypothèse par un passage à la limite. On notera que la condition de courbure ne peut être vérifiée que si φ est plurisousharmonique, car $i \operatorname{Ricci}(X) \leqslant 0$.

Si $\pi_X : NX \longrightarrow X$ est la projection du fibré NX, on définit une application $\sigma : NX \longrightarrow \mathbb{C}^p$ par

$$\sigma(\zeta) = \pi_X(\zeta) + h.\zeta \ , \quad \zeta \in NX \ .$$

Il est clair, d'après le théorème des fonctions implicites, que σ est un isomorphisme d'un voisinage V de la section nulle dans NX, sur un voisinage V' de X dans Ω.

On construit donc une rétraction holomorphe $r : V' \to X$ (c'est-à-dire une application holomorphe $r : V' \to X$ telle que $r(z) = z$ pour tout $z \in X$) en posant

$$r = \pi_X \circ \sigma^{-1} .$$

Il ne nous reste plus qu'à préciser V et V'.

On se donne une fonction $\rho > 0$ sur X telle que pour tout $z \in X$, il existe un polydisque $D(z, \rho(z))$ de centre z, de rayon $\rho(z)$, dans lequel X est un graphe. De façon précise, on suppose :

(40) $D(z, \rho(z))$ est le produit des deux disques $D' \subset T_z X$, $D'' \subset (T_z X)^{\perp}$ de centre z et de rayon $\rho(z)$.

(41) $X \cap D(z, \rho(z))$ est le graphe d'une application holomorphe

$$u_z : D' \to D'' .$$

Si on pose $\varphi_\rho(z) = \sup_{\zeta \in X \cap D(z,\rho(z))} \varphi(\zeta)$, on obtient le résultat général suivant, qui ne fait intervenir que des données géométriques de X.

THÉORÈME 4. - <u>Soit</u> φ <u>une fonction plurisousharmonique sur</u> X <u>telle que</u>

$$id'd''\varphi + i(\varepsilon + 1 + \mathrm{Inf}(2n,p))\mathrm{Ricci}(X) \geqslant 0 \quad (\varepsilon > 0),$$

$$\int_X e^{-\varphi}\, dV < +\infty ,$$

ρ <u>une fonction vérifiant les hypothèses</u> (40), (41), <u>et</u> h <u>le scindage holomorphe du théorème 3.</u>

<u>Alors l'application</u> $\sigma(z, \xi) = z + h(z).\xi$, <u>définie sur</u> NX, <u>est injective sur un "voisinage"</u> V <u>de la section nulle dans</u> NX <u>de la forme</u>

$$V = \{(z, \xi) \in NX \ ; \ |\xi| < C_1 \, e^{-\varphi_\rho(z)} \rho(z)^{2n+1}\} .$$

<u>Il existe une constante</u> $C_2 > 0$ <u>et une rétraction holomorphe</u> $r : U \to X$ <u>sur l'ouvert</u>

$$U = \{\zeta \in \mathbb{C}^p \ ; \ (\exists z \in X) \ |\zeta - z| < C_2 \, e^{-\varphi_\rho(z)} \rho(z)^{2n+1}\}$$

<u>Les constantes</u> $C_1 > 0$, $C_2 > 0$, <u>sont le produit de constantes universelles (ne dépendant que de la dimension</u> p) <u>et de</u>

$$[(1 + \tfrac{1}{\varepsilon}) \int e^{-\varphi}\, dV]^{-1} .$$

Démonstration. Dans toute la suite, on désignera par α_j les constantes ne dépendant que de la dimension p, et on posera

$$C = (1 + \frac{1}{\varepsilon}) \int_X e^{-\varphi}\, dV .$$

On considère en tout point $z \in X$ un système de coordonnées linéaires $(\zeta_1, \ldots, \zeta_p)$ tel que $\frac{\partial}{\partial \zeta_1}, \ldots, \frac{\partial}{\partial \zeta_n}$ soit une base orthonormée de $T_z X$, et $\frac{\partial}{\partial \zeta_{n+1}}, \ldots, \frac{\partial}{\partial \zeta_p}$ une base orthonormée de $(T_z X)^{\perp}$. Les vecteurs $\frac{\partial}{\partial \zeta_{n+1}}, \ldots, \frac{\partial}{\partial \zeta_p}$ définissent un repère local de NX au-dessus de $X \cap D_z$; on note $\xi_{n+1}, \ldots, \xi_p$ les coordonnées correspondantes dans les fibres de NX. La section $h \in \mathrm{Hom}(NX, T\Omega|_X)$ est donc définie dans $X \cap D_z$ par une matrice

$$H_z(\zeta_1, \ldots, \zeta_n) = \left[h_{jk}(\zeta_1, \ldots, \zeta_n)\right]_{\substack{1 \leqslant j \leqslant p \\ n+1 \leqslant k \leqslant p}} ,$$

avec $h_{jk}(z) = \delta_{jk}$, $n+1 \leqslant j,k \leqslant p$.

Soit Δ_z le polydisque de centre z et de rayon $\frac{1}{2}\rho(z)$, contenu dans $D(z, \rho(z))$. Dans Δ_z, $|h|^2$ est équivalent à une constante universelle près à

$$|H|^2 = \sum_{j,k} |h_{jk}|^2$$

(noter que l'application u_z de (41) a ses dérivées bornées dans Δ_z), et on tire de (39), grâce aux inégalités de Cauchy, que pour tout $\zeta \in \Delta_z$,

$$(42) \quad \begin{cases} |H_z(\zeta)| \leqslant \alpha_1 C^{1/2} e^{1/2\varphi_\rho(z)} \rho(z)^{-n} \\ \sup\limits_{1 \leqslant j \leqslant n} \left| \frac{\partial H_z}{\partial \zeta_j}(\zeta) \right| \leqslant \alpha_2 C^{1/2} e^{1/2\varphi_\rho(z)} \rho(z)^{-n-1} . \end{cases}$$

a/ <u>Injectivité de</u> σ <u>sur</u> $V \subset NX$.

De (42) il résulte en particulier pour $\zeta = z$:

$$(43) \qquad C^{1/2} e^{1/2\varphi_\rho(z)} \rho(z)^{-n} \geqslant \alpha_3 ,$$

et si (z, ξ) appartient au "voisinage" V, on a

$$(44) \quad |\sigma(z, \xi) - z| = |h(z).\xi| \leqslant \alpha_4 C_1 C^{1/2} e^{-1/2\varphi_\rho(z)} \rho(z)^{n+1} .$$

Supposons $\sigma(z, \xi) = \sigma(z', \xi')$ pour deux points distincts (z, ξ), (z', ξ') de V, ce qui ne peut se produire que si $z \neq z'$; on a par exemple

$$e^{-1/2\varphi_\rho(z')} \rho(z')^{n+1} \leqslant e^{-1/2\varphi_\rho(z)} \rho(z)^{n+1} .$$

(43) et (44) entraînent donc

(45) $|z' - z| \leqslant 2\alpha_4 C_1 C^{1/2} e^{-1/2\varphi_\rho(z)} \rho(z)^{n+1} \leqslant \alpha_5 C_1 C \rho(z)$,

et $z' \in \Delta_z$ dès que $C_1 C$ est assez petit .

On montre aisément à partir de (41) , (45) que

$$\text{angle}(T_z X ; z'-z) \leqslant \alpha_6 C_1 C^{1/2} e^{-1/2\varphi_\rho(z)} \rho(z)^n ,$$

car les dérivées secondes de u_z sont bornées par $\alpha_7 \rho(z)^{-1}$ dans Δ_z . Ecrivons maintenant

$$0 = \sigma(z', \xi') - \sigma(z, \xi) = z' - z + H_z(z').\xi' - H_z(z).\xi$$

$$= z' - z + (H_z(z') - H_z(z)).\ \xi' + H_z(z).(\xi' - \xi) .$$

D'après (42) et l'inégalité $|\xi'| \leqslant C_1 e^{-\varphi_\rho(z)} \rho(z)^{2n+1}$, on obtient , si $C_1 C$ est assez petit :

$$|(H_z(z') - H_z(z)).\xi'| \leqslant \alpha_8 C_1 C^{1/2} e^{-1/2\varphi_\rho(z)} \rho(z)^n .\ |z' - z|$$

$$\leqslant \alpha_9 C_1 C |z' - z| < |z' - z| ,$$

$$\text{angle } (z' - z ; z' - z + (H_z(z') - H_z(z)).\ \xi') \leqslant \alpha_{10} C_1 C^{1/2} e^{-1/2\varphi_\rho(z)} \rho(z)^n .$$

D'autre part, comme le vecteur non nul $H_z(z).(\xi' - \xi)$ se projette sur le vecteur de coordonnées $\xi' - \xi$ dans $N_z X$, (42) implique

$$\text{angle}(T_z X ; H_z(z).(\xi' - \xi)) \geqslant |\xi' - \xi| .\ |H_z(z).(\xi' - \xi)|^{-1}$$

$$\geqslant \alpha_{11} C^{-1/2} e^{-1/2\varphi_\rho(z)} \rho(z)^n .$$

Les trois évaluations d'angle qui précèdent sont contradictoires dès que $C_1 C$ est assez petit; on a donc démontré l'injectivité de σ sur V .

b/ Existence de l'ouvert U .

Comme nous n'avons fait aucune hypothèse de régularité sur la fonction ρ , l'ensemble V n'est pas nécessairement un véritable voisinage de la section nulle dans NX .

Il nous faut commencer par "régulariser" ρ .

On remarque qu'il existe une constante $\alpha_{12} \in]0,1[$ telle que pour tout $\zeta \in \Delta_z = D(z, \frac{1}{2}\rho(z))$, X soit un graphe dans le polydisque $D(\zeta, \alpha_{12}\rho(z))$ (c'est-à-dire que les hypothèses (40) , (41) relatives à ζ , $D(\zeta, \alpha_{12}\rho(z))$ sont vérifiées).

On peut donc remplacer ρ par la fonction

$$\rho'(\zeta) = \sup_{z\in X} \alpha_{12}(\rho(z) - 2\,|\zeta - z|) \quad , \zeta \in X .$$

ρ' est lipschitzienne de rapport $2\alpha_{12}$, à moins que $\rho' \equiv +\infty$, auquel cas X est le graphe d'une application $\mathbb{C}^n \to \mathbb{C}^{p-n}$, et l'ouvert $U = \mathbb{C}^p$ convient !

Posons pour tout $z \in X$ et $0 < t \leqslant \rho'(z)$:

$$\varphi_t(z) = \sup_{\zeta \in X \cap D(z,t)} \varphi(\zeta) .$$

La plurisousharmonicité de φ entraîne que $\varphi_t(z)$ est continue par rapport à (t,z) dans le domaine $0 < t \leqslant \rho'(z)$.

D'autre part, on a la majoration évidente

$$e^{-\varphi_t(z)} \cdot \frac{\pi^n}{n!}\, t^{2n} \leqslant \int_X e^{-\varphi}\, dV ,$$

de sorte que l'expression $e^{-\varphi_t(z)} \cdot t^{2n+1}$, $t \in]0, \rho'(z)]$, atteint son maximum en un point $t = \hat\rho(z) \in]0, \rho'(z)]$, et que la fonction

$$e^{-\varphi_{\hat\rho}(z)} \quad \hat\rho(z)^{2n+1}$$

est continue sur X . Il résulte du lemme de Schwarz que les conditions (40), (41) (41) sont bien satisfaites pour les polydisques $D(z, \hat\rho(z))$.

L'ensemble $\hat V$ associé à $\hat\rho$ comme dans l'énoncé du th. 4 est donc ouvert, et l'application holomorphe σ , qui est injective sur $\hat V$, est un isomorphisme de $\hat V$ sur l'ouvert $\sigma(\hat V) \subset \mathbb{C}^p$ (on pourrait aussi de façon plus élémentaire utiliser le théorème des fonctions implicites).

L'inégalité évidente qui suit, valable pour $|\zeta - z| < \frac{1}{4}\rho(z)$:

$$\rho'(\zeta) \geqslant \frac{\alpha_{12}}{2}\, \rho(z)$$

entraîne par définition de $\hat\rho$

$$(46) \quad e^{-\varphi_{\hat\rho}(\zeta)} \quad \hat\rho(\zeta)^{2n+1} \geqslant \exp(-\varphi_{\frac{\alpha_{12}}{2}\rho(z)}(\zeta)) \cdot \left[\frac{\alpha_{12}}{2}\, \rho(z)\right]^{2n+1}$$

$$\geqslant \left(\frac{\alpha_{12}}{2}\right)^{2n+1} e^{-\varphi_\rho(z)} \quad \rho(z)^{2n+1} ,$$

car $D(\zeta, \frac{\alpha_{12}}{2}\rho(z)) \subset D(z, \rho(z))$.

Fixons $z \in X$; d'après (46) l'ensemble $\hat V$ contient l'adhérence $\overline{W}$ de l'ouvert

$$W = \{(\zeta, \xi) \in NX \ ; \ |\zeta - z| < \frac{1}{4}\rho(z) \quad \text{et} \quad |\xi| < C_3\, e^{-\varphi_\rho(z)} \quad \rho(z)^{2n+1}\}$$

dès que $C_3 < \alpha_{13}\, C_1$.

On va montrer que $\sigma(W)$ contient la boule de centre z et de rayon $C_2\, e^{-\varphi_\rho(z)}\, \rho(z)^{2n+1}$ lorsque $C_2\, C$ est assez petit, ce qui achèvera la démonstration. Il est clair que le rayon de la plus grande boule incluse dans $\sigma(W)$ est égal à la distance de z au bord $\partial\sigma(W) = \sigma(\partial W)$ de $\sigma(W)$.

Supposons qu'un point $(\zeta, \xi) \in W$ soit tel que

$$(47) \quad |\sigma(\zeta, \xi) - z| = |\zeta - z + h(\zeta).\xi| = C_2\, e^{-\varphi_\rho(z)}\, \rho(z)^{2n+1} .$$

Alors d'après (42), (43) :

$$|\zeta - z| \leqslant \alpha_{14}\, C_3\, C^{1/2}\, e^{-1/2\, \varphi_\rho(z)}\, \rho(z)^{n+1} + C_2\, e^{-\varphi_\rho(z)}\, \rho(z)^{2n+1}$$
$$\leqslant \alpha_{15}(C_2 + C_3) C^{1/2}\, e^{-1/2 \varphi_\rho(z)}\, \rho(z)^{n+1} .$$

On en déduit comme dans la première partie :

$$\text{angle } (\zeta - z \,;\, T_\zeta X) \leqslant \alpha_{16}(C_2 + C_3) C^{1/2}\, e^{-1/2\, \varphi_\rho(z)}\, \rho(z)^n$$
$$\text{angle } (h(\zeta).\xi \,;\, T_\zeta X) \geqslant |\xi| . |h(\zeta).\xi|^{-1} \geqslant \alpha_{17}\, C^{-1/2\, \varphi_\rho(z)}\, \rho(z)^n ,$$

et lorsque $(C_2 + C_3)C$ est assez petit :

$$\text{angle } (\zeta - z, -h(\zeta).\, \xi) \geqslant \frac{1}{2}\, |\xi| .\, |h(\zeta).\xi|^{-1} ,$$

$$|\zeta - z + h(\zeta).\xi| \geqslant \left[\sin(\tfrac{1}{4}|\xi| . |h(\zeta).\xi|^{-1})\right] . \left[\, |\zeta - z| + |h(\zeta).\xi|\right]$$
$$\geqslant \frac{1}{8}\, |\xi| + \alpha_{18}\, C^{-1/2}\, e^{-1/2\, \varphi_\rho(z)}\, \rho(z)^n\, |\zeta - z| ,$$

ce qui est vrai même si $|\xi| = 0$.

Il résulte alors de (47) :

$$|\xi| \leqslant 8\, C_2\, e^{-\varphi_\rho(z)}\, \rho(z)^{2n+1} ,$$
$$|\zeta - z| \leqslant \alpha_{18}^{-1}\, C_2\, C^{1/2}\, e^{-1/2 \varphi_\rho(z)}\, \rho(z)^{n+1} \leqslant \alpha_{19}\, C_2\, C\, \rho(z) \;;$$

lorsque $C_2\, C$ est assez petit, on voit que (ζ, ξ) ne peut appartenir à la frontière ∂W de W , donc la distance de z à $\sigma(\partial W)$ est bien minorée par $C_2\, e^{-\varphi_\rho(z)}\, \rho(z)^{2n+1}$. ∎

Nous allons maintenant transcrire le théorème 4 sous une forme plus exploitable dans la pratique. On suppose que la variété X est définie par des équations

$$F_1 = F_2 = \ldots = F_N = 0 ,$$

telles que le rang du système $(dF_j)_{1 \leqslant j \leqslant N}$ soit égal à codim $X = p - n$ en tout point de X . Calculons la courbure de Ricci de X en un point $z \in X$ où les formes $(dF_j)_{j\in J_o}$ sont indépendantes, $J_o \subset \{1,\dots,N\}$, $|J_o| = p - n$. Si l'on considère dF_j comme une section de N^*X , $\bigwedge_{j\in J_o} dF_j$ définit une section de dét $N^*X \simeq$ dét TX ; par conséquent

$$\text{Ricci}(X) = c(\text{dét } TX) = -d'd'' \text{Log } |\bigwedge_{j\in J_o} dF_j|^2 ,$$

ce qui entraîne

$$i\,\text{Ricci}(X) + id'd'' \text{Log } \sum_J |\bigwedge_{j\in J} dF_j|^2$$

$$= id'd'' \text{Log } \sum_J \left|\frac{\bigwedge_{j\in J} dF_j}{\bigwedge_{j\in J_o} dF_j}\right|^2 \geqslant 0 .$$

D'autre part , par définition de la métrique de $\wedge T^* \mathbb{C}^p$, on a

$$|\bigwedge_{j\in J} dF_j|^2 = \sum_K |\Delta_{J,K}|^2 ,$$

où

$$\Delta_{J,K} = \text{dét } \left[\frac{\partial F_j}{\partial z_k}\right]_{j\in J,\ k\in K} ,$$

$$J \subset \{1,\dots,N\} ,\quad K \subset \{1,\dots, p\} \quad , |J| = |K| = p - n.$$

On a donc

$$i\,\text{Ricci}(X) + id'd'' \text{Log } \sum_{J,K} |\Delta_{J,K}|^2 \geqslant 0 ,$$

et on voit qu'on peut prendre pour poids φ toute fonction

(48) $$\varphi = 2\ell \text{ Log } \Delta + \varphi_1 ,$$

avec les notations

$$\ell = \varepsilon + 1 + \text{Inf}(2n, p) ,\ \Delta^2 = \sum_{J,K} |\Delta_{J,K}|^2$$

et où φ_1 est une fonction plurisousharmonique sur X telle que

$$\int_X \Delta^{-2\ell} e^{-\varphi_1} dV < +\infty .$$

Nous pouvons maintenant montrer de façon très précise l'existence de rétractions holomorphes, déja discutée par C.A.BERENSTEIN et B.A.TAYLOR [1] dans un cadre analogue.

THÉORÈME 5. - <u>Soient</u> $\varphi_1, \varphi_2, \chi$ <u>des fonctions plurisousharmoniques sur l'ouvert pseudoconvexe</u> $\Omega \subset \mathbb{C}^p$, <u>telles que</u>

(49) $$\int_X \Delta^{-2\ell} e^{-\varphi_1} dV < +\infty ,\ \ell = \varepsilon + 1 + \text{Inf}(2n , p) ,$$

(50) $$|F| = \left(\sum_{j=1}^N |F_j|^2\right)^{1/2} \leqslant e^{\varphi_2} ,$$

(51) $z \in \Omega$ *et* $|\zeta - z| < e^{-\chi(z)}$ *impliquent :*

$$\zeta \in \Omega\ ,\ \varphi_1(\zeta) < \varphi_1(z) + A\ ,\ \varphi_2(\zeta) \leqslant \varphi_2(z) + A, \quad \text{et} \quad \chi(\zeta) \leqslant \chi(z) + A.$$

Alors il existe une rétraction holomorphe $r : U \to X$ *définie sur l'ouvert*

$$U = \{z \in \Omega\ ;\ |F(z)| < e^{-\psi(z)}\}\ ,$$

où ψ *est la fonction plurisousharmonique*

(52) $$\psi = \varphi_1 + C_2\,\varphi_2 + C_3\,\chi + C_4 \log \Delta + C_5\ ,$$

avec $C_2 = (2n+2)(p-n)-1,\ C_3 = (2n+2)(p-n)+2n,\ C_4 = 2(\ell-n-1)\ ,$

$$C_5 = \log\left(\int_X \Delta^{-2\ell}\, e^{-\varphi_1}\, dV\right) + \log \frac{1}{\varepsilon} + \alpha\,(1+\varepsilon+A)\ ,$$

et $\alpha > 0$ *une constante ne dépendant que de* N *et* p .

Démonstration. On désignera par β_j les constantes du type $\alpha^{1+\varepsilon+A}$, et

$$C = \left(1 + \frac{1}{\varepsilon}\right) \int \Delta^{-2\ell}\, e^{-\varphi_1}\, dV\ .$$

a/ Montrons que la fonction

(53) $$\rho = \beta_1\, \Delta\, e^{-(p-n)\varphi_2 - (p-n+1)\chi}$$

vérifie les hypothèses (40) , (41).

D'après (50), (51) et les inégalités de Cauchy, les dérivées premières des F_j sont majorées par $\beta_2\, e^{\varphi_2(z)+\chi(z)}$ sur la boule de centre z et de rayon $\frac{1}{2}\, e^{-\chi(z)}$, les dérivées secondes par $\beta_3\, e^{\varphi_2(z)+2\chi(z)}$.

On a donc $|\Delta| \leqslant \beta_4\, e^{(p-n)(\varphi_2+\chi)}$, ce qui permet de choisir β_1 tel que

(54) $$\rho \leqslant e^{-\chi}\ .$$

Fixons un point $z \in X$, que nous prendrons comme origine des coordonnées pour simplifier ; quitte à effectuer une transformation unitaire sur $(F_1,\dots,F_N)$, on peut supposer que les différentielles $d_zF_1,\dots,d_zF_{p-n}$ sont orthogonales, et que $d_zF_{p-n+1} = \dots = d_zF_N = 0$.

On choisit un système de coordonnées $(\zeta_1,\dots,\zeta_p)$ tel que

$$d_zF_1(\zeta) = a_1\zeta_1,\ \dots,\ d_zF_{p-n}(\zeta) = a_{p-n}\zeta_{p-n}\ .$$

On a alors

$$\Delta(z) = |a_1| \dots |a_{p-n}|\ , \quad \text{et}\ |a_j| = |d_zF_j| \leqslant \beta_2 e^{\varphi_2(z)+\chi(z)}\ ,$$

ce qui entraîne

(55) $$|a_j| \geqslant \beta_5\, \Delta(z)\ e^{-(p-n-1)(\varphi_2(z)+\chi(z))}\ .$$

On peut écrire

$$a_j^{-1}(F_j(\zeta) - F_j(z)) = \zeta_j + G_j(\zeta) \ , \ 1 \leqslant j \leqslant p-n \ , \text{ avec}$$

$$|dG_j(\zeta)| \leqslant \beta_6 \ \Delta(z)^{-1} \ \exp((p-n-1)(\varphi_2(z)+ \chi(z)) + \varphi_2(z) + 2\,\chi(z))|\zeta|$$

sur la boule $|\zeta| \leqslant \frac{1}{2} e^{-\chi(z)}$.

On en déduit que l'application $G = (G_1,\ldots,G_{p-n})$ est $\frac{1}{2}$ lipschitzienne sur une certaine boule de centre z et de rayon $\beta_7\Delta(z) \exp(-(p-n)\,\varphi_2(z) - (p-n+1)\chi(z))$; l'assertion relative à ρ résulte alors du théorème des fonctions implicites. Si β_7 est assez petit, on aura de plus dans cette boule

$$(56) \qquad \Delta(\zeta) \geqslant \frac{1}{2} \Delta(z) \ .$$

b/ D'après (48), (51), (54), (56), on a $e^{\varphi(\zeta)} \leqslant \beta_8 e^{\varphi(z)}$ pour $|\zeta - z| < \rho(z)$, donc $e^{\varphi_\rho(z)} \leqslant \beta_8 e^{\varphi(z)}$. Le théorème 4 implique que l'application σ est injective sur le voisinage

$$V = \{(z, \xi) \in NX \ ; \ |\xi| < \beta_9 \ C^{-1} e^{-\varphi(z)} \ \rho(z)^{2n+1}\} ,$$

avec (cf. (43)) $\beta_9 C^{-1} e^{-\varphi(z)} \ \rho(z)^{2n+1} \leqslant \rho(z)$.

Plaçons-nous en un point $z_o \in X$. Pour tout point $(\zeta, \xi) \in \partial V$, avec $|\zeta - z_o| < \beta_{10}\rho(z_o)$ et $|\xi| = \beta_9 C^{-1} e^{-\varphi(\zeta)} \ \rho(\zeta)^{2n+1}$, le raisonnement de a/ montre que

$$|F(\sigma(\zeta, \xi))| \geqslant \operatorname*{Inf}_j |a_j| \times \text{ distance } (\sigma(\zeta,\xi),X) \ .$$

La partie b/ de la démonstration du théorème 4 entraîne dans les mêmes conditions (β_9 et β_{10} assez petits) :

$$\begin{aligned} \text{distance}(\sigma(\zeta,\xi),X) &= \inf_{z\in X, |z-z_o|\leqslant\frac{1}{2}\rho(z_o)} \text{distance}(\sigma(\zeta,\xi),z) \\ &\geqslant \inf_{z\in X, |z-z_o|\leqslant\frac{1}{2}\rho(z_o)} \beta_{11} C^{-1} e^{-\varphi(z)} \rho(z)^{2n+1} \\ &\geqslant \beta_{12} C^{-1} e^{-\varphi(z_o)} \ \rho(z_o)^{2n+1} \ , \end{aligned}$$

et comme $|z_o - \sigma(\zeta,\xi)| < \rho(\sigma(\zeta, \xi))$, on a d'après (51), (53), (54) :

$$e^{-\varphi(z_o)} \ \rho(z_o)^{2n+1} \geqslant \beta_{13} e^{-\varphi(\sigma(\zeta,\xi))} \ \rho(\sigma(\zeta, \xi))^{2n+1} \ .$$

On obtient donc au point $\sigma = \sigma(\zeta,\xi) \in \partial\sigma(V)$:

$$|F(\sigma)| \geqslant \beta_{14} C^{-1} \Delta(\sigma) \ \exp(-(p-n-1)(\varphi_2(\sigma) + \chi(\sigma)))e^{-\varphi(\sigma)}\rho(\sigma)^{2n+1} = e^{-\psi(\sigma)}\rho(\sigma)^{2n+1}$$

(cf. (48),(52),(53),(55),56)).

Il en résulte que l'ouvert V contient toutes les composantes connexes de U qui rencontrent la sous-variété X. On définit la rétraction r par $r = \pi_X \circ \sigma^{-1}$ sur ces composantes, et r = point constant de X sur les autres composantes de U.

REMARQUE 5. On a de plus par construction

$$|r(\zeta) - \zeta| < e^{-\chi(\zeta)}$$

pour tout point ζ de l'une des composantes connexes de U rencontrant X.

REMARQUE 6. Dans les applications, on aura intérêt à remplacer (49) par une condition plus maniable. Si ω est la réunion des boules ouvertes de centre $z \in X$ et de rayon $\rho(z)$, on a

$$\int_X \Delta^{-2} e^{-\varphi_1} dV \leqslant \beta_{15} \int_\omega \Delta^{-2\ell} e^{-\varphi_1} \rho^{-2(p-n)}$$

$$\leqslant \beta_{16} \int_\omega \Delta^{-2(\ell+p-n)} e^{-\varphi_1} e^{2(p-n)[(p-n)\varphi_2+(p-n+1)\chi]}$$

compte-tenu de la définition (53) de ρ.

Supposons maintenant que les conditions suivantes soient réalisées (avec des fonctions plurisousharmoniques $\varphi_2, \varphi_3, \chi$) :

(57) $\Delta \geqslant e^{-\varphi_3}$,

(58) $|F| \leqslant e^{\varphi_2}$,

(59) $z \in \Omega$ et $|\zeta - z| < e^{-\chi(z)}$ impliquent

$$\zeta \in \Omega, \ \varphi_2(\zeta) \leqslant \varphi_2(z) + A, \ \varphi_3(\zeta) \leqslant \varphi_3(z) + A, \ \chi(\zeta) \leqslant \chi(z) + A .$$

Alors on peut choisir

$$\varphi_1 = 2(p-n)[(p-n)\varphi_2 + (p-n+1)\chi] + 2(\ell+p-n)\varphi_3 + (p+\varepsilon)\mathrm{Log}(1+|z|^2),$$

$$\psi = C_2'\varphi_2 + C_3'\varphi_3 + C_4'\chi + C_5' + (p+\varepsilon)\mathrm{Log}(1+|z|^2),$$

avec $\ell = \varepsilon + 1 + \mathrm{Inf}(2n,p)$, et (compte-tenu de ce que $|\Delta| \leqslant \beta_4 e^{(p-n)(\varphi_2+\chi)}$) :

$$C_2' = 2(\ell+p-n)(p-n) - 1, \quad C_3' = 2(\ell+p-n),$$

$$C_4' = 2(\ell+p-n+1)(p-n) + 2n, \quad C_5' = 2\,\mathrm{Log}\frac{1}{\varepsilon} + \alpha(1+\varepsilon+A).$$

Ces dernières estimations précisent et généralisent les résultats antérieurs de C.A.BERENSTEIN et B.A.TAYLOR [1]. Ainsi, soit χ une fonction plurisousharmonique sur Ω vérifiant les conditions suivantes :

(60) $\chi \geqslant 0$, et $\mathrm{Log}(1+|z|) = O(\chi(z))$;

(61) il existe une constante A telle que

$z \in \Omega$ et $|z-\zeta| < \exp(-\chi(z))$ implique que

$$\chi(\zeta) \leqslant \chi(z) + A .$$

On définit l'algèbre $A_\chi(\Omega)$ comme l'ensemble des fonctions holomorphes f sur Ω telles qu'il existe des constantes $A_1, A_2 \geqslant 0$ telles que

(62) $$|f(z)| \leqslant A_1 \exp(A_2\, \chi(z)) .$$

L'hypothèse (61) est généralement exprimée sous une forme un peu plus générale dans la littérature (voir par exemple [8]), mais tous les poids usuels satisfont la condition plus restrictive que nous avons donnée*.

COROLLAIRE 1. - <u>Soit</u> X <u>une sous-variété de dimension</u> n <u>de l'ouvert pseudoconvexe</u> $\Omega \subset \mathbb{C}^p$, <u>définie par les équations</u> $F_1 = F_2 = \ldots = F_N = 0$, <u>avec</u> $F_1,\ldots,F_N \in A_\chi(\Omega)$. <u>On suppose que la quantité</u>

$$\Delta = \left(\sum_{|J| = |K| = p-n} \left| \text{dét}\left[\frac{\partial F_j}{\partial z_k} \right]_{j\in J,\ k\in K} \right|^2 \right)^{1/2}$$

<u>est non nulle, et vérifie une minoration du type</u>

$$\Delta \geqslant \exp(-A_1\, \chi(z) - A_2) \quad , \ \underline{\text{pour tout}} \ \ z \in X .$$

<u>Alors il existe des constantes</u> $A_3, A_4 > 0$ <u>et une rétraction holomorphe</u> $r : U \to X$ <u>définie sur l'ouvert</u>

$$U = \{z \in \Omega \ ; \ |F(z)| < \exp(-A_3\,\chi(z) - A_4)\} .$$

5. Extension des fonctions holomorphes avec contrôle de la croissance.

On se replace tout d'abord dans la situation générale des paragraphes 0,1 et 2 : X désigne une variété kählérienne faiblement pseudoconvexe de dimension n , E, M des des fibrés hermitiens au-dessus de X , M étant de rang N . On considère un sous-ensemble analytique $Y = F^{-1}(0)$ de X , lieu des zéros d'une section holomorphe F de M , et on pose

$$U = \{z \in X \ ; \ |F(z)| < 1\}.$$

* On peut démontrer en outre que les classes d'algèbres $A_\chi(\Omega)$ correspondantes sont les mêmes.

On a dans ce cadre un théorème d'extension , qui généralise le théorème de B.JENNANE [9].

THÉORÈME 6. - Soient η , q deux réels > 0 tels que $|F|^{-2q}$ ne soit localement sommable en aucun point de Y (en général q sera un entier $\geqslant 1$, par exemple $q = \sup_{z \in Y} \operatorname{codim} Y_z$, ou $q = \operatorname{Inf}(n,N)$) .

On suppose que Y est X-négligeable (cf. §3, définition 2)*, et que la forme de courbure de E satisfait à l'inégalité

$$(63) \qquad ic(E) \geqslant_N \left(\frac{\eta}{1+|F|^2} + \frac{q}{|F|^2} \right) < ic(M).F,F > -i \operatorname{Ricci}(X) .$$

Alors pour toute section g de E au-dessus de U , telle que $\int_U |g|^2 dV < +\infty$, il existe une section G de E au-dessus de X, coïncidant avec g sur Y, et telle que

$$\int_X \frac{|G|^2 dV}{(1+|F|^2)^{q+\eta}} \leqslant C(q,\eta) \int_U |g|^2 dV ,$$

avec $C(q,\eta) = 1 + \frac{(q+1)^2}{\eta}$ si $q \geqslant 1$, $= \frac{1}{2^q-1} + \frac{(q+1)^2}{\eta}$ si $0 < q < 1$.

Démonstration. On cherche l'extension G sous la forme

$$G = \lambda(|F|^2)g - u ,$$

où λ est une fonction réelle de classe C^∞ à support dans $]-\infty, 1[$, telle que $\lambda = 1$ au voisinage de 0, et $u \in C^\infty(X,E)$, $u = 0$ sur Y .

L'analyticité de G équivaut à

$$(64) \qquad D''u = v ,$$

avec $v = D''(\lambda(|F|^2)g) = \lambda'(|F|^2) < F , D'F > g$.

Comme toute solution de l'équation (64) est de classe C^∞ , il suffit d'imposer que $|u|^2 \, |F|^{-2q}$ soit localement sommable pour assurer l'annulation de u sur Y .

Soit Z une partie fermée de X contenant Y , telle que $X \setminus Z$ satisfasse les hypothèses de la définition 2. On résout l'équation (64) dans $X \setminus Z$, après avoir multiplié la métrique de E par $(1+|F|^2)^{-\eta}|F|^{-2q}$.

Pour cette nouvelle métrique, la forme de courbure $c'(E)$ du fibré E est donnée par

$$c'(E) = c(E) + \eta \left[\frac{<D'F,D'F>}{(1+|F|^2)^2} + \frac{|F|^2 <D'F,D'F> - <D'F,F> \wedge <F,D'F>}{(1+|F|^2)^2} - \frac{<c(M).F,F>}{1+|F|^2} \right]$$

$$+ q \left[\frac{|F|^2 <D'F,D'F> - <D'F,F> \wedge <F,D'F>}{|F|^4} - \frac{<c(M).F,F>}{|F|^2} \right] .$$

* Comme pour le théorème 2, cette hypothèse est en fait superflue.

Soit $K = \det(T^*X)$ le fibré canonique de X . Comme la (1,1)-forme

$i(|F|^2 \langle D'F , D'F\rangle - \langle D'F,F\rangle \wedge \langle F, D'F\rangle)$ est $\geqslant 0$, il résulte de l'hypothèse (63) que

$$ic'(K^* \otimes E) \geqslant_N \eta . i \frac{\langle D'F, D'F\rangle}{(1+|F|^2)^2} \otimes \mathrm{Id}_E .$$

La forme $v \in C^\infty_{0,1}(X,E) = C^\infty_{n,1}(X, K^* \otimes E)$ vérifie clairement les inégalités

$$|v|^2 \leqslant \lambda'^2 . |F|^2 \, |D'F|^2 \, |g|^2$$

$$\leqslant 1/\eta \; \lambda'^2 . |F|^2 (1 + |F|^2)^2 \, |g|^2 \langle ic'(K^*\otimes E)\wedge v, v\rangle \quad ,$$

grâce au lemme 1, § 3, ligne (21). D'après H.SKODA [13] , théorème 2 et remarques consécutives (comparer aussi avec la proposition du § 2), il existe une solution $u \in C^\infty_{n,o}(X\setminus Z, K^* \otimes E) = C^\infty(X\setminus Z, E)$ de l'équation (64) , telle que

$$\int_{X\setminus Z} \frac{|u|^2}{(1+|F|^2)^\eta |F|^{2q}} dV \leqslant \int_{X\setminus Z} \frac{1/\eta \; \lambda'^2 . |F|^2 (1 + |F|^2)^2 |g|^2}{(1 + |F|^2)^\eta |F|^{2q}} dV$$

$$= \frac{1}{\eta} \int_U \frac{(\lambda'(|F|^2))^2}{|F|^{2q-2}} (1 + |F|^2)^{2-\eta} |g|^2 \, dV \; ; \tag{65}$$

comme la fonction $\frac{(\lambda'(|F|^2))^2}{|F|^{2q-2}} (1 + |F|^2)^{2-\eta}$ est bornée, la dernière intégrale du second membre est bien finie.

La section $G = \lambda(|F|^2)g - u$ est donc holomorphe dans $X\setminus Z$ et localement L^2 dans X , par conséquent G se prolonge en une section holomorphe de E sur X (hypothèse de la définition 2) ; on voit que u est de classe C^∞ dans X , et que $u = 0$ sur Y d'après (65). On obtient :

$$|G|^2 \leqslant \left(1 + \frac{1}{(1 + 1/|F|^2)^q - 1}\right) \lambda^2 \, |g|^2 + \left(1 + \frac{1}{|F|^2}\right)^q |u|^2 ,$$

$$\frac{|G|^2}{(1+|F|^2)^q} \leqslant \frac{\lambda^2 \, |g|^2}{(1+|F|^2)^q - |F|^{2q}} + \frac{|u|^2}{|F|^{2q}} ,$$

$$\int_X \frac{|G|^2 \, dV}{(1+|F|^2)^{q+\eta}} \leqslant \int_U \frac{g^2}{(1+|F|^2)^\eta} \left[\frac{\lambda^2}{(1+|F|^2)^q - |F|^{2q}} + \frac{1}{\eta} \frac{\lambda'^2 . (1+|F|^2)^2}{|F|^{2q-2}}\right] .$$

On fait tendre convenablement λ vers la fonction λ_o définie par

$$\lambda_o(t) = 1 - t^{q+1/2} \quad \text{pour} \quad 0 \leqslant t \leqslant 1 ,$$

$$= 0 \quad \text{pour} \quad t \geqslant 1 \; ;$$

le prolongement G de g va tendre vers une section encore notée G

telle que

$$\int_X \frac{|G|^2\, dV}{(1+|F|^2)^{q+\eta}} \leq \int_U \frac{|g|^2}{(1+|F|^2)^{\eta}} \left[\frac{\lambda_o^2}{(1+|F|^2)^q - |F|^{2q}} + \frac{1}{\eta}\, \frac{\lambda_o'^2 \cdot (1+|F|^2)^2}{|F|^{2q-2}}\right],$$

avec $\dfrac{[\lambda_o'(|F|^2)]^2 (1+|F|^2)^2}{|F|^{2q-2}} = (\frac{q+1}{2})^2\ (1+|F|^2)^2 \leq (q+1)^2$ dans U, et

$(1+|F|^2)^q - |F|^{2q} \geq \mathrm{Inf}(1\ ,\ 2^q-1)$ dans U, car la fonction $(1+x)^q - x^q$ est monotone sur $[0,1]$.

On peut donc prendre $C(q,\eta) = \mathrm{Sup}(1\ ,\ \frac{1}{2^q-1}) + \frac{(q+1)^2}{\eta}$. ∎

On remplace désormais X par un ouvert pseudoconvexe Ω de $\mathbb{C}^p$, et on suppose que $E = \mathbb{C}$, $M = \mathbb{C}^N$ sont des fibrés triviaux, dont les métriques sont données respectivement par les poids $e^{2q\psi - \varphi}$, $e^{2\psi}$ (φ , ψ fonctions plurisousharmoniques de classe C^∞) . On a donc

$\mathrm{Ricci}(\Omega) = 0$, $c(E) = d'd''\varphi - 2q\, d'd''\psi$, $c(M) = -\, 2d'd''\psi \otimes \mathrm{Id}_M$, de sorte que la condition (63) est vérifiée.

COROLLAIRE 2. - <u>Soient</u> g <u>une fonction holomorphe dans l'ouvert</u>

$$U = \{z \in \Omega\ ;\ |F(z)| < e^{-\psi(z)}\}\ ,$$

<u>telle que</u> $\int_U |g|^2\, e^{2q\psi - \varphi} < +\infty$, <u>et</u> η <u>un réel</u> > 0 .
<u>Alors il existe une fonction holomorphe</u> G <u>qui coïncide avec</u> g <u>sur l'ensemble analytique</u> $X = F^{-1}(0)$, <u>et telle que</u>

$$\int_\Omega \frac{|G|^2\, e^{2q\psi - \varphi}\, dV}{(1+|F|^2\, e^{2\psi})^{q+\eta}} \leq C(q,\eta) \int_U |g|^2\, e^{2q\psi - \varphi}\, dV .$$

Par un passage à la limite évident, le corollaire 1 s'étend au cas où φ est plurisousharmonique quelconque, et ψ localement minorée.*

Reprenons maintenant les notations et les hypothèses du théorème 5 : X est la sous-

*On améliore ainsi les estimations de B.JENNANE [9] , grâce au choix de poids plus naturellement adaptés au problème posé.
Il peut paraître surprenant que le corollaire 1 fasse intervenir un poids non plurisousharmonique $2q\psi - \varphi$, mais cette situation s'explique par le fait qu'on a "récupéré de la plurisousharmonicité" en jouant sur la négativité du fibré M . Lorsque $F(z) = z = (z_1,\ldots,z_p)$, $q = p$, et ψ = constante, le corollaire 1 redonne le théorème d'HÖRMANDER-BOMBIERI sous une forme optimale, utile pour la théorie des nombres (cf. H. SKODA [16]).

variété lisse de l'ouvert $\Omega \subset \mathbb{C}^p$ définie par les équations $F_1 = \ldots = F_N = 0$.
On pose $n = \dim X$, $\Delta^2 = \sum_{|J|=|K|=p-n} \left|\det\left[\frac{\partial F_j}{\partial z_k}\right]_{j\in J,\, k\in K}\right|^2_n$, $F = (F_1,\ldots,F_N)$, $|F|^2 = |F_1|^2 + \ldots + |F_N|^2$, et on désigne par $dV_X = \frac{\omega^n}{n!}\Big|_X$ l'élément de volume canonique de X ; on suppose que ψ est la fonction plurisousharmonique donnée par le théorème 5 ou la remarque 6, et que l'inégalité $|\zeta - z| < e^{-\chi(z)}$ entraîne de plus $\varphi(\zeta) \leqslant \varphi(z) + A$.

COROLLAIRE 3. - <u>Pour toute fonction holomorphe</u> g <u>sur</u> $X = F^{-1}(0)$ <u>et tout réel</u> $\eta > 0$, <u>il existe une fonction holomorphe</u> G <u>dans</u> Ω <u>qui prolonge</u> g, <u>telle que</u>

$$\int_\Omega \frac{|G|^2\, e^{-\varphi - 2\eta\psi}}{(|F|^2 + e^{-2\psi})^{p-n+\eta}}\, dV \leqslant \alpha^{1+A}\left(1+\frac{1}{\eta}\right)\int_X |g|^2\, \Delta^{-2}\, e^{-\varphi}\, dV_X ,$$

<u>où</u> α <u>est une constante ne dépendant que de</u> p <u>et</u> N.

<u>Démonstration</u>. On choisit $q = \operatorname{codim} X = p-n$; si $r : U \to X$ est la rétraction du théorème 5, on étend g à U en posant $\tilde{g} = g \circ r$ sur les composantes de U qui rencontrent X, et $\tilde{g} = 0$ sur les autres composantes.

Réexaminons maintenant les arguments utilisés dans la démonstration du théorème 5, en conservant les mêmes notations.

Les composantes connexes de U qui rencontrent X forment un voisinage tubulaire de X, dont la coupe suivant le plan normal $(T_zX)^\perp$ constitue approximativement un polydisque de multirayon $(|a_1|^{-1}e^{-\psi}, \ldots, |a_{p-n}|^{-1}e^{-\psi})$.

Un tel polydisque est contenu par construction dans la boule de centre z et de rayon $\rho(z) \leqslant e^{-\chi(z)}$, et son volume est donné par :

$$\pi^{p-n}\, |a_1|^{-2} \ldots |a_{p-n}|^{-2}\, e^{-2(p-n)\psi} \approx \pi^q\, \Delta^{-2}\, e^{-2q\psi} .$$

On en déduit visiblement d'après la remarque 5 que

$$\int_U |\tilde{g}|^2\, e^{2q\psi - \varphi}\, dV \leqslant \alpha^{1+A} \int_X |g|^2\, \Delta^{-2}\, e^{-\varphi}\, dV ,$$

et la conclusion résulte du corollaire 2.

Le corollaire 3 paraît pratiquement optimal, en dehors du fait que l'on souhaiterait pouvoir prendre $\eta = 0$.

On observera que les constantes C_2, C_3, C_4, C_5 qui interviennent dans la définition de ψ, et qui ne sont probablement pas les meilleures possibles, n'auront en général aucune importance dans les applications du corollaire 3, puisqu'on peut les "tuer" en choisissant η assez petit, et qu'en pratique ψ sera $\geqslant 0$.

REMARQUE 7. Explicitons le corollaire 3 en termes plus familiers, sous les hypothèses suivantes :

$$|g| \leqslant e^{\gamma} \quad , \; |F| \leqslant e^{\varphi_2} \quad , \quad \Delta \geqslant e^{-\varphi_3} \quad ,$$

dans lesquelles on suppose que γ, φ_2, φ_3, χ sont des fonctions plurisousharmoniques vérifiant toutes l'analogue de (51) ; dans ces conditions, on peut choisir comme fonction ρ (cf. (40) , (41) et le théorème 5) la fonction

$$\rho = \beta_4 \, \Delta e^{-(p-n)\varphi_2 - (p-n+1)\chi} \quad .$$

On obtient alors, en posant $\omega = \bigcup_{z \in X} D(z, \rho(z))$ et en remplaçant φ par 2φ :

$$\int_X |g|^2 \, \Delta^{-2} \, e^{-2\varphi} \, dV_X \leqslant \beta_{17} \int_\omega e^{2\gamma + 2\varphi_3 - 2\varphi} \rho^{-2(p-n)} \, dV \; ;$$

on choisit donc

$$\varphi = \gamma + (p-n+1)\,\varphi_3 + (p-n)\left[(p-n)\varphi_2 + (p-n+1)\chi\right] + (p+\eta)\mathrm{Log}(1+|z|) \; ,$$

ce qui donne

$$\int_\Omega |G|^2 \; \frac{e^{-2\varphi - 2\eta\psi}}{(e^{2\varphi_2} + e^{-2\psi})^{p-n+\eta}} \, dV \leqslant \beta_{18}\left(1 + \frac{1}{\eta}\right)^2 ,$$

d'où $|G| \leqslant \beta_{19}\left(1 + \frac{1}{\eta}\right) e^{\varphi + \eta\psi + p\chi} \, . \, (e^{\varphi_2} + e^{-\psi})^{p-n+\eta}$.

COROLLAIRE 4. - <u>Sous les hypothèses du corollaire</u> 1 [<u>voir</u> (60), (61), (62) ; X <u>est définie par</u> $F_1, F_2, \ldots, F_N \in A_\chi(\Omega)$, <u>et on suppose que</u> $\Delta \geqslant \exp(-A_1\chi - A_2)$] , <u>une fonction holomorphe</u> g <u>sur</u> X <u>se prolonge en une fonction</u> $G \in A_\chi(\Omega)$ <u>si et seulement si</u> g <u>vérifie la condition</u> :

$$|g(z)| \leqslant \exp(A_3\chi(z) + A_4) \quad , \text{ pour tout } \; z \in X \; .$$

BIBLIOGRAPHIE

[1] BERENSTEIN (C.A.) and TAYLOR (B.A.). - Interpolation problems in $\mathbb{C}^n$ with applications to harmonic analysis, à paraître au Journal d'Analyse Math. de Jérusalem.

[2] DEMAILLY (J.-P.). - Relations entre les différentes notions de fibrés et de courants positifs, à paraître.

[3] DEMAILLY (J.-P.) et SKODA (H.). - Relations entre les notions de positivités de P.A.Griffiths et de S.Nakano pour les fibrés vectoriels, Séminaire P.Lelong-H.Skoda (Analyse), 19e année, 1978-1979, Lecture Notes (à paraître).

[4] DOUADY (A.) et VERDIER (J.-L.). - Séminaire de Géométrie analytique, E.N.S., 1972-1973, Différents aspects de la positivité, Astérisque 17, 1974, Société Mathématique de France.

[5] GRIFFITHS (P.A.). - Hermitian differential geometry, Chern classes and positive vector bundles, Global analysis, Princeton University Press, p. 185-251, 1969.

[6] HÖRMANDER (L.). - L^2 estimates and existence theorems for the $\bar{\partial}$ operator, Acta Math., 113, p. 89-152, 1965.

[7] HÖRMANDER (L.). - An introduction to Complex analysis in Several Variables, Princeton, Van Nostrand Company, 1966, 2e édition, 1973.

[8] HÖRMANDER (L.). - Generators for some rings of analytic functions, Bull. Amer. Math. Soc. 73, p. 943-949, 1967.

[9] JENNANE (B.). - Extension d'une fonction définie sur une sous-variété avec contrôle de la croissance, Séminaire P.Lelong-H.Skoda (Analyse), 17e année, 1976-1977, Lecture Notes n° 694, Springer, Berlin, Heidelberg, New York, 1978.

[10] NAKANO (S.) - Vanishing theorems for weakly 1-complete manifolds II, Publ. RIMS, Kyoto University, vol. 10, p. 101, 1974.

[11] SKODA (H.). - Application des techniques L^2 à la théorie des idéaux d'une algèbre de fonctions holomorphes avec poids, Annales scient. de l'Ecole Normale Supérieure, 5, p. 545-579, 1972.

[12] SKODA (H.). - Formulation hilbertienne du Nullstellensatz dans les algèbres de fonctions holomorphes, paru dans "l'Analyse harmonique dans le domaine complexe". Lecture Notes in Mathematics, n° 336, Springer, Berlin,Heidelberg,New York,1973.

[13] SKODA (H.). - Morphismes surjectifs et fibrés linéaires semi-positifs, Séminaire P.Lelong-H.Skoda (Analyse), 17e année, 1976-1977, Lecture Notes n° 694, Springer, Berlin,Heidelberg,New York, 1978.

[14] SKODA (H.). - Morphismes surjectifs de fibrés vectoriels semi-positifs, Ann. Scient. de l'Ecole Normale Supérieure, 4e série, t. 11, p. 577-611, 1978.

[15] SKODA (H.). - Relèvement des sections globales dans les fibrés semi-positifs, Séminaire P.Lelong-H.Skoda (Analyse), 19e année, 1978-1979, Lecture Notes (à paraître).

[16] SKODA (H.). - Estimations L^2 pour l'opérateur $\overline{\partial}$ et applications arithmétiques. Séminaire P.Lelong (Analyse), 16e année, 1975-1976, Lecture Notes n° 578, Springer, Berlin-Heidelberg,New York, 1977.

INTEGRALES DE COURBURE ET POTENTIELS SUR LES HYPERSURFACES ANALYTIQUES DE $\mathbb{C}^n$

par BERNARD GAVEAU

A Monsieur P.LELONG

Introduction. L'objet de ce travail est d'obtenir des conditions géométriques nécessaires satisfaites par l'intégrale de la courbure scalaire d'un diviseur d'une fonction holomorphe ayant une certaine croissance. Dans [2] un tel résultat avait été obtenu pour le cas d'un diviseur de la boule de $\mathbb{C}^2$, et la démonstration donnée peut en être simplifiée et généralisée aux classes de Hardy de la boule. Ce que nous faisons au §3 .

Nous donnons aussi une estimée plus fine en utilisant la théorie du potentiel le long du diviseur même, appliquée à une équation du type de Monge-Ampère satisfaite par la courbure (ou par le déterminant de la courbure de Ricci) ; les résultats obtenus concernent les diviseurs de fonctions holomorphes bornées ou de classe de Hardy dans la boule unité de $\mathbb{C}^n$ ou les diviseurs de fonctions entières de type exponentiel dans $\mathbb{C}^n$.

1. Cas des ensembles analytiques dans $\mathbb{C}^2$: calculs préliminaires.

Soit f une fonction holomorphe dans un domaine $D \subset \mathbb{C}^2$ que nous spécifierons ultérieurement. Commençons par calculer la courbure des surfaces de niveau de f . Soit $V_\zeta = f^{-1}(\zeta)$. Nous supposerons V_ζ sans singularités.

Lemme 1 : la courbure K de V_ζ pour la métrique euclidienne est définie par

$$K = |Q(f)|^2 (\|\nabla f\|^2)^{-3} \tag{1}$$

où $Q(f) = f_{zz} f_w^2 - 2 f_{zw} f_z f_w + f_{ww} f_z^2$ et $\|\nabla f\|^2 = |f_z|^2 + |f_w|^2$ est le carré du gradient de f pour la métrique euclidienne.

Preuve : supposons que nous utilisions z comme paramètre local le long de V_ζ . La restriction de la métrique euclidienne $ds^2 = |dz|^2 + |dw|^2$ à V_ζ est alors

$$ds^2|_{V_\zeta} = \frac{|f_z|^2+|f_w|^2}{|f_w|^2}|d\bar{z}|^2 \equiv g|dz|^2$$

(Cela supposons qu'en soit sur des points où z peut être pris comme paramètre local de V_ζ donc $f_w \neq 0$).

Alors la courbure K est donnée par le laplacien de $\log g$ le long de V_ζ

$$K = \frac{1}{g}\frac{\partial^2}{\partial z\partial\bar{z}} \log g .$$

Un calcul immédiat (voir par exemple [1]) donne le résultat.

Nous avons alors en notant Δ le laplacien euclidien de V_ζ pour la métrique eiclidienne de V_ζ

Lemme 2 : $\Delta(\log\|\nabla f\|^2) = K$

Preuve : plaçons nous là où $f_w \neq 0$: alors

$$K = \Delta \log \frac{|f_z|^2+|f_w|^2}{|f_w|^2} = \Delta\log \|\nabla f\|^2 \qquad \text{car} \qquad \Delta \log |f_w|^2 = 0$$

Corollaire : la courbure K satisfait la relation suivante

(2) $\Delta(\log K) = -3K+2 \sum_{\rho_i\in\theta} \delta_{\rho_i}$ où θ désigne l'ensemble des points ombilics de V_ζ ie des points où K s'annule.

2. Cas d'un diviseur de fonction de la classe de Hardy dans la boule.

Considérons maintenant la boule $B(0,1)$ de centre 0 et rayon 1 de $\mathbb{C}^2$ ensemble de (z_1,z_2) avec $|z_1|^2+|z_2|^2 <1$ et soit f une fonction holomorphe de la classe de Hardy H^p pour un $p \geq 1$ dans la boule.

Théorème 1 : 1) Soit $\varepsilon > 0$ fixé quelconque. $d(z,\partial B)$ la distance de $z\in B$ à ∂B . Pour presque tout $\zeta\in\mathbb{C}$, on a

(3) $$\int_{V_\zeta} d(z, B)^{2+\varepsilon} K(z)\, dv(z) < +\infty$$

où K est la courbure euclidienne de V_ζ et dv l'élément d'aire euclidienne.

2) <u>si</u> $g_\zeta^{(R)}(z_o,z)$ <u>est la fonction de Green euclidienne de la composante de</u> $V_\zeta \cap B(0,R)$ <u>contenant</u> z_o , <u>alors pour tout</u> $\zeta \in \mathbb{C}$, <u>le potentiel de la courbure satisfait</u>

(4) $$\int_{V_\zeta \cap B(0,R)} g_\zeta^{(R)}(z_o,z)\, K(z)\, d\sigma(z) \leqslant C \log\left(\frac{1}{1-R}\right)$$

<u>où</u> C <u>est une constante absolue.</u>

3) <u>de plus (5)</u> $\Sigma\, g_\zeta^{(R)}(z_o,z_i) \leqslant C \log \frac{1}{1-R}$ <u>où la somme porte sur les points ombilicaux de</u> V_ζ

Preuve : 1) notons $P_R(z)$ **la fonction** $R^2 - |z|^2$ restreinte à V_ζ . Calculons d'abord $\Delta P_R^{2+\varepsilon}$ on a

(5) $$\Delta P_R^{2+\varepsilon} = (2+\varepsilon)\,(P_R)^{\varepsilon}\,(-P_R + (1+\varepsilon)\,\|\partial P_R\|^2)$$

où $\|\partial P_R\|^2$ est le carré de la longueur du gradient de P_R (le long de l'ensemble analytique V_ζ) ; ∂P_R est donc la projection orthogonale de $\nabla|z|^2$ (ie du vecteur radial) sur l'ensemble analytique. Son carré de longueur est donc $\leqslant \cos^2\theta$ où θ et l'angle entre la tangente à l'ensemble analytique et la normale à la sphère de centre 0 passant par le point considéré. Appliquons maintenant à $V_\zeta \cap B(0,R)$ la formule de Green aux fonctions $u = \log \|\nabla f\|^2$ et $v = P_R^{2+\varepsilon}$ (on suppose V_ζ sans singularités, donc u est C^∞).

(6) Alors $$\int_{V_\zeta \cap B(0,R)} (u\,\Delta v - v\Delta u)\, d\sigma = \int_{V_\zeta \cap S(0,R)} \left(u\,\frac{\partial v}{\partial n} - v\,\frac{\partial u}{\partial n}\right) ds\ .$$

Ici v est telle que v et $\frac{\partial v}{\partial n}$ s'annulent sur $V_\zeta \cap S(0,R)$, donc le $2^{\underline{nd}}$ membre est nul. D'où compte-tenu du lemme 2

$$\int_{V_\zeta \cap B(0,R)} P_R^{2+\varepsilon}\, K d\sigma = \int_{V_\zeta \cap B(0,R)} \log \|\nabla f\|^2 \Delta P_R^{2+\varepsilon}\, d\sigma$$

(7) $$\leqslant (2+\varepsilon)\int_{V_\zeta \cap B(0,R)} -P_R^{1+\varepsilon} \log \|\nabla f\|^2\, d\sigma\ +$$

$$+ (2+\varepsilon)(1+\varepsilon) \int_{V_\zeta \cap B(0,R)} P_R^{\varepsilon} \log^+ (\|\nabla f\|^2) \cos^2\theta\, d\sigma\ .$$

Maintenant comme f est de la classe de Hardy

$$\log^+ \|\nabla f\|^2 \leqslant C \log^+ \left(\frac{1}{R-|z|^2}\right)$$

De plus $(R^2 - |z|^2)^\varepsilon \log^+\left(\frac{1}{1-|z|^2}\right) \leqslant C'$ constante ne dépendant que de ε .

Donc le second terme du second membre de (7) est contrôlé par

$$C'' \int_{V_\zeta} \cos^2\theta \, d\sigma$$

et l'on sait que cette intégrale est finie si f est de la classe de Névanlinna (critère de Malliavin, voir [4]) . Pour étudier le premier terme du second membre de (7), majorons-le en remplaçant $-\log \|\nabla f\|^2$ par $\log^- \|\nabla f\|^2$ et $+P_R^{1+\varepsilon}$ par 1 et intégrant en ζ

$$\int_{\mathbb{C}} d\zeta d\bar\zeta \int_{V_\zeta} \log^- \|\nabla f\|^2 \, d\sigma = \int_{B(0,1)} \|\nabla f\|^2 \log^- \|\nabla f\|^2 \, dv(z)$$

où $dv(z) = \frac{d\zeta d\bar\zeta}{\|\nabla f\|^2} d\sigma(z)$ est précisément le volume euclidien. Comme $x \log^- x$ est borné et que la boule est de volume borné, on conclut que l'intégrale dans le 1er terme du 2nd membre de (7) est bornée pour presque tout $\zeta \in \mathbb{C}$.

2) résulte de l'application de la formule de Green (6) avec

$$u = \log \|\nabla f\|^2 \quad , \quad v = g_\zeta^{(R)}(z_o, z) \quad : \text{on a}$$

$$(8) \quad -\log \|\nabla f\|^2(z_o) = \int_{V_\zeta \cap B(0,R)} g_\zeta^{(R)}(z_o,z) \, K(z) \, d\sigma(z) - \int_{V_\zeta \cap S(0,R)} \log \|\nabla f\|^2 \, d\mu$$

où $d\mu$ est la mesure harmonique z_o ; alors

$$(9) \quad \int_{V_\zeta \cap S(0,R)} \log \|\nabla f\|^2 \, d\mu \leqslant \int \log^+ \|\nabla f\|^2 \, d\mu \leqslant C \log \frac{1}{(1-R^2)}$$

car f est bornée. Par conséquent (8) et (9) donnent la formule (4) attendue.

<u>Remarque</u>: initialement le théorème 1 avait été énoncé pour des fonctions bornées: H.SKODA m'a signalé que l'estimée logarithmique des dérivées se généralisait aux classes de Hardy.

3. Diviseur des fonctions d'une classe de Hardy de la boule de $\mathbb{C}^2$.

Nous allons compléter le résultat obtenu dans [2] en le généralisant aux classes de Hardy et en même temps simplifier un peu sa démonstration en utilisant la théorie de Paley-Littlewood non linéaire développée dans [1] .

Soit B la boule de $\mathbb{C}^2$, f holomorphe dans B ; posons

$$\varphi_f(\zeta) = \int_{V_\zeta} d^3(z,\partial B)\, K(z)\, d\sigma\,(z)$$

φ_f étant finie ou non sur $\mathbb{C}$; nous nous proposons d'étudier les moments de la mesure $\varphi_f(\zeta)\, d\zeta\, d\bar{\zeta}$.

Théorème 3 : soit $p = 4q+2$ (q entier) ou $p = 0$ ou $p = +\infty$.
Si f est dans la classe $H^{p+2}(B)$, alors

$$\int_{\mathbb{C}} |\zeta|^p\, \varphi_f(\zeta)\, d\zeta d\bar{\zeta} < +\infty \;.$$

et particulier $\varphi_f(\zeta) < +\infty$ pp dans $\mathbb{C}$.

Démonstration : on a $dv(z) = \dfrac{d\zeta d\bar{\zeta}}{\|\nabla f\|^2}\, d\sigma_\zeta(z)$

où dv(z) désigne le volume euclidien. Comme

$$K(z)\|\nabla f\|^2 = \frac{|Q(f)|^2}{(\|\nabla f\|^2)^2} \leq \|\nabla^2 f\|^2(z)$$

où $\|\nabla^2 f\|^2$ désigne la somme des dérivées secondes des carrés de f, on a :

$$\int_{\mathbb{C}} |\zeta|^p\, \varphi_f(\zeta)\, d\zeta d\bar{\zeta} \leq \int_{B(0,1)} d^3(z,\partial B)\, |f|^p \|\nabla^2 f\|^2\, dv(z) \tag{10}$$

Introduisons alors pour f holomorphe dans B , les deux fonctions de Paley-Littlewood de f définies par

$$(g_2(f)(\sigma))^2 = \int_0^1 (1-r)^3\, \|\nabla^2 f\|^2(r\sigma)\, dr$$

et

$$(g_{1,4}(f)(\sigma))^4 = \int_0^1 (1-r)^3\, \|\nabla f\|^4(r\sigma)\, dr \;.$$

où $\sigma \in S(0,1)$ $r\sigma$ est le point de $B(0,1)$ de coordonnées polaires (r,σ) .

$g_2(f)$ est fonction de Paley-Littlewood linéaire des dérivées $2^{\underline{\text{ndes}}}$; on a

$$\|g_2(f)\|_{L^2(S(0,1))}^{2} \leqslant \|f\|_{H^2}$$

$g_{1,4}(f)$ est fonction de Paley-Littlwood non linéaire des dérivées premières de f , on a

$$\|g_{1,4}(f)\|_{L^4(S(0,1))} \leqslant C\,\|f\|_{H^4}$$

(voir [1] pour les démonstrations et on fait des résultats plus précis utilisant la théorie du potentiel de la métrique de Bergmann).

Si $p = 2q'$ $(q' = 2q+1)$, on a

$$\partial_{z_i}\partial_{z_j} f^{q'+1} = (q'+1)\, f^{q'}\, \partial_{z_i}\partial_{z_j} f + q'(q'+1) f^{q'-1}\, \partial_{z_i} f\, \partial_{z_j} f$$

de sorte que

$$|f|^p\, \|\nabla^2 f\|^2 \leqslant C\,(\|\nabla^2 (f^{q'+1})\|^2 + |f^{q'-1}|^2\, \|\nabla f\|^4)$$

or $q' = 2q+1$, $f^{\frac{q'+1}{2}}$ est donc holomorphe et $\|\nabla(f^{\frac{q'+1}{2}})\|^4 = |f^{q'-1}|^2\, \|\nabla f\|^4$

d'où

$$|f|^p\, \|\nabla^2 f\|^2 \leqslant C(\|\nabla^2 (f^{q'+1})\|^2 + \|\nabla(f^{\frac{q'+1}{2}})\|^4)$$

Le $2^{\underline{\text{nd}}}$ membre de (10) se contrôle par

(12) $$C\,(\|g_2(f^{q'+1})\|^2_{L^2(S)} + \|g_{1,4}(f^{\frac{q'+1}{2}})\|^4_{L^4(S)}$$

et donc par les résultats rappelés plus haut par

$$C\,(\|f^{q'+1}\|^2_{H^2} + \|f^{\frac{q'+1}{2}}\|^4_{H^4})$$

donc est fini si $f \in H^{p+2}$

Remarque : dans [1] , ces résultats de théorie de Paley-Littlewood sont utilisés pour majorer les volumes des images de domaines admissibles par une application holomorphe de la boule à valeur dans $\mathbb{C}^2$.

4. Diviseur des fonctions entières de $\mathbb{C}^2$

Soit f une fonction entière de type exponentiel. Posons

$$M(f,r) = \sup_{B(0,r)} |f(z)|$$

$$\text{Ord } f = \limsup_{r \to +\infty} \frac{\log \log M(f,r)}{\log r}$$

que nous supposerons donc fini. On a alors

Proposition : on a pour tout $\beta >$ Ord f , si $p_R = R^2 - |z|^2$

$$\int_{V_\zeta \cap B(0,R)} P_R^2 \, K \, d\sigma \leqslant \varphi_R(\zeta) + CR^\beta \, \sigma(\zeta,R)$$

où
$$\sigma(\zeta,R) = \int_{V_\zeta \cap B(0,R)} d\sigma$$

et $\varphi_R(\zeta)$ est une fonction telle que

$$\int \varphi_R(\zeta) \, d\zeta d\bar{\zeta} \leqslant C \, R^6 \; .$$

où C est une constante absolue.

Preuve : appliquons toujours la formule de Green (6) avec $u = \log\|\nabla f\|^2$ et $v = P_R^2$: on obtient toujours la formule (7) mais cette fois $R \to +\infty$: on a donc

$$\int_{V_\zeta \cap B(0,R)} P_R^2 \, K \, d\sigma \leqslant 2 \int_{V_\zeta \cap B(0,R)} -P_R \log\|\nabla f\|^2 \, d\sigma + 2\int_{V_\zeta \cap B(0,R)} \log^+\|\nabla f\|^2 \cos^2\theta d\sigma$$

Soit $\varphi_R(\zeta)$ la $1^{\underline{ere}}$ intégrale : elle se majore en moyenne comme précédemment

$$\int d\zeta d\bar{\zeta} \int_{V_\zeta \cap B(0,R)} -P_R \log\|\nabla f\|^2 d\sigma \leqslant \int_{B(0,R)} \|\nabla f\|^2 \log^-\|\nabla f\|^2 \, P_R \, dv$$

et cette dernière intégrale est majorée par CR^6. La seconde intégrale se majore alors par

$$CR^{\beta}\ \sigma(\zeta,R)$$

en majorant le $\cos^2\theta$ par 1 et $\log^+\|\nabla f\|^2$ par CR^{β}.

<u>Théorème 2</u> : <u>pour presque tout</u> $\zeta \in \mathbb{C}$, <u>on a</u> $\int_{V_\zeta \cap B(0,R)} P_R^2\, K\, d\sigma \leqslant CR^{\sup(6,2\beta+2)}$

<u>si</u> $R \longrightarrow +\infty$.

Démonstration : d'après [4] , l'aire euclidienne $\sigma(S,R) = \bar{\sigma}(\zeta,R)R^2$ où $\bar{\sigma}$ est l'aire projective. Comme cette dernière est contrôlée par R^{β} (voir [4] et plus généralement [3]) .

5. <u>Cas des hypersurfaces de $\mathbb{C}^3$: calculs préliminaires.</u>

Soit $f(z_1,z_2,z_3) = \zeta$ une hypersurface V_ζ de $\mathbb{C}^3$. Utilisons les coordonnées locales (z_1,z_2) pour représenter V_ζ . Soit $g = \det g_{i\bar{j}}$ le déterminant de la métrique euclidienne sur V_ζ .

<u>Lemme 3 : on a</u>

$$g_{i\bar{j}} = \begin{pmatrix} 1+\dfrac{|f_{z_1}|^2}{|f_{z_3}|^2} & \dfrac{f_{z_1}\bar{f}_{z_2}}{|f_{z_3}|^2} \\ \\ \dfrac{\bar{f}_{z_1}f_{z_2}}{|f_{z_3}|^2} & 1+\dfrac{|f_{z_2}|^2}{|f_{z_3}|^2} \end{pmatrix}$$

et

$$g = \det g_{i\bar{j}} = \frac{\|\nabla f\|^2}{|f_{z_3}|^2} .$$

Preuve : calcul de routine.

Soit $$R_{i\bar{j}} = \partial_i\bar{\partial}_j \log g_i = \frac{\partial_i\bar{\partial}_j g}{g} - \frac{\partial_i g\bar{\partial}_j g}{g^2} .$$

C'est la courbure de Ricci de la métrique induite.

Lemme 4 : on a

$$R_{i\bar{j}} = \frac{1}{(\|\nabla f\|^2)^2} \sum_{k=1}^{3} B_i^k \bar{B}_j^k$$

$$B_i^k = f_{z_\ell} \partial_i f_{z_n} - f_{z_n} \partial_i f_{z_\ell}$$ où (k,ℓ,n) est permutation circulaire de $(1,2,3)$.

Preuve : on calcule

$$\partial_i g = \partial_i \left(\frac{\|\nabla f\|^2}{|f_{z_3}|^2}\right) = \frac{\partial_i \|\nabla f\|^2}{|f_{z_3}|^2} - \|\nabla f\|^2 \frac{\partial_i f_{z_3}}{f_{z_3}^2 \bar{f}_{z_3}}$$

$$= \frac{1}{f_{z_3}^2 \bar{f}_{z_3}} [\, f_{z_3} (\partial_i f_{z_1} \bar{f}_{z_1} + \partial_i f_{z_2} \bar{f}_{z_2} + \partial_i f_{z_3} \bar{f}_{z_3}) - \partial_i f_{z_3} \|\nabla f\|^2 \,]$$

$$= \frac{1}{f_{z_3}^2 \bar{f}_{z_3}} [\, \bar{f}_{z_1} B_i^2 + \bar{f}_{z_2} B_i^1 \,]$$

$$\bar{\partial}_j \partial_i g = \frac{1}{f_{z_3}^2} [\, \frac{\overline{\partial_j f_{z_1}}}{\bar{f}_{z_3}} B_i^2 - \frac{\overline{\partial_j f_{z_3}}\, \bar{f}_{z_1}}{\bar{f}_{z_3}^2} B_i^2 + \frac{\overline{\partial_j f_{z_2}}}{\bar{f}_{z_3}} B_i^1 - \frac{\overline{\partial_j f_{z_3}}}{\bar{f}_{z_3}^2} \bar{f}_{z_2} B_i^1 \,]$$

$$= \frac{1}{|f_{z_3}|^4} [\, B_i^1 \bar{B}_j^1 + B_i^2 \bar{B}_j^2 \,]$$

Alors

$$R_{i\bar{j}} = \frac{\partial_i \bar{\partial}_j g}{g} - \frac{\partial_i g \,\bar{\partial}_j g}{g^2} = \frac{1}{|f_{z_3}|^2 (\|\nabla f\|^2)^2} [\, \|\nabla f\|^2 (B_i^2 \bar{B}_j^2 + B_i^1 \bar{B}_j^1) -$$

$$- (|f_{z_1}|^2 B_i^2 \bar{B}_j^2 + \bar{f}_{z_1} f_{z_2} B_i^2 \bar{B}_j^1 + \bar{f}_{z_2} f_{z_1} B_i^1 \bar{B}_j^2 + |f_{z_2}|^2 \bar{B}_j^1 B_i^1)]$$

$$= \frac{1}{|f_{z_3}|^2(\|\nabla f\|^2)^2}[\,|f_{z_2}|^2 B_i^2 \bar{B}_j^2 + |f_{z_i}|^2 B_i^1 \bar{B}_j^1 - \bar{f}_{z_1} f_{z_2} B_i^2 \bar{B}_j^1 -$$

$$- \bar{f}_{z_2} f_{z_1} B_i^1 \bar{B}_j^2 + |f_{z_3}|^2 (B_i^2 \bar{B}_j^2 + B_i^1 \bar{B}_j^1)]$$

Alors il apparaît

$$f_{z_2} B_i^2 \overline{(f_{z_2} B_j^2 - f_{z_1} B_j^1)} + f_{z_1} B_i^1 \overline{(f_{z_1} B_j^1 - f_{z_2} B_j^2)} = \overline{(f_{z_2} B_j^2 - f_{z_1} B_j^1)}(f_{z_2} B_i^2 - f_{z_1} B_i^1)$$

ou $f_{z_2} B_j^2 - f_{z_1} B_i^1 = f_{z_2}(\partial_i f_{z_1} f_{z_3} - \partial_i f_{z_3} f_{z_1}) - f_{z_1}(\partial_i f_{z_2} f_{z_3} - \partial_i f_{z_3} f_{z_2})$

$$= f_{z_3}(f_{z_2} \partial_i f_{z_1} - f_{z_1} \partial_i f_{z_2}) = f_{z_3} B_i^3$$

Pour continuer le calcul supposons que f s'écrive

$$- \zeta + f(z_1, z_2, z_3) = \varphi(z_1, z_2) - z_3$$

Dans ce cas $\quad B_i^1 = -f_{z_3} \partial_i f_{z_2} = \varphi_{2i}$

$$B_i^2 = \varphi_{1i}$$

$$B_i^3 = \varphi_1 \varphi_{2i} - \varphi_2 \varphi_{1i}$$

(ici $\varphi_{ij} = \partial_i \partial_j \varphi$) $\varphi_i = \partial_i \varphi$) .

Alors on a

$$R_{i\bar{j}} = \frac{1}{(1+\|\nabla\varphi\|^2)^2} ((\varphi_1 \varphi_{2i} - \varphi_2 \varphi_{1i})\overline{(\varphi_1 \varphi_{2j} - \varphi_2 \varphi_{1j})} + \varphi_{2i} \bar{\varphi}_{2j} + \varphi_{1i} \bar{\varphi}_{1j})$$

Définissons la fonction det $R_{i\bar{j}}$ sur V_ζ par l'égalité :

$$(R_{ij}\, dz^i \wedge d\bar{z}^j)^{\wedge 2} = (\det R_{i\bar{j}}) dv_\zeta$$

où dv_ζ est le volume euclidien de $V_\zeta \subset \mathbb{C}^3$.

Lemme 5 : on a $\det R_{i\bar{j}} = \dfrac{|\varphi_{21}^2 - \varphi_{11}\varphi_{22}|^2}{(1+\|\nabla\varphi\|^2)^4}$

Démonstration: par définition de $dv_\zeta = (1+\|\nabla\varphi\|^2)dz_1 \wedge d\bar{z}_1 \wedge dz_2 \wedge d\bar{z}_2$,

on a

$$\det R_{i\bar{j}} = (R_{1\bar{1}}R_{2\bar{2}} - R_{1\bar{2}}R_{2\bar{1}})(1+\|\nabla\varphi\|^2)^{-1} \equiv D\,.\,(1+\|\nabla\varphi\|^2)^{-1}\,.$$

Donc d'après le lemme 4

$$D = \frac{1}{(1+\|\nabla\varphi\|^2)^4}[\,(|\varphi_1\varphi_{21} - \varphi_2\varphi_{11}|^2 + |\varphi_{21}|^2 + |\varphi_{11}|^2)(|\varphi_1\varphi_{22} - \varphi_2\varphi_{12}|^2 + |\varphi_{21}|^2 + |\varphi_{22}|^2) -$$

$$- |(\varphi_1\varphi_{21} - \varphi_2\varphi_{11})\overline{(\varphi_1\varphi_{22} - \varphi_2\varphi_{12})} + \varphi_{21}\bar{\varphi}_{22} + \varphi_{11}\bar{\varphi}_{12}|^2\,]$$

$$= \frac{1}{(1+\|\nabla\varphi\|^2)^4}[\,|\varphi_1\varphi_{21} - \varphi_2\varphi_{11}|^2(|\varphi_{21}|^2 + |\varphi_{22}|^2) + |\varphi_1\varphi_{22} - \varphi_2\varphi_{12}|^2(|\varphi_{21}|^2 + |\varphi_{11}|^2)$$

$$+ |\varphi_{21}|^4 + |\varphi_{11}|^2|\varphi_{22}|^2 - ((\varphi_1\varphi_{21} - \varphi_2\varphi_{11})\overline{(\varphi_1\varphi_{22} - \varphi_2\varphi_{12})}(\bar{\varphi}_{21}\varphi_{22} + \bar{\varphi}_{11}\varphi_{12})$$

$$+ \varphi_{21}\bar{\varphi}_{22}\bar{\varphi}_{11}\varphi_{12} \quad + \text{conjugué de ces 2 derniers termes})]$$

$$= \frac{1}{(1+\|\nabla\varphi\|^2)^4}[\,|\varphi_{21}^2 - \varphi_{11}\varphi_{22}|^2 + (\varphi_1\varphi_{21} - \varphi_2\varphi_{11})\{\overline{(\varphi_1\varphi_{21} - \varphi_2\varphi_{11})}(|\varphi_{21}|^2 + |\varphi_{22}|^2) -$$

$$- \overline{(\varphi_1\varphi_{22} - \varphi_2\varphi_{12})}(\bar{\varphi}_{21}\varphi_{22} + \bar{\varphi}_{11}\varphi_{12})\} +$$

$$+(\varphi_1\varphi_{22} - \varphi_2\varphi_{12})\{\overline{(\varphi_1\varphi_{22} - \varphi_2\varphi_{12})}(|\varphi_{21}|^2 + |\varphi_{11}|^2) - \overline{(\varphi_1\varphi_{21} - \varphi_2\varphi_{11})}(\varphi_{21}\bar{\varphi}_{22} + \varphi_{11}\bar{\varphi}_{12})\}]$$

$$= \frac{1}{(1+\|\nabla\varphi\|^2)^4}\left[|\varphi_{21}^2 - \varphi_{11}\varphi_{22}|^2 + (\varphi_1\varphi_{21} - \varphi_2\varphi_{11})\{\varphi_{22}(\overline{-\varphi_2(\varphi_{11}\varphi_{22} - \varphi_{12}^2)}) + \varphi_{21}(\overline{\varphi_1(\varphi_{21}^2 - \varphi_{11}\varphi_{22})})\} + \right.$$

$$\left. + (\varphi_1\varphi_{22} - \varphi_2\varphi_{12})\{\varphi_{21}\overline{\varphi_2(\varphi_{11}\varphi_{22} - \varphi_{12}^2)} + \varphi_{11}\overline{\varphi_1(\varphi_{11}\varphi_{22} - \varphi_{12})}\}\right]$$

$$= \frac{|\varphi_{21}^2 - \varphi_{11}\varphi_{22}|^2}{(1+\|\nabla\varphi\|^2)^3}$$

<u>Lemme 6 : Le long de l'ensemble analytique</u> V_ζ , <u>on a</u> $\partial_i\bar{\partial}_j \log(\|\nabla f\|^2) = R_{i\bar{j}}$.

<u>De plus</u>

$$\partial_i\bar{\partial}_j \log(\text{dét } R_{i\bar{j}}) = -4R_{i\bar{j}} + 2T_{i\bar{j}}$$

<u>où</u> $(T_{i\bar{j}})$ <u>est le courant d'intégration le long du diviseur ensemble des zéros du déterminant de la courbure de Ricci.</u>

Preuve : comme $g = \dfrac{\|\nabla f\|^2}{|f_{z_3}|^2}$, on a là où $f_{z_3} \neq 0$

$$R_{i\bar{j}} = \partial_i\bar{\partial}_j \log g = \partial_i\bar{\partial}_j \log \|\nabla f\|^2 .$$

Par ailleurs vu le lemme 5 , on a

$$\partial_i\bar{\partial}_j \log \text{dét } R_{i\bar{j}} = \partial_i\bar{\partial}_j \log |\varphi_{21} - \varphi_{11}\varphi_{22}|^2 - 4\,\partial_i\bar{\partial}_j \log (1+\|\nabla\varphi\|^2)^3 .$$

Le premier terme est exactement $2T_{i\bar{j}}$ d'après le théorème de Poincaré et le 2nd terme est exactement $-4R_{i\bar{j}}$ d'après la première formule.

<u>Corollaire</u> : <u>en dehors des zéros de</u> dét $R_{i\bar{j}}$, log (dét $R_{i\bar{j}}$) <u>est une fonction plurisurharmonique sur l'ensemble analytique</u> V_ζ . <u>De plus elle satisfait l'équation</u>

de Monge-Ampère complexe

$$\text{dét}\ (\partial_i \bar{\partial}_j \log (\text{dét}\ R_{i\bar{j}})) = 16\ \text{dét}\ R_{i\bar{j}} \quad .$$

Enfin , si Δ est le laplacien euclidien le long de V_ζ on a

$$\Delta \log \|\nabla f\|^2 = K$$

$$\Delta \log(\text{dét}\ R_{i\bar{j}}) = -4K + 2\ d\sigma$$

où K est la courbure scalaire de V_ζ et $d\sigma$ l'aire euclidienne de l'ensemble des zéros du déterminant de la courbure de Ricci.

Evidemment ces résultats se généralisent dans le cas d'une hypersurface de $\mathbb{C}^n$.

6. Diviseur de fonction bornée dans la boule.

Le théorème analogue au théorème 1 mais pour $\mathbb{C}^n$ est alors

Théorème 1' : 1) si K désigne la courbure scalaire de $V_\zeta = f^{-1}(\zeta)$ avec f holomorphe bornée , pour tout $\varepsilon > 0$, pour presque tout $\zeta \in \mathbb{C}$, on a

$$(3)' \qquad \int_{V_\zeta} d(z,\partial B)^{2+\varepsilon}\ K(z)\ d\sigma(z) < +\infty \ .$$

2) On a de même pour tout $\zeta \in \mathbb{C}$

$$(4)' \qquad \int_{V_\zeta \cap B(0,R)} g_\zeta^{(R)}(z_0,z)\ K(z)\ d\sigma(z) \leqslant C \log \frac{1}{1-R} \ .$$

7. Cas d'une fonction entière de $\mathbb{C}^n$

Le résultat analogue au théorème 2 énonce exactement de la même façon dans $\mathbb{C}^n$.

Théorème 2' : soit f entière d'ordre fini Ord f et $\beta >$ Ord f . Pour presque $\zeta \in \mathbb{C}$, on a

$$\int_{V_\zeta \cap B(0,R)} P_R^2\ K\ d\sigma \leqslant C\ R^{\max(2n+2,\ 2\beta+2n-2)} \quad \text{si} \quad R \longrightarrow +\infty \ .$$

8. Remarques.

a) Le critère de la $2^{\underline{nde}}$ partie du théorème 1 ou 1' fait intervenir la fonction de Green de la métrique euclidienne le long du diviseur même ; cette fonction est un être intrinsèquement lié au diviseur (plongé dans $\mathbb{C}^n$) et le problème majeur est d'en avoir des estimées asymptotiques. Dans [3] , sont obtenues de telles estimées pour les variétés algébriques.

Signalons seulement le résultat suivant

Théorème 3 : *si* V *est un diviseur de fonction holomorphe définie au voisinage de la boule de* $\mathbb{C}^2$, *alors la fonction de Green de* $V \cap B(0,1)$ *satisfait*

$$C' \, d(z,\partial B) \leqslant g(z_o,z) \leqslant C \, d(z,\partial B)$$

lorsque $z \longrightarrow \partial B$, *où* C *et* C' *sont des constantes indépendantes de* z .

b) Par ailleurs dans les théorèmes 1 ou 1' ,nous avons encore qu'une version statistique : pour presque tout ζ une certaine intégrale sur V_ζ est bornée. Pour avoir une version individuelle (ie pour tout ζ) , il faudrait démontrer que pour tout ζ

$$\int_{V_\zeta} d(z,\partial B)^{1+\varepsilon} \log^- \|\nabla f\|^2 \, d\sigma < +\infty \ .$$

(C'est en effet pour majorer cette intégrale que nous avons besoin d'intégrer en $\zeta \in \mathbb{C}$) . Une telle estimée est vraisemblable, puisque $\int_{V_\zeta} d(z,\partial B) d\sigma < +\infty$ d'après le critère de Malliavin [4] pour les diviseurs des fonctions de la classe de Névanlinna.

Références

[1] B. GAVEAU : Valeurs frontières des fonctions harmoniques ou holomorphes Cas de la boule CR Acad Sci Paris, 288, 1979, 403-406 et article détaillé Proc. Conf. Analytic functions, Pologne, 1982.

[2] B. GAVEAU, P. MALLIAVIN : Courbure des surfaces de niveau d'une fonction holomorphe bornée CR Acad Sci Paris 1981.

[3] B. GAVEAU, J. VAUTHIER : Répartition des zéros des fonctions de type exponentiel sur une variété algèbrique lisse CR Acad Sci Paris 283 1976 635 - 638 .

[4] P. LELONG : Fonctions entières (n variables) et fonctions plurisousharmoniques d'ordre fini dans $\mathbb{C}^n$ J Analyse Jerusalem 1963 365-407.

[5] P. MALLIAVIN : Fonctions de Green d'un ouvert strictement pseudoconvexe et classe de Névanlinna CR Acad Sci Paris 1974.

Cet article est la version détaillée d'une Note aux Comptes Rendus de l'Académie des Sciences de Paris, Octobre 1981.

Le 14 septembre 1981

Tour Béryl
40 Avenue d'Italie
75013 Paris

Exposé fait en Juin 1980.

FONCTIONS HOLOMORPHES ET PARTICULE CHARGÉE DANS UN CHAMP MAGNÉTIQUE UNIFORME

par

B.GAVEAU et G.LAVILLE

Introduction.

L'exposé qui suit a pour but de comparer les équations apparaissant dans l'étude d'une particule chargée dans un champ magnétique constant en mécanique quantique (électron de Landau) et les équations apparaissant dans l'étude des valeurs au bord de fonctions holomorphes. Pour rendre cette étude plus lisible par les mathématiciens ne connaissant pas le côté physique de ce problème,nous avons au début rappelé brièvement des calculs bien connus .

Le lien entre l'électron de Landau et l'oscillateur harmonique à deux dimensions est bien connu, son lien avec l'oscillateur à une seule dimension est étudié dans ce qui suit. D'autre part, il apparaît ici une nouvelle symétrie à ce problème celle du groupe d'Heisenberg.

1. - Hamiltonien de la particule.

Considérons une particule de charge q , de masse m placée dans un champ magnétique $\vec{B}$ constant .

Soient $\vec{A}$ le potentiel vecteur : $\vec{\nabla} \times \vec{A} = \vec{B}$, $\vec{r}$ et $\dot{\vec{r}}$ la position et la vitesse de la particule . Le Lagrangien s'écrit :

$$(1) \qquad \mathscr{L}(\vec{r},\dot{\vec{r}},t) = \frac{1}{2} m \dot{\vec{r}}^2 + q \dot{\vec{r}}.\vec{A}(\vec{r},t) .$$

Soit $\vec{p}$ l'impulsion, par définition :

$$\vec{p} = \left(\frac{\partial \mathscr{L}}{\partial \dot{x}_1} , \frac{\partial \mathscr{L}}{\partial \dot{x}_2} , \frac{\partial \mathscr{L}}{\partial \dot{x}_3} \right) = m \dot{\vec{r}} + q \vec{A}(\vec{r},t)$$

d'où l'Hamiltonien

$$(2) \qquad \mathscr{H}(\vec{r},\vec{p},t) = \vec{p}.\dot{\vec{r}} - \mathscr{L}(\vec{r},\dot{\vec{r}},t) = \frac{1}{2m} \vec{p}^2 - \frac{q}{m} \vec{p}.\vec{A} + q^2 \vec{A}^2 .$$

Quantifions cet Hamiltonien : P et R étant les observables d'impulsion et de position : $P = \frac{\hbar}{i} \vec{\nabla}$, $\vec{R} = (x_1,x_2,x_3)$, l'Hamiltonien total vaut :

(3) $$\frac{1}{2m}[\vec{P} - q\,\vec{A}(R,t)]^2 .$$

Prenons une base de R^3 telle que le champ $\vec{B}$ soit dirigé suivant l'axe des x_3 : $\vec{B} = B_o\,\vec{e}_3$, choisissons comme jauge $\vec{A} = -\frac{x_2}{2} B_o\,\vec{e}_1 + \frac{x_1}{2} B_o\,\vec{e}_2$.

Remarquons que cette jauge respecte la symétrie cylindrique. En séparant le mouvement dans le plan $(\vec{e}_1,\vec{e}_2)$ du mouvement dans la direction $\vec{e}_3$, l'Hamiltonien plan s'écrit :

(4) $$H_1 = \frac{-\hbar^2}{2m}\left[\Delta - \frac{id\,B_o}{\hbar}\left(x_2\frac{\partial}{\partial x_1} - x_1\frac{\partial}{\partial x_2}\right) - (x_1^2 + x_2^2)\left(\frac{q\,B_o}{2\hbar}\right)^2\right] .$$

Remarquons que l'Hamiltonien total dans R^3 s'écrit $P_3/_{2m} + H_1$, P_3 étant l'impulsion dans la direction $\vec{e}_3$.

2. - Transformée de Fourier dans le groupe d'Heisenberg.

Soit G_H le groupe d'Heisenberg de degré 1 , c'est-à-dire $G_H = \{(z,t) \in \mathbb{C}\times\mathbb{R}$ avec la loi de groupe

$$(z,t)(z',t') = (z+z',\ t+t' + 2\ \mathrm{Im}\ z\,\overline{z}')$$

prenons pour base de son algèbre de Lie les champs de vecteurs invariants à gauche :

$$X = \frac{\partial}{\partial x} + 2y\frac{\partial}{\partial t} \quad ; \quad Y = \frac{\partial}{\partial y} - 2x\frac{\partial}{\partial t} \quad ; \quad T = \frac{\partial}{\partial t}$$

avec $z = x + iy$. Introduisons le Laplacien incomplet

$$\Delta_K = X^2 + Y^2$$

(5) $$= \frac{\partial^2}{\partial x^2} + \frac{\partial^2}{\partial y^2} + 4\frac{\partial}{\partial t}\left(y\frac{\partial}{\partial x} - x\frac{\partial}{\partial y}\right) + 4(x^2 + y^2)\frac{\partial^2}{\partial t^2} .$$

Sur le groupe G_H , on peut envisager deux types de transformée de Fourier :

a/ la transformée de Fourier "totale" : les classes d'équivalence des représentations unitaires irréductibles peuvent s'indexer par $\lambda \in R$; π^λ étant une telle représentation , prenons $f \in L^2(\mathbb{R})$,

$$\pi^\lambda(x,y,t)f(\xi) = \exp[i\,\lambda\,(t + 4\,\xi y - 2xy)]\ f(\xi - x)$$

(6) $$d\pi^\lambda(X) = -\frac{\partial}{\partial \xi} \quad ; \quad d\,\pi^\lambda(Y) = 4i\lambda\xi \quad ; \quad d\pi^\lambda(T) = i\,\lambda .$$

Soient X_h et P_h les observables de position et d'impulsion de l'oscillateur harmonique quantique. Il est bien connu que l'Hamiltonien de cet oscillateur est

(7) $$H_2 = \frac{P_h^2}{2m} + \frac{1}{2} m \omega^2 X_h^2 = - \frac{\hbar^2}{2m} \frac{d^2}{d\xi^2} + \frac{1}{2} m \omega^2 \xi^2$$

On s'aperçoit que

$$d \pi^{\lambda}(i \hbar X) = P_h$$
$$d \pi^{\lambda}(\frac{-i}{4} Y) = X_h$$

(8) $$d\pi^{\lambda} \left|\frac{-\hbar^2}{2m} (X^2 + Y^2)\right| = \frac{1}{2m} (P_h^2 + 16 \lambda^2 \hbar^2 X_h^2)$$

Prenons $\omega = \frac{4\lambda\hbar}{m}$.

L'Hamiltonien H_2 est le transformé de l'opérateur $\frac{-\hbar^2}{2m}(X^2 + Y^2)$.

Ce résultat , bien connu, est à l'origine du nom "groupe d'Heisenberg".

b/ la transformée de Fourier partielle (cf. [3]) (x,y) étant fixé, effectuons une transformée de Fourier relativement à la variable t :

Notons $\mathcal{F}$ la transformée de Fourier partielle :

$$\mathcal{F}(X) = \frac{\partial}{\partial x} + 2 i y \tau$$
$$\mathcal{F}(Y) = \frac{\partial}{\partial y} - 2 i x \tau$$
$$\mathcal{F}(T) = i \tau .$$

D'après (5), on a :

(9) $$\mathcal{F}(\Delta_K) = \frac{\partial^2}{\partial x^2} + \frac{\partial^2}{\partial y^2} + 4 i \tau (y \frac{\partial}{\partial x} - x \frac{\partial}{\partial y}) - 4(x^2 + y^2) \tau^2 .$$

Comparons maintenant les formules (4) et (9) , elles se correspondent avec :

$$\tau = \frac{-q B_o}{4\hbar} ; x_1 = x ; x_2 = y .$$

D'où :

(10) $$\mathcal{F}(- \frac{\hbar^2}{2m} \Delta_K) (x,y, \frac{-q B_o}{4\hbar}) = H_1 .$$

Cette formule donne le lien entre l'Hamiltonien H_1 et le Laplacien incomplet Δ_K.

Mais d'après (8)

(11) $$d\pi^{\lambda} \left[\frac{\hbar^2}{2m} \Delta_K \right] = H_2 .$$

Les formules (10) et (11) montrent immédiatement le lien entre deux situations physiques : l'oscillateur harmonique à une seule dimension et le mouvement de la particule. Le lien entre ce dernier problème et l'oscillateur à deux dimensions est bien connu.

Utilisons la formule de Plancherel pour le groupe d'Heisenberg ; φ étant une fonction régulière :

$$-\frac{\hbar^2}{2m}\Delta_K \varphi(x,y,t) = \int_{\mathbb{R}} \operatorname{tr}[\pi^\lambda(x,y,t)\, H_2\, \pi^\lambda(\varphi)]\, \frac{|\lambda|\, d\lambda}{(2\pi)^2}$$

d'après (10) nous obtenons finalement :

(12) $$H_1\, \varphi(x,y,\frac{-q\,B_o}{2\hbar}) = \frac{1}{(2\pi)^2}\int_{-\infty}^{+\infty}\int_{-\infty}^{+\infty} \operatorname{tr}[\pi^\lambda(x,y,t)H_2\, \pi^\lambda(\varphi)]e^{it\, q\, B_o/2\hbar}\, dt|\lambda|d\lambda.$$

3. - Fonctions de Cauchy-Riemann et fonctions propres de l'Hamiltonien.

Considérons l'opérateur $\Delta_K + i\,\alpha\,T$ sur le groupe G_H, avec α réel (cf. [1]). Dans le cas $\alpha = 1$ les fonctions $f : H \to \mathbb{C}$ qui satisfont à l'équation $(\Delta_K + iT)f = 0$ sont les fonctions de Cauchy-Riemann

$$\mathcal{F}(\Delta_K + i\,\alpha T) = \frac{\partial^2}{\partial x^2} + \frac{\partial^2}{\partial y^2} + 4\, i\, \tau(y\frac{\partial}{\partial x} - x\frac{\partial}{\partial y}) - 4(x^2+y^2)\tau^2 - \tau\,\alpha$$

$$= \frac{-2m}{\hbar^2} H_1 + \alpha\frac{q\,B_o}{4\hbar} = \frac{-2m}{\hbar^2}(H_1 - \alpha\frac{q\,B_o\hbar}{8m})$$

or, nous savons (cf. [1]) que l'opérateur $\Delta_K + i\,\alpha\,T$ est inversible si et seulement si $\alpha \neq 4(2n+1)$, $n \in \mathbb{N}$. Dans le cas contraire, l'ensemble des fonctions de carré intégrable sur G_H et annulant $\Delta_K + i\,4(2n+1)T$ est de dimension infinie (fonctions de Cauchy-Riemann dans le cas $n = 0$).

Il est bien connu que les valeurs propres de la particule chargée sont $(2n+1)\frac{q\,\hbar\,B_o}{2m}$, ce qui correspond aux valeurs trouvées. Dans le cas du niveau fondamental, les fonctions propres sont les transformées de Fourier partielles des fonctions de Cauchy-Riemann.

Pour le calcul effectif d'une base de fonctions propres de carré intégrable, la méthode classique (cf. [2]) consiste à prendre l'Hamiltonien H_1, et à effectuer une séparation des variables ce qui amène l'équation hypergéométrique confluente.

Une autre méthode consiste à suivre les idées de [3] . Partant de l'égalité (4) mais écrite sous forme complexe : en posant $z = x_1 + ix_2$, (4) devient :

$$(13)\qquad H_1 - \frac{q\hbar B_o}{2m} = \frac{-\hbar^2}{2m}\left(2\frac{\partial}{\partial z} - \frac{qB_o}{2\hbar}\bar{z}\right)\left(2\frac{\partial}{\partial \bar{z}} + \frac{qB_o}{2\hbar}z\right).$$

Donc, les fonctions propres correspondant à la première valeur propre $\frac{q\hbar B_o}{2m}$ peuvent se calculer en cherchant $f(z, B_o)$ telle que :

$$(14)\qquad \frac{\partial f}{\partial z} + \frac{qB_o z}{4\hbar} f = 0.$$

Remarquons que nous avons équivalence entre $H_1 f = \frac{q\hbar B_o}{2m} f$ et l'équation (14).

Ceci peut se voir, par exemple en remarquant que par transformée de Fourier $(\Delta_K + 4iT)\varphi = 0$ équivaut à φ fonction de Cauchy-Riemann.

L'équation (14) possède les solutions évidentes :

$$(15)\qquad f(z, B_o) = k(z, B_o)\, e^{\frac{qB_o}{4\hbar}|z|^2}$$

où $k(z, B_o)$ est une fonction holomorphe en la variable z .

Les fonctions propres correspondant aux autres valeurs propres s'obtiennent par la méthode classique (en introduisant les opérateurs de création et d'annihilation).

Remarque: si l'on ajoute un potentiel $V \geqslant 0$, l'égalité (13) s'écrit

$$(16)\qquad H_1 + V - \frac{q\hbar B_o}{2m} = \frac{-\hbar^2}{2m}\left(2\frac{\partial}{\partial z} - \frac{qB_o}{2\hbar}\bar{z}\right)\left(2\frac{\partial}{\partial z} + \frac{qB_o}{2\hbar}z\right) + V.$$

Soient μ et φ valeur et vecteur propres

$$(H_1 + V)\varphi = \mu\varphi$$

en utilisant (16) on trouve :

$$\frac{\hbar^2}{2m}\int \left|2\frac{\partial\varphi}{\partial z} + \frac{qB_o}{2\hbar} z\varphi\right|^2 dz\, d\bar{z} + \int V|\varphi|^2 dz\, d\bar{z} = \mu\int|\varphi|^2\, dz\, d\bar{z} - \frac{q\hbar B_o}{2m}\int |\varphi|^2\, dz\, d\bar{z}$$

d'où , par positivité :

$$\frac{q\hbar B_o}{2m} \leqslant \mu$$

nous avons donc la plus petite valeur propre possible pour tous les potentiels positifs (sous réserve d'existence d'états liés).

4. - Lien entre les valeurs propres de H_1 et H_2.

Choisissons une fonction $\varphi_n \in L^2(G_H)$, normalisée telle que

(17) $$[\Delta_K + i\,4(2n+1)T]\varphi_n = 0$$

avec $n \in \mathbb{N}$. Supposons de plus que $\Delta_K \varphi_n$ et $T\varphi_n$ appartiennent à $L^2(G_H)$.

Par transformée de Fourier totale :

$$\pi^\lambda\left(-\frac{\hbar^2}{2m}\Delta_K \varphi_n\right) - \frac{\hbar^2}{2m} 4(2n+1)\,\pi^\lambda(i\,T\varphi_n) = 0$$

(18) $$H_2\,\pi^\lambda(\varphi_n) = -\frac{\hbar^2}{m} 2(2n+1)\pi^\lambda(\varphi_n)$$

(comme $\lambda = \frac{\omega m}{4\hbar}$, on retrouve les valeurs propres de l'oscillateur $(n + 1/2)\hbar\omega$) .

Effectuons une transformée de Fourier partielle

$$\mathcal{F}(\Delta_K \varphi_n) + i\,4(2n+1)\,\mathcal{F}(T\varphi_n) = 0$$

d'après (10) :

$$H_1 \mathcal{F}\varphi_n\left(x,y,-\frac{qB_o}{4\hbar}\right) = -\frac{\hbar^2}{m} 2(2n+1)\tau\mathcal{F}(\varphi_n) = (n+1/2)\frac{q\hbar B_o}{m}\mathcal{F}(\varphi_n).$$

Utilisons la formule de Plancherel en prenant une fonction $\psi_n \in L^2$ ainsi que ses dérivées jusqu'à l'ordre 2 (ceci afin d'assurer les convergences dans les intégrales qui suivent) :

$$\int_{\mathbb{R}} \operatorname{tr}\left[\pi^\lambda(\psi_n)^* \left(H_2 + \frac{\hbar^2}{m} 2(2n+1)\lambda\right)\pi^\lambda(\psi_n)\right]\frac{|\lambda|\,d\lambda}{(2\pi)^2}$$

$$= -\frac{\hbar^2}{2m}\int_{G_H} \overline{\psi}_n(\Delta_K + i\,4(2n+1)T)\,\psi_n\,d\sigma$$

(19) $$= \frac{1}{2\pi}\int_{\mathbb{R}^3} \mathcal{F}(\overline{\psi}_n)\,\mathcal{F}\left(-\frac{\hbar^2}{2m}\Delta_K - i\frac{\hbar^2}{m}2(2n+1)T\right)\mathcal{F}(\psi_n)\,d\sigma$$

$$= \frac{1}{2\pi}\int_{\mathbb{R}^3} \mathcal{F}(\overline{\psi}_n)\left(H_1 + \frac{\hbar^2}{m}2(2n+1)\tau\right)\mathcal{F}(\psi_n)\,d\sigma\,.$$

Soient $E_n^{(1)}$ et $E_n^{(2)}(\lambda)$ les valeurs propres de H_1 et H_2 l'égalité (19) peut encore s'écrire :

$$\int_{\mathbb{R}^3}\mathcal{F}(\overline{\psi}_n)\,H_1\mathcal{F}(\psi_n)\,d\sigma - E_n^{(1)}\int_{\mathbb{R}^3}\mathcal{F}(\overline{\psi}_n)\,\mathcal{F}(\psi_n)\,d\sigma = \int_{\mathbb{R}}\operatorname{tr}[\pi^\lambda(\psi_n)^*(H_2 - E_n^{(2)}(\lambda))\pi^\lambda(\psi_n))\frac{|\lambda|d\lambda}{2\pi}.$$

Si on utilise (12) on voit que :

(20) $$E_n^{(1)} = \int_{\mathbb{R}} E_n^{(2)}(\lambda)\operatorname{tr}[\pi^\lambda(\psi_n)^*\,\pi^\lambda(\psi_n)]\frac{|\lambda|d\lambda}{2\pi}$$

sous réserve que ψ_n soit normalisée.

L'intérêt de la formule (20) est que si l'on ajoute un potentiel $V(x_1,x_2)$ à l'Hamiltonien H_1, les calculs précédents restent valables à condition de remplacer l'Hamiltonien de l'oscillateur par :

$$\frac{P_h^2}{2m} + \frac{1}{2} m\omega^2 X_h^2 + d\pi^\lambda(V)$$

$d\pi^\lambda(V)$ jouant le rôle de potentiel. La formule (20) relira encore les deux valeurs propres.

5. - Remarque : Spin et champ électrique.

Les considérations précédentes sont encore valables si la particule possède un spin et si l'on est en présence d'un champ électrique constant. Dans ce cas, la formule (3) donnant l'Hamiltonien doit être remplacée par l'Hamiltonien

$$(21) \qquad \frac{1}{2m} [\vec{P} - q \vec{A}(\vec{R}, t)]^2 + q U(\vec{R}, t) - \frac{q\hbar}{2m} (\sigma_x B_x + \sigma_y B_y + \sigma_z B_z) .$$

Ce nouvel Hamiltonien agissant maintenant sur le produit tensoriel des états de spin et des états de spin et des états de position, σ_x, σ_y, σ_z étant les matrices de Pauli .

Comme $\vec{B} = B_o \vec{e}_3$, $B_x = B_y = 0$, $\sigma_z = \begin{pmatrix} 1 & 0 \\ 0 & -1 \end{pmatrix}$ pour le spin, nous aurons seulement à considérer un terme $- \frac{q \hbar B_o}{2m} \begin{pmatrix} 1 & 0 \\ 0 & -1 \end{pmatrix}$ ce qui montre que l'on se ramène à deux deux équations découplées, donc on se ramène au cas sans spin, quitte a ajouter le terme $\pm \frac{q \hbar B_o}{2m}$, nous prendrons donc, comme Hamiltonien (en prenant par exemple le signe +)

$$(22) \qquad \frac{1}{2m} [\vec{P} - q \vec{A}(\vec{R}, t)]^2 + q U(\vec{R}, t) + \frac{q \hbar B_o}{2m}$$

$U(\vec{R}, t)$ représentant le potentiel électrique (potentiel scalaire). Supposons que ce potentiel ne dépende que de x_1 et x_2 les équations (10) et (11) deviendront

$$\mathcal{F}(- \frac{\hbar^2}{2m} \Delta_K + q U + \frac{q \hbar B_o}{2m}) = H_1 + q U + \frac{q \hbar B_o}{2m}$$

$$d\pi^\lambda(- \frac{\hbar^2}{2m} \Delta_K + q U + \frac{q \hbar B_o}{2m}) = H_2 + \frac{q \hbar B_o}{2m} + q\, d\, \pi^\lambda(U).$$

On trouve l'oscillateur harmonique avec un potentiel perturbateur du type

$$\frac{q \hbar B_o}{2m} + q\, d\, \pi^\lambda(U).$$

Si l'on note $\alpha_n^{(1)}$ les valeurs propres de $H_1 + q\,U + \frac{q\,\hbar\,B_o}{2m}$ et $\alpha_n^{(2)}(\lambda)$ les valeurs propres de l'oscillateur perturbé , le calcul effectué précédemment pourra se faire de même et l'on aura encore une formule analogue à (20)

$$\alpha_n^{(1)} = \int_{\mathbb{R}} \alpha_n^{(2)}(\lambda)\,\mathrm{tr}[\pi^{\lambda}(\varphi_n)^{*}\,\pi^{\lambda}(\varphi_n)]\,\frac{|\lambda|\,d\lambda}{2\pi}\ .$$

BIBLIOGRAPHIE

[1] FOLLAND (G.), STEIN (E.). - Estimates for the $\overline{\partial}_b$-complex and analysis on the Heisenberg group. Comm. on pure and appl. Math., vol. 27, pp. 429-522, 1974.

[2] LANDAU (L.), LIFSCHITZ (E.). - Mécanique quantique . Ed. Mir, Moscou.

[3] LAVILLE (G.). - Formule de représentation intégrale dans le bord d'un domaine de C^n. C.R.Acad.Sc., Paris, t. 287, pp. 129-130, 1978.

[4] POGOSYAN (G.) and TER-ANTONYAN (V.) .- Lien entre les fonctions d'onde polaires et cartésiennes d'une particule chargée nonrelativiste dans un champ magnétique homogène. Teoreticheskaya i Matematiches Fizika, vol. 40, n° 1, pp. 140-143, Juillet 1979.

Université Pierre et Marie Curie
L.A. au C.N.R.S., n° 213
4, Place Jussieu 75230-PARIS CEDEX 05

INTÉGRALE DE DIRICHLET SUR UNE VARIÉTÉ COMPLEXE I

par Bernard GAVEAU et Julian ŁAWRYNOWICZ

Introduction

En une variable complexe, l'intégrale de Dirichlet joue un rôle fondamental dans la définition des mesures harmoniques, fonction de Green, capacités etc., et dans la démonstration des théorèmes fondamentaux, par exemple le théorème de Riemann. Ce rôle important est dû au principe variationnel associé à l'intégrale de Dirichlet et au fait que l'intégrale de Dirichlet est un invariant conforme.

En analyse complexe à plusieurs variables, l'intégrale de Dirichlet peut être définie lorsqu'on s'est donnée une métrique kählérienne, mais son défaut fondamental est de ne plus être un invariant biholomorphe. Pour obtenir des invariants biholomorphes du type "potentiel", une première approche étudiée depuis plusieurs années, est l'équation de Monge-Ampère complexe (voir [7] , [3] , [22], [30] et [14]) . Le laplacien est alors remplacé par l'équation non linéaire de Monge-Ampère ; dans l'interprétation "contrôle optimal" étudiée par le premier auteur [11], l'équation de Monge-Ampère apparaît comme un mélange de toutes les théories du potentiel associées à toutes les métriques kählériennes possibles et en cela, elle semble donc être un substitut plausible pour la théorie du potentiel de plusieurs variables.

Nous avons proposé dans [12] une autre possibilité , plus strictement liée au potentiel, pour définir une intégrale de Dirichlet invariant biholomorphe, que nous allons détailler dans cet article. Décrivons brièvement l'idée que nous utilisons.

Dans [5], Beurling et Deny ont introduit la notion d'espace de Dirichlet et de forme de Dirichlet générale , à partir desquelles ils peuvent définir un balayage, une capacité et en général toutes les notions usuelles de théorie du potentiel ; cette théorie a été reprise très récemment par Fukushima [10] d'un point de vue d'analyse fonctionnelle et de théorie des processus de Markov abstraits, mais malheureusement avec peu d'applications. Plus précisement, dans notre contexte à chaque fonction plurisousharmonique G , dans un domaine D de $\mathbb{C}^n$, on peut associer une intégrale de Dirichlet

$$I[u,dd^cG] = \int_D du \wedge d^cu \wedge (dd^cG)^{n-1} .$$

Lorsque G est convenablement normalisée, par exemple $0 \leqslant G \leqslant 1$ partout, alors la norme $\|u\|_D = (\sup\{I[u,dd^cG] : 0 \leqslant G \leqslant 1\})^{1/2}$ est bien définie sur les u à supports compacts de classe C^1 et donc, par définition, est un invariant biholomorphe. Par rapport à cette norme, nous développons une théorie du balayage qui conduit en particulier à une nouvelle définition de la capacité ; cette capacité est le supremum des capacités définie par les intégrales de Dirichlet $I[.u,dd^cG]$ et de plus, pour chaque compact K est associé une fonction G"extrêmale", comme il résulte du théorème de von Neumann de théorie de jeux [25] appliqué dans notre contexte (tout au moins en deux variables complexes). Ce résultat, le calcul d'Okada [26], joints à la théorie de Fukushima permettent de montrer l'existence d'un point-selle d'une certaine fonctionnelle ce qui donne ainsi une définition satisfaisante du balayage sur un compact.

Nous donnons quelques applications, en particulier à une théorie non linéaire des particules élémentaires dûe au second auteur et à L. Wojtczak [23] , [24] .

Cette approche est donc assez différente de celle de l'équation de Monge-Ampère complexe. L'article présenté ici la première partie d'un travail en cours d'élaboration, en particulier concernant des généralisations à $\mathbb{C}^n$, une définition de mesure harmonique et l'allure au bord des fonctions de plusieurs variables.

1. Propriétés de la norme $|||u|||_p(D,q)$

Soit D un domaine de $\mathbb{C}^n$ et G une fonction de la classe $P_2(D)$ des fonctions $C^2(D)$-plurisousharmoniques avec $0 \leq G \leq 1$ partout. Considérons D comme la variété kählérienne munie de la mesure $d\nu$ et de la distance géodésique ds^2 associées à sa métrique g :

(1) $$g_{j\bar{k}} = G_{|j\bar{k}} \equiv (\partial^2/\partial z_j \partial \bar{z}_k)G \ .$$

Alors

$$d\nu = \text{dét } g \, d\nu_{eucl}, \quad ds^2 = g_{j\bar{k}} dz_j \otimes d\bar{z}_k \ ,$$

(2) $$g_{j\bar{k}} \, dz_j \wedge d\bar{z}_k \equiv (1/2i) \, dd^c G \equiv \partial\bar{\partial} G \ .$$

Si u est une fonction de la classe $C^1(D)$, $g : D \longrightarrow \mathbb{R}^+$ est continue et $p \geq 1$, on pose

(3) $$|||u|||_p(D,q) = (\sup_{G \in P_2(D)} \int_D q \|\nabla u\|^p_{ds^2} \, d\nu)^{1/p} \ .$$

Notons aussi $P_o(D)$ la classe des fonctions plurisousharmoniques $C^o(D)$ avec $0 \leq G \leq 1$ partout et $C_c^\infty(D)$ la classe des fonctions $C^\infty(D)$ à support compact. Nous avons besoin de

LEMME 1. Pour $u \in C_c^\infty(D)$ et $1 \leq p \leq 2$ on a

(4) $\|u\|_p(D,q) < +\infty$,

(5) $\|u\|_p^p(D,q) \equiv [\|u\|_p(D,q)]^p = \sup\limits_{G\in P_o(D)} \int_D q\|\nabla u\|^p_{ds^2}\, d\nu$.

Si de plus $n=p=2$, on peut remplacer aussi le supremum par le supremum sur la classe $P_2(D)$ des fonctions plurisousharmoniques dans D , comprises entre 0 et 1 et pas nécessairement continues.

Preuve . Les relations (4) et (5) résultent d'une régularisation de G et du lemme suivant (Chern, Levine et Nirenberg [8]) : Soit $\Delta(0;r)$ un polydisque ouvert de rayon $r=(r_j)$ et de centre 0 dans $\mathbb{C}^n$ et soit $G:\Delta(0;r)\longrightarrow(0;1)$ une fonction plurisousharmonique de classe $C^2(\Delta(0;r))$. Alors, pour chaque $\tilde{r}=(\tilde{r}_j)$, $0<\tilde{r}_j<r_j$, $j=1,\ldots,n$, il existe une constante A , indépendante de G et dépendante seulement de r et $\tilde{r}$, telle que pour tous les indices $j_1,\ldots,j_L,k_1,k_\ell,\ldots,k_\ell$ $1\leq j_1<\ldots<j_\ell\leq n$, $1\leq k_1<\ldots<k_\ell\leq n$, on a l'estimation

$$\int_{\Delta(0;z)} \{\left|\begin{vmatrix} G_{|j_1\bar{k}_1}\cdots G_{|j_i\bar{k}_\ell} \\ \cdots\cdots\cdots\cdots \\ G_{|j_Lk_1}\cdots G_{|j_L\bar{k}_\ell}\end{vmatrix}\right| + \sum_{k=1}^{n} |G_{|k}|^2\}d\nu_{eucl}\leq A \quad .$$

Dans $\mathbb{C}^2$, $(dd^cG)^{n-1} = dd^cG$ donne un courant défini au sens de de Rham et il n'y a pas de problème de définition.

Remarque 1. Pour $p=2$ on a

(6) $\|u\|_2(D,q) = [\sup\limits_{G\in P_2(D)} \int_D q\,du\wedge d^cu\wedge(dd^cG)^{n-1}$.

On note $\mathcal{H}^1_p(D,q)$ l'espace de $u\in C^1(D)$ avec $\|u\|_p(D,q)<+\infty$.

On a

LEMME 2. $\mathcal{H}^1_p(D,q)$ avec la norme (3) est complet.

Preuve. Soit (u_n) une suite de Cauchy pour $|||\ |||_p(D,q)$; si nous prenons pour $G(z)=|z|^2$, on voit que (u_n) est suite de Cauchy pour la norme usuelle de Sobolev dans $\mathbb{C}^n$, donc elle converge vers u. Extrayons de (u_n) une suite (v_k), $v_k=u_{n_k}$, $k=1,2,\ldots,$ telle que

$$|||v_k-v_{k-1}|||_p(D,q)\leqslant 2^{-k} .$$

On a alors

$$u-v_k=\sum_{n=k}^{+\infty}(v_{n+1}-v_n),\text{ d'où } |||u-v_k|||_p(D,q)\leqslant\sum_{n=k}^{+\infty}2^{-n}=2^{-k+1}$$

et, par conséquent, $u_n\longrightarrow u$ au sens de l'espace de Sobolev.

On notera $\mathcal{H}^1_{o,p}(D,q)$ l'adhérence des $u\in C^\infty_c(D)$ dans $\mathcal{H}^1_p(D,q)$.

A cause des lemmes 1 et 2, $\mathcal{H}^1_{o,p}(D,q)$ est complet lorsqu'on le considère avec la norme définie par $P_2(D)$ ou $P_o(D)$.

La norme (3) est un invariant biholomorphe :

THEOREME 1. Si $f:D\longrightarrow f[D]$ est biholomorphe, où $f[D]\subset\mathbb{C}^n$ et $u\in\mathcal{H}^1_{o,p}(f[D],q)$, alors

$$u\circ f\in\mathcal{H}^1_{o,p}(D,q\circ f)$$

et

$$|||u\circ f|||_p(D,q\circ f)=|||u|||_p(f[D],q) .$$

Preuve. Pour $G\in P_2(f[D])$ on écrit, d'après (2) et (1),

$$\int_D(q\circ f)\|\nabla(u\circ f)\|^p_{ds^2}\,d\nu$$

$$=\int_D(q\circ f)[g^{j\bar k}(u\circ f)_{|j}(u\circ f)_{|\bar k}]^{\frac12 p}\,d\nu$$

$$=\int_D(q\circ f)[g^{jk}(u_{|r}\circ f)(u_{|\bar s}\circ f)f^r_{|j}\bar f^{\bar s}_{\bar k}]^{\frac12 p}\,d\nu .$$

Soit $h = (f^{-1})^* g$ de sorte que

$$h^{r\bar{s}} \circ f = g^{j\bar{k}} f^{r}_{|j} f^{\bar{s}}_{|\bar{k}} \quad .$$

Alors $\tilde{h} = \partial\partial(G\circ f^{-1})$ car $g = \partial\bar{\partial}G$ et on obtient

$$\int_D (q\circ f)\|\nabla(u\circ f)\|^p_{ds^2} d\nu$$

$$= \int_D (q\circ f)[(h^{r\bar{s}}\circ f)(u_{|r}\circ f)(u_{|\bar{s}}\circ f)]^{\frac{1}{2}p} d\nu$$

$$= \int_{f[D]} q[h^{r\bar{s}} u_{|r} u_{|\bar{s}}]^{\frac{1}{2}p} dV \quad ,$$

où dV est le volume de la métrique h .

Mais considérons $q\circ f^{-1}$, alors elle est dans $P_2(D)$ car f est holomorphe, d'où

$$\|u\circ f\|_p(D, q\circ f) \leq \|u\|_p(f[D], q)$$

et par biholomorphisme on a la réciproque.

2. Exemples

Nous avons besoin de

COROLLAIRE 1. Pour $u \in \mathcal{H}^1_2(B,1)$, où B est la boule unité de $\mathbb{C}^2$, il existe une constante C telle que pour tout Φ convexe croissante à valeurs dans $[-1;0]$, si on pose $P(z) = |z|^2 - 1$, on a :

(7) $$\int_D (\Phi'\circ P)\|\nabla u\|^2_{eucl} d\nu_{eucl} + \int_B (\Phi''\circ P)\|\partial_b u\|^2_{eucl} d\nu_{eucl} \leq C \ ,$$

où $(\partial_b u)_{eucl} = \bar{z}_2 u_{|1} - \bar{z}_1 u_{|2}$,

Preuve. Soit $G = \Phi \circ P$. Alors

(8) $$-4\|\nabla u\|^2_{eucl} d\nu_{eucl} = \partial\bar{\partial}P \wedge \partial u \wedge \bar{\partial}u \quad ,$$

$$(9)\ -4\|\partial_b u\|^2_{eucl}\, d\nu_{eucl} = -4(z_2\bar{z}_2 u_{|1} u_{|\bar{1}} - z_1\bar{z}_2 u_{|1} u_{|\bar{2}} - z_2\bar{z}_1 u_{|2} u_{|1} + z_1\bar{z}_1 u_{|2} u_{|\bar{2}})\, d\nu_{eucl} =$$

$$= (\bar{z}_1 dz_1 + \bar{z}_2 dz_2)\wedge(z_1 d\bar{z}_1 + z_2 d\bar{z}_2)\wedge(u_{|1} dz_1 + u_{|2} dz_2)\wedge(u_{|\bar{1}}\, d\bar{z}_1 + u_{|\bar{2}}\, d\bar{z}_2) =$$

$$= \partial P\wedge\bar{\partial}P\wedge\partial u\wedge\bar{\partial}u \quad .$$

Par conséquent on a

$$-4[\,(\Phi' o P)\|\nabla u\|^2_{eucl}\, d\nu_{eucl} + (\Phi'' o P)\|\partial_b u\|_{eucl}\, d\nu_{\ eucl}]$$

$$= [\,(\Phi' o P)\partial\bar{\partial}P + (\Phi'' o P)\partial P\wedge\bar{\partial}P]\wedge\partial u\wedge\bar{\partial}u$$

$$= \partial[\,(\Phi' o P)\bar{\partial}P\ ]\wedge\partial u\wedge\bar{\partial}u = \partial\bar{\partial}(\Phi o P)\wedge\partial u\wedge\bar{\partial}u$$

$$= \partial\bar{\partial}G\wedge\partial u\wedge\bar{\partial}u = -4\|\nabla u\|^2_{ds^2}\, d\nu \quad ,$$

où ds^2 est déterminé par le (2). D'où le résultat par le lemme 1.

Exemple 1 (de fonctions dans $\mathcal{H}^1_2(D,1)$ si D est la boule unité B de $\mathbb{C}^2$). Prenons dans le (7) Φ telle que $\Phi'(s) \geqslant C_1/|s|$, où $C_1 > 0$. Alors, pour une boule quelconque $B(0;r)$ de $\mathbb{C}^2$,où $0 < r < 1$, on obtient pour $-1 < s < 0$:

$$(10)\quad 0 < \int_s^0 \psi_u(\zeta)(1+\zeta)^{n-\frac{1}{2}}\, d\zeta \leqslant C_2|s| \ ,\text{où}\ \Psi_u(\zeta) = \int_{C_\zeta} \|du\|^2\, d\zeta \ ,$$

où $C_\zeta = B(0;(1+\zeta)^{\frac{1}{2}}$ et $d\zeta$ est l'aire euclidienne de C_ζ .

En effet, la première intégrale dans le (7) s'exprime par

$$\int_0^1 \Phi'(r^2-1)\Psi_u(r^2-1)r^{2n-1}\, dr = \int_{-1}^0 \Phi'(\zeta)\bar{\Psi}_u(\zeta)(1+\zeta)^{n-\frac{1}{2}}\, d\zeta \ .$$

Mais Φ' croît et est ≥ 0 , d'où

$$\int_B (\Phi' o P) \| \nabla u \|^2_{eucl} d\nu_{eucl} \geq \int_s^o \Phi'(\zeta)\, \Psi_u(\zeta)(1+\zeta)^{n-\frac{1}{2}} d\zeta \geq \Phi'(s) \int_s^o \Psi_u(\zeta)(1+\zeta)^{n-\frac{1}{2}} d\zeta .$$

Par le corollaire 1 et l'estimation $\Phi'(s) \geq C_1/|s|$, où $C_1 > 0$, on obtient

$$C \geq (C_1/|s|) \int_o^s \Psi_u(\zeta)(1+\zeta)^{n-\frac{1}{2}} d\zeta > 0$$

c'est-à-dire (10).

La condition (10) n'est pas catastrophique, car $|\Psi_u(\zeta)|$ est intégrable pour r voisin de 0 . La deuxième intégrale dans le (7) est au contraire catastrophique. En effet, prenons Φ de sorte que $\Phi'(s) \simeq 0$ si $s \leq s_o$ et $\Phi'(s) = 1/s_o$ si $s_o + \varepsilon \leq s \leq 0$, où $\varepsilon > 0$. On aura alors une condition $\Phi''(s) \simeq 1/s_o\varepsilon$ si $s_o \leq s \leq s_o + \varepsilon$, d'où

$$(11) \qquad 0 < \int_s^{s_o+\varepsilon} \Psi_u^b(\zeta)(1+\zeta)^{n-1} d\zeta \leq C_3 |s_o| \varepsilon \text{ , où } \Psi_u^b(\zeta) = \int_{C_\zeta} \|\partial_b u\|^2 \, d\zeta$$

où $\partial_b u = \bar{z}_2 u_{|1} - \bar{z}_1 u_{|2}$. Cela indique que $\|\partial_b u\|^2$ tend au sens de $L^2(S^3)$ vers 0 (où S^3 désigne la sphère 3-dimensionnelle de $\mathbb{R}^4$) , parce que $\bar{\Psi}_u^b(s_o) \leq C_4 |s_o|$, où $C_4 > 0$, pour $s_o \longrightarrow 0$.

<u>Exemple 2</u> (de fonctions dans $\mathcal{H}_2^1(D,1)$ si D est le polydisque unité Δ de $\mathbb{C}^2$) . Pour $u \in \mathcal{H}_2^1(\Delta,1)$ il existe une constante C_5 telle que si F^r est une fonction de la classe $C_c^\infty(\Delta)$, valant 1 pour $|z| \leq r$ et 0 pour $|z| > r + \frac{1}{2}(1-r)$, $0 < r < 1$, avec $F^r(z) = F^r(z^1)F^r(z^2)$ pour $z = (z^1, z^2)$, et si $\Delta(0;r)$ est le polydisque de centre 0 et de rayon r , on a

$$(12)\quad \|u\|_{\Delta}^{2} \leqslant C_s \, /\!/u/\!/_{\Delta}^{2} + \limsup_{r\to 1-} \sum_{j,k=1}^{2} \int_{\mathbb{C}^2\setminus\Delta(0;2)} |F^r_{|j}|^2 |F^r_{|\bar{k}}|^2 dz\, d\bar{z} \;,$$

où $/\!/u/\!/_D$ est la norme de Sobolev d'ordre 2 .

En effet, on étudie la forme (8), c'est-à-dire

$$(g_{2\bar{2}}u_{|1}u_{|\bar{1}} - g_{2\bar{1}}u_{|1}u_{|\bar{2}} - g_{1\bar{2}}u_{|2}u_{|\bar{1}} + g_{1\bar{1}}u_{|\bar{2}}u_{|\bar{2}}) \wedge dz_1 \wedge d\bar{z}_1 \wedge dz_2 \wedge d\bar{z}_2 \quad .$$

Evidemment, par (1) et $-1 \leqslant G \leqslant 0$, on a

$$\int_{\Delta(0;2)} g_{2\bar{2}}u_{|1}u_{|\bar{1}}\, dzd\bar{z} \leqslant \int_{\Delta} (F^r)^2 g_{2\bar{2}} |u_{|1}|^2\, dzd\bar{z}$$

$$\leqslant \int_{\Delta} \frac{g_{2\bar{2}}}{-G}(F^r|u_{|1}|)^2 dzd\bar{z} \leqslant \int_{\Delta} \left(\frac{g_{2\bar{2}}}{-G} + \left|\frac{G_{|2}}{-G}\right|^2\right)(F^r|u_{|1}|)^2\, dzd\bar{z}$$

$$= \int_{\Delta} \frac{1}{G} G_{|2} \frac{\partial}{\partial\bar{z}_2}(F^r|u_{|1}|)^2\, dzd\bar{z} - \int_{\Delta} \frac{\partial}{\partial\bar{z}_2}\left[\frac{1}{G} G_{|2}(F^r|u_{|1}|)^2\right] dzd\bar{z}$$

$$\leqslant \int_{\Delta} \frac{1}{G} G_{|2} \frac{\partial}{\partial\bar{z}_2}(F^r|u_{|1}|)^2\, dzd\bar{z}$$

$$= \int_{\Delta} \frac{1}{G} G_{|2}(F^r)^2 \frac{\partial}{\partial\bar{z}_2}|u_{|1}|^2 dzd\bar{z} + \int_{\Delta} \frac{2}{G} G_{|2}\, F^r\, F^r_{|\bar{2}} |u_{|1}|^2 dzd\bar{z} \;.$$

Mais $2ab \leqslant \varepsilon^2 a^2 + \varepsilon^{-2} b^2$ pour $a,b,\varepsilon \in \mathbb{R}$, $\varepsilon \neq 0$, d'où la deuxième intégrale se contrôle par

$$\varepsilon^2 \int_{\Delta} \left|\frac{1}{G} G_{|2}\, F^r u_{|1}\right|^2 dzd\bar{z} + \frac{1}{\varepsilon^2} \left|F^r_{|z} u_{|1}\right|^2 dzd\bar{z} \quad .$$

La première intégrale se contrôle par

$$\eta^2 \int_{\Delta} \left|\frac{1}{G} G_{|2} F^r u_{|1}\right|^2 dzd\bar{z} + \frac{1}{2\eta^2}\int_{\Delta} |F^r u_{|\bar{1}\bar{2}}|^2\, dzd\bar{z} + \frac{1}{2\eta^2}\int_{\Delta} |F^r u_{|\bar{1}\bar{2}}|^2\, dzd\bar{z}$$

pour $\eta \in \mathbb{R}$, $\eta \neq 0$. Ainsi on obtient

$$\int_\Delta [\frac{g_{2\bar{2}}}{-G} + (1-\varepsilon^2 - \dot{\eta}^2) \; |\frac{G_{|2}}{G}|^2] (F^r|u_{|1}|)^2 \, dzd\bar{z}$$

$$\leqslant \frac{1}{\varepsilon^2} \int_\Delta |F^r_{|2} \, u_{|1}|^2 \, dzd\bar{z} + \frac{1}{2\eta^2}\int_\Delta |F^r u_{|1\bar{2}}|^2 \, dzd\bar{z} + \frac{1}{2\eta^2}\int_\Delta |F^r u_{|\bar{1}2}| \, dzd\bar{z}$$

pour $\eta \in \mathbb{R}$, $\eta \neq 0$. Ainsi on obtient

$$\int_\Delta [\frac{g_{2\bar{2}}}{-G} + (1-\varepsilon^2 - \eta^2) \; |\frac{G_{|2}}{G}|^2] (F^r|u_{|1}|)^2 \, dzd\bar{z}$$

$$\leqslant \frac{1}{\varepsilon^2}\int_\Delta |F^r_{|2}u_{|1}|^2 \, dzd\bar{z} + \frac{1}{2\eta^2}\int_\Delta |F^r u_{|1\bar{2}}|^2 \, dzd\bar{z} + \frac{1}{2\eta^2}\int_\Delta |F^r u_{|\bar{1}2}|^2 \, dzd\bar{z} \; .$$

Prenons $\varepsilon^2 + \eta^2 \leqslant 1$. Alors

$$\int_{\Delta(0;r)} \frac{g_{2\bar{2}}}{-G} (F^r|u_{|1}|)^2 \, dzd\bar{z} \leqslant \frac{1}{\varepsilon^2} \int_{\mathbb{C}\setminus\Delta(0;r)} |F^r_{|\bar{2}}u_{|1}|^2 \, dzd\bar{z} + \frac{1}{\eta^2} // u //^2_{\Delta(0;r)}$$

Comme $2g_{\bar{1}2} \leqslant g_{1\bar{1}} + g_{2\bar{2}}$, on obtient l'estimée avec dérivées mixtes, d'où le résultat (12) .

3. <u>Une variation de métrique kählérienne dans l'intégrale de Dirichlet</u>

Pour la capacité (3) on a :

<u>THEOREME 2. Si G est une fonction stationnaire de (3), elle satisfait l'équation différentielle</u>

(13) $$\frac{\partial^2}{\partial z_j \partial \bar{z}_k} [g(\frac{\partial u}{\partial z_r} \frac{\partial u}{\partial \bar{z}_s})^{\frac{1}{2}p} (g^{rs})^{\frac{1}{2}p-1} (g^{j\bar{k}} g^{r\bar{s}} - \tfrac{1}{2}pg^{r\bar{k}} g^{j\bar{s}}) \det g] = 0 \; .$$

<u>Pour</u> $n = 1$, $p \neq 2$ <u>et</u> $D = \Delta(0;1)$ <u>soit</u> $G(z) \longrightarrow \varphi(z_o)$ <u>et</u>

$$q(z)\,|\,u_z(z)|^{p}(G_{z\bar z})^{-\frac{1}{2}p} \longrightarrow \psi(z_o)$$

pour $z \longrightarrow z_o \in \partial\,\Delta(0;1)$. Puis soit h_φ et h_ψ les prolongements harmoniques de φ et ψ , respectivment, tandis que G_ψ le potentiel de Green correspondant à la fonction $(|u_z|^{p}/h_\psi)^{2/p}$ dans $\Delta(0;1)$. Alors $G = h_\varphi + G_\psi$.

Preuve : On a $d\tilde s^2 = \tilde g_{j\bar k}\, dz_j \otimes d\bar z_k$, où $\tilde g_{j\bar k} = \tilde G_{|j\bar k} = \dfrac{\partial^2 \tilde G}{\partial z_j \partial \bar z_k}$ et

$$(14)\quad \|\nabla u\|^2_{d\tilde s^2} = \tilde g^{j\bar k}\,\frac{\partial u}{\partial z_j}\,\frac{\partial u}{\partial \bar z_k},\; d\tilde v = \text{dét}\,\tilde g\; d\nu_{\text{eucl}} ,$$

où $\tilde g^{r\bar m}\,\tilde g_{\ell\bar m} = \delta^r_{\bar m}$ et $\tilde g = [\,\tilde g_{\ell\bar m}\,]$ (cf. (1) et (2) ; δ désigne la fonction δ de Kronecker) . Alors, par le (3),

$$(15)\quad |\!|\!|u|\!|\!|^p_p\,(D,g) = \sup_{\tilde G \in P_2(D)} \int_D F(z,\bar z,\tilde G,\tilde G',\tilde G'')\,d\nu_{\text{eucl}} ,$$

où

$$F(z,\bar z,\tilde G,\tilde G',\tilde G'') = g\Big(\tilde g^{j\bar k}\,\frac{\partial u}{\partial z_j}\,\frac{\partial u}{\partial \bar z_k}\Big)^{\frac{1}{2}p}\,\text{dét}\,\tilde g\ .$$

Si G est une fonction stationnaire de (15) et

$$F_{\underline G} = F_{\tilde G}(z,\bar z,\tilde G,\tilde G',\tilde G'')\,\Big|_{\tilde G=G(z),\,\tilde G'=G'(z),\,\tilde G''=G''(z)} ,$$

on a

$$F_{\underline G} - \frac{\partial}{\partial z_k} F_{\underline G_{|\underline k}} - \frac{\partial}{\partial \bar z_k} F_{G_{|\bar k}} + \frac{\partial^2}{\partial z_j \partial z_k} F_{\underline G_{|\underline{jk}}} + 2\,\frac{\partial^2}{\partial z_j \partial \bar z_k} F_{\underline G_{|\underline{j\bar k}}} + \frac{\partial}{\partial \bar z_j \partial \bar z_k} F_{\underline G_{|\underline{jk}}} = 0\ .$$

Mais

$$F_{\underline G} = 0,\ \frac{\partial}{\partial z_k} F_{\underline G_{|\underline k}} = 0,\ \frac{\partial}{\partial \bar z_{\bar k}} F_{G_{|\bar k}} = 0,\ \frac{\partial^2}{\partial z_j \partial z_k} F_{\underline G_{|\underline{jk}}} = 0,\ \frac{\partial^2}{\partial \bar z_j \partial \bar z_k} F_{\underline G_{|\underline{jk}}} = 0 ,$$

d'où

$$\frac{\partial^2}{\partial z_j \partial \bar z_k}\, F_{\underline G_{|\underline{j\bar k}}} = 0 ,$$

c'est-à-dire

$$(16) \qquad \frac{\partial^2}{\partial z_j \partial \bar{z}_k}\{ q \left(\frac{\partial u}{\partial z_r} \frac{\partial u}{\partial \bar{z}_s} \right)^{\frac{1}{2}p} \left[\left(g^{r\bar{s}} \right)^{\frac{1}{2}p} \text{dét } g \right]_{G|j\bar{k}} \} = 0 \ .$$

On calcule :

$$(g^{r\bar{m}})_{G|j\bar{k}} \, g_{L\bar{m}} + g^{r\bar{m}}(g_{\ell\bar{m}})_{G|j\bar{k}} = 0 \quad \text{et} \quad (g_{L\bar{m}})_{G|j\bar{k}} = \delta_{j\ell}\,\delta_{\bar{m}\bar{k}} \ ,$$

d'où

$$(g^{r\bar{s}})_{G|j\bar{k}} = (g^{r\bar{m}})_{G|j\bar{k}}\,\delta^{\bar{s}}_{\bar{m}} = (g^{r\bar{m}})_{G|j\bar{k}}\, g_{\ell\bar{m}}\, g^{\ell\bar{s}} = -\delta_{j\ell}\,\delta_{\bar{m}\bar{k}}\, g^{r\bar{m}} g^{\ell\bar{s}} = -\delta_{\bar{m}\bar{k}}\, g^{r\bar{m}} g^{\ell\bar{s}} = -g^{r\bar{k}}\, g^{j\bar{s}}$$

et

$$\left[(g^{r\bar{s}})^{\frac{1}{2}p} \text{détg} \right]_{G|j\bar{k}} = -\frac{1}{2}p (g^{r\bar{s}})^{\frac{1}{2}p-1} g^{r\bar{k}} g^{j\bar{s}} \text{détg} + (g^{r\bar{s}})^{\frac{1}{2}p} g^{j\bar{k}} \text{détg} = \left(g^{j\bar{k}} g^{r\bar{s}} - \frac{1}{2}p g^{r\bar{k}} g^{j\bar{s}}\right)(g^{r\bar{s}})^{\frac{1}{2}p-1} \text{détg}.$$

Par (16) on déduit (13)

Pour $n = 1$ et $p \neq 2$ on obtient

$$\frac{\partial^2}{\partial z \partial \bar{z}} \left[g \left(\frac{\partial u}{\partial z} \frac{\partial u}{\partial \bar{z}} \right)^{\frac{1}{2}p} \left(\frac{\partial^2 G}{\partial z \partial \bar{z}} \right)^{-\frac{1}{2}p} \right] = 0 \ ,$$

d'où

$$g \left(\frac{\partial u}{\partial z} \frac{\partial u}{\partial \bar{z}} \right)^{\frac{1}{2}p} \left(\frac{\partial^2 G}{\partial z \partial \bar{z}} \right)^{-\frac{1}{2}p} = h_\psi \ ,$$

c'est-à-dire $G_{z\bar{z}} = (|u_z|^p / h_\psi)^{2/p}$ dans $\Delta(0,1)$, d'où $G = h_\varphi + G_\psi$.

4. Autres espaces de Dirichlet invariants biholomorphes

On a en fait montré un peu mieux que dans les sections 1 et 2 . Par (14) ,

$$(17) \qquad du \wedge d^c u \wedge dd^c G = \|\nabla u\|^2_{\partial\bar{\partial} G}\, d\nu \ ,$$

(cf. (8) , où $d\nu$ est la mesure associée à la distance géodésique ds^2 correspondante à $\partial\bar{\partial} G$ (déterminée par le (2)) . Alors on déduit :

Remarque 2 . L'estimée (12) est valable pour $G \leq 0$ et on peut remplacer $\| u \|_p(D,g)$ par

(18) $$\| u \|_p(G,q) = [\sup_{G \in P'(D)} \int_D q \, \| \nabla u \|^2_{-\partial\bar{\partial}\log(-G)} \, d\nu]^{1/p} ,$$

où $P'(D)$ dénote la classe de fonctions $C^2(D)$-plurisousharmoniques avec $-1 \leq G \leq 0$ partout.

Dans le supremum (3), on peut prendre une famille de métriques $\partial\bar{\partial}G$ avec G plurisousharmoniques de la classe C^1 sur la fermeture de D strictement pseudoconvexe, et telle que $G(z) \leq 0$ dans D avec la C^1-norme ≤ 1 . En utilisant le théorème de Fefferman [9] (p.2) on peut montrer facilement la proposition suivante :

PROPOSITION 1. Si $f : D \longrightarrow D'$ est une bijection biholomorphe entre deux domaines strictement pseudoconvexes D et D' de $\mathbb{C}^n$, l'application $u \longmapsto u \circ F$ est une bijection bicontinue.

En ce cas particulier on vérifie que la norme est estimée pour la boule unité B de $\mathbb{C}^2$ par la norme de Sobolev d'ordre 2 . Ici on n'a donc plus une condition aussi rigide pour la valeur frontière ; celle-ci n'est pas nécessairement constante comme dans le cas de $\mathcal{H}^1(B)$.

5. Cas d'un domaine strictement pseudoconvexe de $\mathbb{C}^2$-la norme $| u |_p(D,q)$

Soit D_o un domaine strictement pseudoconvexe de $\mathbb{C}^2$ à frontière de classe C^2, $Q(D_o)$ la famille des fonctions G plurisousharmoniques dans D avec $0 \leq G \leq 1$ partout et $\lim_{z \to \partial D_o} \sup G(z) = 1$.

Comme D_o est strictement pseudoconvexe à frontière C^2 , il existe du moins une fonction plurisousharmonique de classe $C^2(D_o)$ dans $Q(D_0)$. Définissons dans ce cas une norme

(17) $$| u |_p(D_o,q) = (\sup_{G \in Q(D_o)} \int_{D_o} q \, | \nabla u |^p_{ds^2} \, d\nu)^{1/p} .$$

Il est clair que $\|u\|_p(D_o,q)$ est plus grande que la norme de Sobolev usuelle. Nous noterons $\hat{\mathcal{H}}^1_{o,p}(D_o,q)$ le complété de $C_c^\infty(D_o)$ pour cette norme. D'après le raisonnement du lemme 2, il s'identifie à un espace de fonctions sur D_o. C'est avec cet espace que nous travaillerons désormais.

6. <u>Une capacité associée</u>

Soit D_o un domaine strictement pseudoconvexe de $\mathbb{C}^2$ à frontière de classe C^2, C_1 un compact de D_o, et $D = D_o \setminus C_1$; on pose

(19) $\mathrm{CAP}_p(D,q) = \inf\{\|u\|_p^p(D_o,q) : u \in C_c^\infty(D_o),\ u|C_1 \geq 1\}$.

On pourrait aussi définir

(20) $\mathrm{CAP}'_p(D,q) = \sup\limits_{G \in Q(D_o)} \inf\{\int_{D_o} q\|\nabla u\|^p_{ds^2}\, d\nu\ ;\ u \in \mathcal{H}^1_{o,p}(D_o,q),\ u|C_1 \geq 1\}$.

Dans le second cas cela signifie que pour chaque $G \in Q(D_o)$ on définit une "norme" de Dirichlet (Cf [5] et [10]) associée au courant positif dd^cG (Cf la remarque 1), qu'on calcule la capacité pour cette norme, qu'on calcule ensuite le supremum des capacités.

<u>THÉORÈME 3. On a</u> $\mathrm{CAP}'_2(D,q) = \mathrm{CAP}_2(D,q)$ <u>pour chaque</u> D <u>et</u> q, <u>et le supremum dans</u> $\mathrm{CAP}'_2(D,q)$ <u>est atteint pour un</u> G <u>de</u> $Q(D_o)$.

<u>Preuve</u> : Apriori, on a $\mathrm{CAP}'_2(D,q) \leq \mathrm{CAP}_2(D,q)$. L'inégalité opposée résulte du théorème fondamental de von Neumann [25] sur la théorie des jeux, ou, plus précisément, d'une variante de ce théorème. Notons X la famille des fonctions $u \in \hat{\mathcal{H}}^1_{o,2}(D,q)$ avec $u \geq 1$ sur C_1 ; alors X est un convexe. Notons Y la famille des courants dd^cG pour $G \in Q(D)$, alors Y est un convexe compact pour la topologie faible. Soit

(21) $$I[u,dd^cG] = \int_{D_o} q\, du \wedge d^cu \wedge dd^cG$$

définie sur $X \times Y$; alors $dd^cG \longrightarrow I[u,dd^cG]$ est linéaire continue sur Y pour tout $u \in X$ fixé . De plus, $u \longmapsto I[u, dd^c G]$ est convexe, or toute fonctionnelle convexe sur un espace de Banach est faiblement semi continue inférieurement. Alors le théorème de von Neumann dit que le jeu (X,Y,I) a une valeur, ce qui signifie précisément le résultat du théorème (Cf la remarque 1) .

Exemple 3 (d'une boule quelconque $B(0;r_o)$ dans la boule unité $B = B(0;1)$ de $\mathbb{C}^2$) . Soit $D_o = B$ et C_1 la fermeture de $B(0;r_o)$. D'après (20) , on pose

$u(z) = U(|z|^2)$ et $G(z) = P(|z|^2)$ pour $z \in B$ et on calcule

$$CAP_2'(D,1) = \sup\{\inf\{\int_o^1 U'^2(r)P'(r)r^2 dr : U(r) > 1 \text{ pour } r < r_o\} : 0 \leq P(r) \leq 1 \text{ pour}$$

$$0 < r < 1 \ , \ P(1) = 1 \text{ et } z \longmapsto P(|z|^2) \text{ plurisousharmonique dans } B\} \ ,$$

doù $U(r) = 1$ pour $0 < r < r_o$ et

$$\frac{d}{dr}[r^2 P'(r)U'] = 0 \ , \text{ or } U(r) = [\int_{r_o}^1 \frac{ds}{s^2P'(s)}]^{-1} \int_r^1 \frac{ds}{s^2P'(s)} \quad \text{pour } r_o < r < 1$$

Mais

$$\int_{r_o}^1 \frac{1}{P'+\delta P'} \frac{dr}{r^2} = \int_{r_o}^1 \frac{1}{P'} \frac{dr}{r^2} - \frac{\delta P(1)}{P'^2(1)} + \frac{\delta P(z_o)}{r_o^2 P'^2(r_o)} + \int_{r_o}^1 \frac{d}{dr}[\frac{1}{r^2P'^2(r)}]\, \delta P dr \ ,$$

or

$$1/P^2(1) = 0 \text{ ou } 1/P'^2(1) \neq 0 \ , \ \delta P(1) = 0 \ , \ \delta P(r_o) = 0$$

et

$$\frac{d}{dr}[\frac{1}{r^2P'}] = 0 \ , \text{ c'est-à-dire } G'' + r^{-1}G' = 0 \ ,$$

d'où $P(r) = c + c' \log r$ pour $r_o \leq r \leq 1$. Ici $c = 1$ et $c' = 1/\log r_o^{-1}$

7. Propriétés de la capacité associée - une généralisation du Lemme de Schwarz

Une modification évidente du théorème 1 donne

THÉORÈME 4. *Si* $f : D_o \to f[D_o]$ *est holomorphe, où* D_o *est un domaine strictement pseudoconvexe de* $\mathbb{C}^2$ *à frontière de classe* C^2, $F[D_o] \subset \mathbb{C}^2$ *et* $u \in \mathcal{H}^1_{o,p}(f[D_o], q)$, *alors*

$$u \circ f \in \mathcal{H}^1_{o,p}(D_o, q \circ f)$$

et

(22) $\|u \circ\|_p(D_o, q\circ f) \geq \|u\|_p(f[D_o], q)$

et de plus, pour tout compact C_1 *de* D_o, *si on pose* $D = D_o \setminus C_1$, *on a*

$$CAP_p(f[D], q) \leq CAP_p(D, q\circ f),$$

(23)

$$CAP'_p(f[D], q) \leq CAP'_p(D, q\circ f).$$

Si f *est biholomorphe, on a égalité dans les inégalités (22) et (23).*

De théorème 3 on déduit facilement (C [30]) :

PROPOSITION 2. *Pour chaque ensemble* $\mathbb{C}^2$ *- polaire* $C_1 = D_o \setminus D$ *de* D_o

(24) $CAP'_2(D, q) = CAP_2(D, q) = 0$.

Remarque 3. La propriété réciproque semble être vraie, mais nous n'avons pas de démonstration rigoureuse.

Théorème 5. *La fonctionnelle* $(u,g) \longmapsto I[u, dd^c G]$, *définie par* (21), *où* $q = const$, *admet un point-selle* (u_o, G_o) *dans l'ensemble des* (u,G) *de* $\mathcal{H}^1_{o,2}(D_o, q) \times Q(D_o)$ *avec* $u \mid C_1 \geq 1$. *De plus, le* u_o *est unique et est la solution plurisousharmonique de l'équation de Monge-Ampère complexe*

(25) $dd^c u \wedge dd^c u = 0$

dans $D = D_o \setminus C_1$, *valant* 1 *sur* C_1 *et* 0 *sur* $C_o = \mathbb{C}^2 \setminus D_o$; *on peut prendre*

$G_o = 1 - u_o$ <u>(potentiel capacitaire extrêmal)</u> .

<u>Preuve</u>. L'existence de G_o a déjà été démontrée dans le théorème 3. Le calcul d'Okada [26] montre alors que la norme de Dirichlet, définie par le courant dd^cG_o , définit un espace de Dirichlet au sens de [5] et [10] . On peut donc appliquer la théorie du balayage et obtenir ainsi u_o (Cf [4] et [6]) . Pour démontrer le dernier point du théorème, appelons u_G le potentiel capacitaire de la métrique associée au courant positif dd^cG . D'après une procédure standard d'approximation nous nous bornons au cas de C_1 à frontière de classe C^2 . On a par définition, pour toute u de $\hat{\mathcal{H}}^1_{o,2}(D_o,q)$ avec $u|C_1 = 1$,

(26) $$\int_{D_o} du_G \wedge d^cu_G \wedge dd^cG \leq \int_{D_o} du \wedge d^cu \wedge dd^cG \quad .$$

Posons $I_G[u] = \int_D du \wedge d^cu \wedge dd^c(G-1)$; on a

$$I_G[u] = \int_D d[du \wedge d^cu \wedge dd^c(G-1)] + \int_D du \wedge dd^cu \wedge d^c(G-1)$$

or

$$\int_D du \wedge dd^cu \wedge d^c(G-1) = -\int_D d^cu \wedge dd^cu \wedge d(G-1) = \int_D d[(d^cu \wedge dd^cu)(G-1)] - \int_D (dd^cu \wedge dd^cu)(G-1) \quad ,$$

d'où

$$I_G[u] = \int_{\partial D} du \wedge d^cu \wedge d^c(G-1) + \int_{\partial D} (d^cu \wedge dd^cu)(G-1) - \int_D (dd^cu \wedge dd^cu)(G-1)$$

Dans (26) , prenons $u = v$,où v est la solution pluri<u>sur</u>harmonique de (25) valant 1 sur C_1 et 0 sur C_o . Alors, comme v est constante sur ∂C_1 et ∂C_o , on a $dv \mid \partial D = 0$; de plus $G \mid \partial C_o = 1$, d'où

$$I_G[v] = -\int_{\partial C_1} (G-1) d^cv \wedge dd^cv \quad .$$

Considérons maintenant $I_{-v}[v]$ pour la même raison

$$I_{-v}[v] = -\int_{\partial C_1} (-v)\, d^c v \wedge dd^c v \quad .$$

Par ailleurs, nous avons

$$\text{(27)} \qquad I_G[v] - I_{-v}[v] = -\int_{\partial C_1} [(G-1)+v]\, d^c(-v) \wedge dd^c(-v) \quad .$$

Or $G - 1 \leqslant -v$ (propriété de la solution de l'équation de Monge-Ampère complexe) et $dd^c(-v) \geqslant 0$ et il est clair, puisque v est constante sur ∂C_1, que $d^c(-v)$ ne contribue à l'intégrale que par la dérivée normale externe à ∂D ; or $-v \mid \partial C_1 = 1$ et $-1 \leqslant v \mid D \leqslant 0$, donc cette dérivée normale est $\leqslant 0$. Par suite, dans (27),

$$I_G[v] - I_{-v}[v] \leqslant 0$$

et d'après (26) nous avons

$$\int_{D_o} dv \wedge d^c v \wedge dd^c(1-v) = \int_{D_o} dv \wedge d^c v \wedge dd^c(1-v) \quad .$$

Or $1 - v$ est dans $Q(D_o)$ et comme v satisfait l'équation de Monge-Ampère complexe $dd^c v \wedge dd^c(1-v) = 0$, ce qui signifie que v est le potentiel u_{1-v}. D'où

$$\int_{D_o} dv \wedge d^c v \wedge dd^c(1-v) = \sup_{G \in Q(D_o)} \int_{D_o} du_G \wedge d^c u_G \wedge dd^c G$$

$$= \sup_{G \in Q(d_o)} \inf\{ \int_{D_o} du \wedge d^c u \wedge dd^c G \ ;\ u \in \hat{\mathcal{R}}^1_{o,2}(D_o, q),\ u)C_1 \geqslant 1\} = CAP_2'(D,q) \quad .$$

Donc le couple $(u_o, G_o) = (v, 1-v)$ réalise bien un point-selle de la fonctionnelle I pour $q = \text{const.}$

8. Les analogues pour variation du potentiel et des courants

Soit $\mathbb{M}$ une variété complexe de dimension complexe n (éventuellement, un domaine de $\mathbb{C}^n$) muni d'une métrique hermitienne h et d'un C^1-champ de tenseurs

H de type $(1,1)$, en particulier $H = J$ (la structure complexe de $\mathbb{M}$). Puis soit D un condensateur de $\mathbb{M}$: un <u>condensateur</u> (D,C_o,C_1) de $\mathbb{M}$ est défini par une partition de $\mathbb{M}$ en deux fermés C_o,C_1 et en un ouvert D ; pour simplifier, nous écrirons D au lieu de (D,C_o,C_1). Soit $g : \mathbb{M} \longrightarrow \mathbb{C}$ une fonction continue (la fonction de <u>non</u>-homogénéité) et p un nombre réel $\geqslant 1$. Nous allons considérer la classe $\mathrm{Adm}\, D$ des fonctions C^2-plurisousharmoniques u sur $c\ell D$ (la fermeture de D) satisfaisant les conditions $0 < u(z) < 1$ pour $z \in D$, $u|\partial C_o = 0$ et $u|\partial C_1 = 1$. Soit [18] :

$$(28)\quad \mathrm{cap}_p(D,q) = \inf_{\tilde{u} \in \mathrm{Adm} D} \left| \int_D q \|\nabla\tilde{u}\|^{p-2}_{ds^2} \,\mathrm{d\acute{e}t}\, H\, d\tilde{u} \wedge d^c\tilde{u} \wedge (dd^c\tilde{u})^{n-1} \right| .$$

Soit ensuite Γ une classe d'homologie de D, dont les coefficients sont réels et $\dim \Gamma = r$. Nous allons considérer tous les courants de τ au sens de de Rham et un recouvrement localement fini $\mathcal{U} = \{U_j : j \in I\}$ de $\mathbb{M}$. La classe des fonctions plurisousharmoniques C^2 sur $U_j \cap D$, définies dans chaque élément du recouvrement et satisfaisant les conditions :

(i) l'oscillation de la fonction u_j dans $U_j \cap D$ est plus petit que 1,

(ii) $du_j = du_k$ dans $U_j \cap U_k \cap D \neq \emptyset$,

étant notée par $\mathrm{adm}(D,\mathcal{U})$. Soit [19] :

$$(29)\quad \mathrm{cap}_p(D,g,\Gamma,\mathcal{U}) = \sup_{\tilde{u} \in \mathrm{adm}(D,\mathcal{U})} \ \inf_{T \in \Gamma} \left| T\left[g \|\nabla\tilde{u}\|^{p-2}_{dS^2}\, \mathrm{d\acute{e}t}\, H\, D^r \tilde{u} \right] \right| ,$$

où

$$D^r u = \begin{cases} d^c u \wedge (dd^c u)^{\frac{1}{2}r - \frac{1}{2}} & \text{pour } r \text{ impair,} \\ du \wedge d^c u \wedge (dd^c u)^{\frac{1}{2}r - 1} & \text{pour } r \text{ pair .} \end{cases}$$

Les seminormes (28) et (29) sont invariantes biholomorphiques (voir le théorème 6 ci-dessous). Pour l'étude d'un analogue complexe du principe de Dirichlet (Cf [27] et [28] une procédure naturelle est de prendre dans (29) pour Γ la classe (2n-1)-dimensionnelle d'homologie des hypersurfaces de niveau $\{z \in c\ell\, D : u(z) = \text{const}\}$. (Par analogie, pour un principe convenable du type de Thomson (Cf [27] et [18] on doit prendre dans le (29) pour Γ la classe 1-dimensionnelle orthogonale d'homologie). On peut espérer l'identité des deux capacités (29) et (28) si nous supposons des conditions convenables, en particulier si nous prenons dans le (29) pour fonctions admissibles seulement les fonctions u définies globalement et si elles vérifient $0 < u(z) < 1$ pour $z \in D$; nous notons $u \in \text{adm} D$ et $\text{cap}_p(D,q,\Gamma)$ au lieu de $\text{cap}_p(D,q,\Gamma,\mathcal{U})$.

Cette idée et les deux définitions s'appliquent pour $H = J$ (la structure complexe de $\mathbb{M}$), $p = 2$ et $q = \text{const}$ de Chern, Lévine et Nirenberg [8] , mais la réponse positive est connue seulement dans certains cas particuliers. Alors pour que la fonctionnelle minimisée ait la valeur minimale pour $\tilde{u} = u$, il faut et il suffit que u satisfasse

(30) $(dd^c u)^n = 0$ (<u>l'équation de Monge-Ampère complexe</u>).

Si

$$(dd^c u)^n = 4^n n!\ \det[(\partial^2/\partial z_j \partial \bar{z}_k)u]\,(\tfrac{1}{2}i)^n\, dz_1 \wedge d\bar{z}_1 \wedge \ldots \wedge dz_n \wedge d\bar{z}_n \quad ,$$

l'équation (30) est l'analogue complexe de

(31) $\det\left[\dfrac{\partial^2}{\partial x_j \partial x_k} u\right] = 0$ (<u>l'équation de Monge-Ampère réelle</u>).

L'équation (31) est un cas particulier des équations de Monge-Ampère complexes généralisées

(32) $dd^c(Fu) \wedge (dd^c u)^{n-1} = 0 \quad , \quad F \in C^2(c\ell\, D)$

ou

(33) $d(G d^c u) \wedge (dd^c u)^{n-1} = 0 \quad , \quad G \in C^1(c\ell D) \quad ,$

qui sont les analogues de (30) pour la capacité générale (28),où

(34) $$d^c(Fu) = Gd^c u \ , \ G = q \, \|\nabla u\|^{p-2}_{ds^2} \, \text{dét} \, H \ .$$

Dans le cas général il faut remplacer la fonction u par un système de fonctions qui satisfont la condition (ii).

Les équations (32) et (33) sont les analogues d'équation (13) pour les capacités (28) et (29) . Pour $n = 1$ on obtient de (32) et (33) :

$$d(\, q \, \|\nabla u\|^{p-2}_{ds^2} \, \text{dét} \, H \, d^c u \,) = 0 \ ,$$

d'où

$$\text{div}\,[\, q \, (q^{11})^{\frac{1}{2}p-1} \, \| \, \text{grad} \, u \, \|^{p-2}_{\text{eucl}} \, H^{1\bar{1}} \, \text{grad} \, u \,] = 0 \ .$$

Cette équation conduit en principe à des applications quasi conformes [21] .

APPENDICE : APPLICATION A UNE THEORIE NON LINEAIRE DE PARTICULES ELEMENTAIRES

par J.LAWRYNOWICZ

9. Quelques propriétés locales et globales de l'équation de Monge-Ampère complexe généralisée.

Etant donné un domaine borné D de $\mathbb{C}^n$, une fonction $f \geqslant 0$ sur D et une fonction φ sur ∂D, on cherche une fonction u plurisousharmonique dans D satisfaisant

$$(dd^c u)^n = f^n \quad \text{dans } D \;, \quad u = \varphi \quad \text{sur} \quad \partial D \;.$$

Lorsque $f = 0$ ce problème est le problème introduit par Bremermann [17] ; dans le cas général ce problème a été traité par Bedford et Taylor [3] . Pour cela, on donne un sens à $(dd^c u)^n$ pour toute fonction u plurisousharmonique continue sur D, ceci est une mesure $\geqslant 0$ (Cf aussi le lemme 1) . Pour traiter le problème de Monge-Ampère complexe, Bedford et Taylor généralisent la méthode des enveloppes supérieures de Bremermann de la façon suivante : ils introduisent la classe $\mathcal{B}(f,\varphi)$ des fonctions v plurisousharmoniques sur D, telles que

$$\lim_{z \to z_0} \sup_{\in \partial D} v(z) \leqslant \varphi(z_0) \quad \text{pour} \quad z \in \partial D \;, \quad \Phi(v) \equiv (\det [\varphi_{j\bar{k}}])^{1/n} \geqslant f \;,$$

où $\varphi_{j\bar{k}}\, dv_{eucl} + v_{j\bar{k}} = d_j d_k^c u$ avec $v_{j\bar{k}}$ étrangère à dv et $\varphi_{j\bar{k}} \in L^1_{loc}$, et ils démontrent le théorème suivant : Si $\mathcal{B}(f,\varphi)$ n'est pas vide, si $f \in L^1(D)$ et si $w = \sup \mathcal{B}(f,\varphi)$ est continu, alors w est solution au sens généralisé du problème de Monge-Ampère, i.e. $\Phi(w) = f$ sur D et $w = \varphi$ sur ∂D.

Pour étudier l'équation de Monge-Ampère complexe généralisée (voir [1] et [2]) nous avons besoin du

<u>LEMME 4. Soit</u> $\mathcal{F}$ <u>un faisceau de groupes abéliens sur un espace topologique</u> X <u>et soit</u> $\mathcal{U} = \{U_j : j \in A\}$ <u>un recouvrement quelconque de</u> X. <u>Alors l'application canonique</u>

$$r_o : H^o(\mathcal{U},\mathcal{F}) \longrightarrow H^o(X,\mathcal{F})$$

<u>des groupes de cohomologie est un isomorphisme et l'application</u>

$$r_1 : H^1(\mathcal{U},\mathcal{F}) \longrightarrow H(X,\mathcal{F})$$

<u>est injective.</u>

<u>Preuve</u>. La première partie de ce lemme est justifiée en principe par la définition du faisceau, par l'hypothèse on conclut, que

$$0 \longrightarrow H^o(\mathcal{U},\mathcal{F}) \xrightarrow{\ i\ } C^o(\mathcal{U},\mathcal{F}) \xrightarrow{\ \delta\ } C^1(\mathcal{U},\mathcal{F})$$

est une suite exacte. Ici $C^j(\mathcal{U},\mathcal{F})$ désigne le groupe des chaînes d'ordre j sur $\mathcal{U}$ à valeurs dans $\mathcal{F}$, $j = 0, 1$ et δ est l'opérateur cobord.

Pour démontrer la deuxième partie il suffit de prouver la proposition suivante : Soit $\mathcal{V} \subset \mathcal{U}$ un raffinement du recouvrement $\mathcal{U}$, $\mathcal{V} = \{V_{ji} j \in I\}$ et soit $r : I \longrightarrow A$ la fonction de raffinement: $V_j \subset U_{r(j)}$. Ensuite, soit $\mathcal{F} \in Z^1(\mathcal{U},\mathcal{F})$ un cocycle d'ordre un sur $\mathcal{U}$. Supposons que si $r_* f \in B^1(\mathcal{V},\mathcal{F})$ est un cobord, alors aussi f est un cobord, i.e. $f \in B^1(\mathcal{U},\mathcal{F})$.

Pour $a,b,c \in A$ on a

$$f_{ab} + f_{bc} + f_{ca} = 0 \quad \text{sur} \quad V_a \cap V_b \cap V_c .$$

D'après la supposition,

$$f_{r(j)\,r(k)} = g_k - g_j \quad \text{sur} \quad V_j \cap V_k \ ; \quad g_j \in \Gamma(V_j, \mathcal{F}) .$$

Ainsi, pour chaque $a \in A$,

$$f_{ar(k)} - f_{ar(j)} = g_k - g_j \quad \text{sur} \quad U_a \cap V_j \cap V_k \quad ,$$

i.e.,

$$\varphi_a = g_j - f_{ar(j)} = g_k - f_{ar(k)}$$

est un élément bien défini de l'ensemble de sections $\Gamma(U_a, \mathcal{F})$. Nous affirmons que

$$f_{ab} = \varphi_b - \varphi_a \quad \text{sur} \quad U_a \cap U_b \quad .$$

En effet,

$$f_{ab} \mid U_a \cap U_b \cap U_{r(j)} = f_{ar(j)} - f_{br(j)}$$

$$= f_{ar(j)} - g_j - [f_{br(j)} - g_j] = \varphi_b - \varphi_a \quad \text{sur} \quad U_b \cap U_b \cap U_j \quad .$$

Car ce résultat est valable pour chaque V_j et comme $\mathcal{V} = \{V_j \ ; \ j \in I\}$ est un recouvrement de X , nous devons obtenir

$$f_{ab} = \varphi_b - \varphi_a \quad \text{sur} \quad U_a \cap U_b$$

et le lemme en résulte aussitôt.

On a donc

<u>Théorème 7. Supposons</u>

(a_1) F <u>pluriharmonique sur</u> $\mathbb{M}$ <u>(non nécessairement munie d'une métrique hermitienne</u> h)

(b_1) <u>l'équation (32) est résoluble localement dans</u> $\mathcal{U} = \{U_j \ : \ j \in I\}$,

(c_1) <u>les solutions</u> u_j <u>de</u> (32) <u>satisfont</u> $du_j = du_k$ <u>dans</u> $U_j \cap U_k \neq \emptyset$,

(d_1) $H^1(\mathbb{M}, \mathbb{C}) = 0$ <u>(par exemple, c'est valable pour</u> $\mathbb{M}$ <u>simplement connexe)</u>.

<u>Alors il existe une solution globale de</u> (32) <u>sur</u> $\mathbb{M}$, <u>égale localement à</u> $u_j - c_j$, <u>où</u> c_j <u>sont constantes.</u>

<u>Preuve</u> . De (b_1) et (c_1) on déduit que $u_j - c_j = c_{jk}$ dans $U_j \cap U_k$, où c_{jk} sont les constantes et $c_{jk} + c_{kl} + c_{lj} = 0$.

Ainsi les éléments c_{jk} engendrent le groupe de cohomologie $H^1(\mathcal{U}, \mathbb{C})$ et l'application canonique $H^1(\mathcal{U}, \mathbb{C}) \longrightarrow H^1(\mathbb{M}, \mathbb{C})$ est injective par le lemme 4. Par le même lemme l'application $H^1(\mathcal{U},\mathbb{C}) \longrightarrow H^1(\mathbb{M},\mathbb{C})$ est un isomorphisme, alors d'après (d_1) il existe des constantes $c_j, j \in I$, telles que

$$u_j - u_k = c_j - c_k, \text{ i.e. } u_j - c_j = u_k - c_k \quad \text{dans} \quad U_j \cap U_k .$$

Cela définit une fonction globale u de la classe C^2 sur la variété $\mathbb{M}$ entière et cette fonction est égale localement à $u_j - c_j$. Alors

$$(35) \qquad (dd^c u)^{n-1} = (dd^c u_j)^{n-1} \quad \text{dans } U_j$$

et

$$(36) \qquad dd^c(Fu) = dd^c(Fu_j) - c_j dd^c F \quad \text{dans } U_j .$$

Car, par (a_1), F est plurisousharmonique sur $\mathbb{M}$, on peut écrire (36) sous la forme

$$dd^c(Fu) = dd^c(Fu_j) \quad \text{dans } U_j ,$$

d'où, par (39), on obtient la relation

$$dd^c(Fu) \wedge (dd^c u)^{n-1} = dd^c(Fu_j) \wedge (dd^c u_j)^{n-1} \quad \text{dans } U_j .$$

D'après (b_1) et l'hypothèse en (c_1) que les u_j sont des solutions de (32) dans U_j, cela achève la preuve du théorème 7.

On a aussi les résultats suivants.

PROPOSITION 3. Supposons :

(a_2) u définie globalement et ne s'annulant pas sur $\mathbb{M}$,

(b_2) G satisfait la condition d'intégrabilité $\partial G \wedge \bar{\partial} u = 0$ sur $\mathbb{M}$,

(c_2) $H^1(\mathbb{M}, \mathcal{O}) = 0$, où $\mathcal{O}$ désigne le faisceau de germes de fonctions holomorphes sur $\mathbb{M}$ (par exemple, c'est valable pour $\mathbb{M}$ de Stein).

Alors il existe une solution globale de $\bar{\partial}(Fu) = G\bar{\partial}u$, G réel, sur $\mathbb{M}$, compatible avec la solution locale donnée.

Théorème 8. Supposons :

(a_3) $G : \mathbb{M} \longrightarrow \mathbb{C}$ est une fonction C^1 quelconque où

$$G = q[h(d^c u, d^c u)]^{\frac{1}{2}p-1} \det H \equiv q\|\nabla u\|^{p-2}_{ds^2} \det H ,$$

(b_3) l'équation (33) est résoluble localement dans $\mathcal{U} = \{u_j : j \in I\}$,

(c_3) les solutions u_j de (33) satisfont $du_j = du_k$ dans $U_j \cap U_k \neq 0$,

(c_4) $H^1(\mathbb{M}, \mathbb{C}) = 0$.

Alors il existe une solution globale de (33) sur $\mathbb{M}$, compatible avec $\{u_j : j \in I\}$.

THEOREME 9. Supposons :

(a_4) D possède un bord C^1-régulier par morceaux et sa fermeture compacte,

(b_4) $n \geqslant 2$, $p = 2$ et q appartient à la classe C^1,

(c_4) u appartient à Adm D et satisfait (33), où $G = q \det H$,

(d_4) $d(Gd^c u) = f dd^c u$,où f appartient à C^1 et $f \geqslant -(n-1)^{-1} G$.

Alors l'infimum en (28) est atteint pour u en question.

Preuve. Posons

$$(37) \qquad \mathcal{J}[u] = \int_D G du \wedge d^c u \wedge (dd^c u)^{n-1} \quad ,$$

où $u \in \text{Adm } D$ et $G = q \det H$ .D'après (28), on a

$$(38) \qquad \text{Cap}_p(D,q) = \inf_{u \in \text{Adm } D} |\mathcal{J}[u]| \quad .$$

Nous calculerons $(d^k/dt^k)\mathcal{J}[u+th]$, $u+h \in \text{Adm } D$, $0 \leqslant t \leqslant 1$, $k = 1,2,$ où les dérivées en $t = 0$ et 1 sont les dérivées à droite ou à gauche. Le problème est bien posé, car Adm D est convexe et $t \longmapsto \mathcal{J}[u+th]$ est un polynôme du degré $n+1$. Notons $u_t = u+th$. On a donc

$$\frac{d}{dt}\mathcal{J}[u_t] = \int_D G\,[dh \wedge d^c u_t \wedge (dd^c u_t)^{n-1} + du_t \wedge d^c h \wedge (dd^c u_t)^{n-1} + (n-1) du_t \wedge d^c h \wedge \varphi_t(n)] \quad ,$$

où $\varphi_t(1) = 0$, $\varphi_t(n) = (dd^c u_t)^{n-2}$ pour $n > 1$ et de plus

$$dh \wedge d^c h \wedge (dd^c u_t)^{n-1} = du_t \wedge d^c h \wedge (dd^c u_t)^{n-1}$$

d'où, par (b_4) , (a_4) et le théorème de Stokes,

$$\frac{d}{dt}\mathcal{J}[u_t] = \int_{\partial D} \{G\,[2hd^c u_t \wedge dd^c u_t + (n-1) du_t \wedge d^c u_t \wedge d^c h + (n-1) hd^c u_t \wedge dd^c u_t] \wedge (dd^c u)^{n-2}\}$$
$$- (n+1) \int_D hd(Gd^c u_t) \wedge (dd^c u_t)^{n-1} \quad .$$

Comme h et du_t sont nulles sur ∂C_o et ∂C_1 , on obtient

(39) $$\frac{d}{dt}\mathcal{J}[u_t] = -(n+1)\int_D hd(Gd^c u_t)\wedge(dd^c u_t)^{n-1} .$$

Respectivement,

$$\frac{d^2}{dt^2}\mathcal{J}[u_t] = -(n+1)\int_D [h(Gd^c h)\wedge dd^c u_t+(n-1)dd^c h\wedge d(Gd^c u_t)]\wedge(dd^c u_t)^{n-2}$$

$$-(n+1)\int_{\partial D} hd^c h\wedge[Gdd^c u_t+(n-1)d(Gd^c u_t)]\wedge(dd^c u_t)^{n-2}$$

$$+(n+1)\int_D dh\wedge d^c h\wedge[Gdd^c u_t+(n-1)d(Gd^c u_t)]\wedge(dd^c u_t)^{n-2} ,$$

où la première intégrale à droite est nulle. Alors

(40) $$\frac{d^2}{dt^2}\mathcal{J}[u_t]=(n+1)\int_D dh\wedge d^c h\wedge[Gdd^c u_t+(n-1)d(Gd^c u_t)]\wedge(dd^c u_t)^{n-2} .$$

A son tour (d_4) implique l'égalité de l'expression en les parenthèses carrées à droite de (40) et de $[G+(n-1)f]dd^c u_t$, et l'estimation $G+(n+1)f \geq 0$ partout. D'autre part, comme $u_t \in \mathrm{Adm}\, D$, elle est plurisousharmonique C^2 , ainsi pour toute fonction $h : D \to \mathbb{C}$ de la classe C^1 on a

$$dh\wedge d^c h\wedge(dd^c u_t)^{n-1} \geq 0$$

partout. Par conséquent, $(d^2/dt^2)\mathcal{J}[u_t] \geq 0$, d'où, par (38),(39) et (c_4), l'infimum en (28) est atteint pour u en question.

Remarque 4. Le théorème 9 est aussi valable pour $n=1$. Raisonnons par analogie et voyons que la condition (d_4) est superflue.

THEOREME 10. Supposons :

(a_5) D possède un bord C^1-régulier par morceaux et sa fermeture compacte,

(b_5) $n \geq 2$, $p=2$ et q appartient à la classe C^2 ,

(c_5) l'infimum en (28) est atteint pour une fonction admissible u ,

(d_5) $d(Gd^c u) = f\, dd^c u$, où $G = q\, \det H$, f est continu et $f \geq 0$.

Alors u satisfait (33) .

Remarque 5. Le théorème 10 est aussi valable pour $n=1$; alors la condition (d_5) est superflue.

10. L'influence de la nonhomogénéité (des milieux physiques)

Les analogues du lemme 1, du théorème 1 et du corollaire 1 sont suivants (voir [16], [17] et [19]).

LEMME 3. Si q appartient à la classe C^1 et pour chaque $u \in \mathrm{Adm}\, D$ il existe une C^1-solution F et $D \longrightarrow \mathbb{C}$ de (34), alors

$$0 \leqslant \mathrm{cap}_p(D, q, \Gamma) \leqslant \mathrm{cap}_p(D, q, \Gamma, \mathcal{U}) < +\infty .$$

THEOREME 6. Soit $f : \mathbb{M} \longrightarrow \mathbb{M}'$ une application biholomorphe, f_* l'homomorphisme induit sur les classes d'homologie et $q : \mathbb{M}' \to \mathbb{C}$ une fonction continue, où $\mathbb{M}'$ est une variété hermitienne munie :

(i) de la métrique h' donnée, en coordonnées locales, par les relations

$$h'^{\ell\bar{m}} = (h^{j\bar{k}} \circ f^{-1})(f^{\ell}_{:j} \circ f^{-1})(\bar{f}^{m}_{:\bar{k}} \circ f^{-1})$$

ou, de manière équivalente, par

$$h'_{\ell\bar{m}} = (h_{j\bar{k}} \circ f^{-1}) f^{-1\, j}_{:\ell} \bar{f}^{-1\, k}_{:\bar{m}} ,$$

où $u_{:j} = (u \circ z^{-1})_{|j} \circ z$ et $z = (z_j)$ est une coordonnée application biholomorphe dans la variété,

(ii) du champ de tenseurs H' du type $(1,1)$ donnée convenablement, en coordonnées locales, par les relations

$$H'^{\ell}_{m} = (H^{j}_{k} \circ f^{-1})(f^{\ell}_{:j} \circ f^{-1}) f^{-1\, k}_{:m} .$$

Alors

$$\mathrm{cap}_p(f[D], q, f_*\Gamma) = \mathrm{cap}_p(D, q\circ f, \Gamma),$$

$$\mathrm{cap}_p(f[D], q, f_*\Gamma, f[U]) = \mathrm{cap}_p(D, q\circ f, \Gamma, \mathcal{U}).$$

COROLLAIRE 2. Soit $\mathbb{M}$ compacte de dimension complexe un, Γ-la classe d'homologie de D, dont les coefficients sont réels, représentée par une courbe de Jordan, que sépare C_0 et C_1, $1 \leqslant p \leqslant 3$,

$$\Phi(\varphi) = q\,[h(\varphi,\varphi)]^{\frac{1}{2}p-1}\, H^1_1 \quad \underline{\text{et}} \quad h(\varphi,\varphi) = h^{1\bar{1}}\,\varphi_1\,\bar{\Psi}_1$$

pour toutes les formes différentielles C^1-dérivables d'ordre un. Puis, pour chaque fonction $r : \mathbb{M} \longrightarrow C$ de classe C^1, soit Ψ une forme C^1-dérivable d'ordre un sur D, qui pour chaque $\gamma \in \Gamma$ possède la propriété

$$\int\!\!\int_D r\Phi(\varphi)\varphi \wedge \bar{\Psi}^c = \int_\gamma r\Phi(\varphi)\varphi \quad ,\text{où} \quad \bar{\Psi}^c = -i(\overline{\Psi_{\bar{1}}}\, dz - \overline{\Psi}_1 d\bar{z})$$

et φ est une forme quelconque C^1-dérivable d'ordre un sur D. Si

$$[1/r\Phi(\Psi)]\{h[r\Phi(\Psi)],\Psi) + h(\Psi, d[r\Phi(\Psi)])\} \leqslant 0 \quad ,$$

alors pour un certain C^1-champ de tenseurs H du type (1,1) sur $\mathbb{M}$ il existe des constantes positives C et $C(a)$, $0<a\leqslant 1$, indépendantes de D (C_o et C_1 inclus) et dépendantes seulement de $\mathbb{M}$ (h inclus), p et q, telle que

$$[1/C(a)][\mathrm{cap}_p(D,|q|)]^a \leqslant \mathrm{cap}_p(D,q,\Gamma,\mathfrak{U}) \leqslant C[\mathrm{cap}_{\tilde{p}}(D,\tilde{g})]^{\frac{1}{2}} \quad , \quad 0<a\leqslant 1 \quad .$$

où $\tilde{p} = 2/(p-1)$ et $\tilde{q} = |q|^{2/(p-1)}$.

Une certaine interprétation physique est étudiée ci-dessous ; c'est une théorie non linéaire des particules élémentaires.

11. Une interprétation physique : la structure des particules élémentaires

Nous décrivons la structure d'une particule élémentaire dans un cadre de géométrie différentielle. Au modèle supposé la particule est située dans l'espace d'observation considérée comme une variété pseudoriemannienne non riemannienne $\mathbb{M}$ ou , plus précisément, comme une fibration $\mathcal{B}_{\mathbb{M}}$, où $\mathbb{M}$ est l'espace de fibre [23] . Les propriétés de l'espace d'observation sont déterminées par l'existence de la particule considérée et les propriétés de sa symétrie. D'autre part, nous pouvons parler de la déformation de l'espace d'observation introduite par la particule décrite de la façon analogue comme dans le cas du champ gravitationnel quand l'existence de la masse change la forme de courbure de l'espace.

Les propriétés quantiques de la particule élémentaire sont exprimées à l'aide de l'équation ondulatoire décrite dans l'espace propre considérée comme une variété riemannienne $\mathbb{N}$ ou, plus précisément, comme une fibration $\mathcal{B}_{\mathbb{N}}$, où $\mathbb{N}$ est l'espace de fibre. D'autre part, nous devrons considérer deux espaces $\mathbb{N}_e$ et $\mathbb{N}_n$. Ici l'espace $\mathbb{N}_e$ est associé à son champ électromagnétique extérieur, tandis que l'espace $\mathbb{N}_n$ est associé à son champ nucléaire. Dans le cadre de la théorie présentée les valeurs de la charge électrique, du moment magnétique, de la charge nucléaire, du moment nucléaire et de la masse sont les paramètres déterminables par la forme de courbure des espaces propres dans lesquels l'équation ondulatoire a la même forme, indépendant du choix de l'espace de particule. L'image de la particule dans l'espace d'observation est alors une image de la particule dans le champ de forces extérieures, où la nature de ces forces nous semble être analogue au cas classique des forces de Coriolis.

Soit $\mathbb{R}$ l'espace de base de $\mathcal{B}_{\mathbb{M}}$ et aussi de $\mathcal{B}_{\mathbb{N}}(\mathbb{N} = \mathbb{N}_e$ ou $\mathbb{N}_n)$, $\mathbb{M}_{\#}$ resp. $\mathbb{N}_{\#}$ la fibre typique de $\mathcal{B}_{\mathbb{M}}$ resp. $\mathcal{B}_{\mathbb{N}}$. Considérons les espaces de Hilbert $\mathbb{H}(\mathcal{B}_{\mathbb{M}})$ resp. $\mathbb{H}(\mathcal{B}_{\mathbb{N}})$ correspondant à l'équation ondulatoire dans les espaces $\mathbb{M}$ resp. $\mathbb{N}$. Nous supposons l'application invertible $\mathbb{M}_{\#}\times\mathbb{R} \ni (x',t') \rightarrow (x,t) \in \mathbb{N}_{\#}\times\mathbb{R}$, qui conduit à la transformation

(41) $$V(x,t) = \exp\{(i/\hbar)\mathcal{H}(x,t)\}$$

de $\mathbb{H}(\mathcal{B}_{\mathbb{M}})$ dans $\mathbb{H}(\mathcal{B}_{\mathbb{N}})$ et à la transformation de $\mathcal{H}(x,t)$ déterminée dans $\mathbb{H}(\mathcal{B}_{\mathbb{N}})$ à l'application $\mathcal{H}'(x',t')$ déterminée dans $\mathbb{H}(\mathcal{B}_{\mathbb{M}})$ où

$$\exp\{(i/\hbar)\mathcal{H}(x,t)\} = \mathbb{1} + \frac{i}{\hbar}\mathcal{H}'(x',t') + \frac{i^2}{2!\hbar^2}[\mathcal{H}'(x',t')]^2 + \ldots$$

et $\mathbb{1}$ est l'opérateur unité dans $\mathbb{H}(\mathcal{B}_{\mathbb{M}})$. Physiquement, le choix de la fonction exp est déterminée par l'unitarité d'application (41).

L'application $V(x,t)$ nous permet de transformer l'équation ondulaire

(42) $$H \mid t > = i\hbar(\partial/\partial t) \mid t >$$

valable pour l'espace $\mathbb{N}$ dans l'équation

(43) $$\underline{H}(V \mid t >) = i\hbar(\partial/\partial t)(V \mid t >)$$

valable pour l'espace $\mathbb{M}$, où

(44) $$\underline{H} = V[H - (\partial/\partial t)\mathcal{H}]V^{-1} .$$

La plus petite valeur propre de H doit être égale à l'énergie au repos. Ensuite, nous pouvons établir une relation entre la forme de courbure Ω^2 d'une fibration principale sur une variété presque complexe (en particulier , complexe) $\mathbb{L}$, dérivée de $\mathbb{N}_e$ et $\mathbb{N}_n$, et l'espace-temps courbé $\mathbb{M}$ par les applications $v_e : \mathbb{N}_e \longrightarrow \mathbb{M}$ et $v_n : \mathbb{N}_n \longrightarrow \mathbb{M}$, de classe C^2 [24] . Ces applications sont induites par V_e et V_n respectivement. La forme Ω^2 est définie par les champs de mesons M et de baryons B , c'est-à-dire Ω^2 = M+iB .

Ainsi nous avons l'interprétation suivante de la procédure proposée :

avant la réaction nucléaire	$\mathbb{M}$ (une variété pseudoriemannienne non-riemannienne-l'espace d'observation d'un système des particules élémentaires)	$\mathbb{N}_e$(une variété riemannienne-l'espace du système des particules élémentaires associé à son champ électro-magnétique extérieur)	$\mathbb{L}$,en particulièr $\mathbb{N}_e \times \mathbb{N}_n$ (le produit Cartésien des variétés complexifie)	$\mathbb{N}_n$(une variété riemannienne-l'espace du système des particules élémentaires associé à son champ nucléaire)
la réaction	↓ F(une application quasi conforme hyperbolique)	↓ U(une application quasi conforme (elliptique))	↓ (une application biholomorphe)	↓ V(une application quasi conforme (elliptique))
après la réaction	$\mathbb{M}'$	$\mathbb{N}'_e$	$\mathbb{L}'$,en particulier,$\mathbb{N}'_e \times \mathbb{N}'_n$	$\mathbb{N}'_n$

Le domaine de biholomorphie de f correspond aux dimensions du système; l'invariance biholomorphe des capacités (19), (20) et (29), et la théorie de l'équation de Monge-Ampère complexe offrent un outil de recherche [20] .

Dans l'espace d'observation $\mathbf{M}$ l'équation ondulatoire (37) peut être écrite sous la forme de l'équation quantique dans l'espace avec la métrique arbitraire g qui correspond à l'application V . Dans le cas d'électron ou de proton l'équation (37) est donnée par l'équation de Dirac écrite sous la forme covariante générale (cf [28] et [29]) :

$$\{\gamma^k [(\partial / \partial x^k) + \Gamma_k] + (mc/\hbar)\} \Psi = 0 \quad ,$$

où m est la masse au repos de la particule, γ^k sont les matrices de Dirac qui satisfont les règles de commutativité

$$\gamma^j \gamma^k + \gamma^k \gamma^j = 2g^{jk}$$

et Γ_k sont les connections spinorielles (i.e. les symboles généralisés de Christoffel) déterminées par

$$\Gamma_k = \frac{1}{4} \gamma^j (\gamma_{j|k} - \{{}^{\ell}_{jk}\} \gamma^{\ell}) - \frac{1}{32} \Gamma r(\gamma \gamma^j \gamma_{j|k}) \gamma$$

avec

$$\gamma = \frac{1}{24} \varepsilon_{jk\ell m} \gamma^j \gamma^k \gamma^{\ell} \gamma^m \; ,$$

où $\varepsilon_{jk\ell m}$ désigne le tenseur complètement antisymétrique de Levi-Civita.

Quand nous considérons le mouvement de la particule dans l'espace -temps de Minkowski, nous pouvons voir que l'équation généralisée de Dirac a des termes additionnels déterminés par le métrique g jouant le rôle du potentiel effectif dans lequel le mouvement de la particule considérée est discuté [24] . En effet, nous obtenons que l'énergie propre de la particule peut être corrigée par l'existence elle-même dans l'espace et son influence sur la déformation de cet espace.

Enfin, nous voudrons ajouter un résultat très intéressant obtenu sur la base de l'équation généralisée de Dirac, c'est-à-dire, on peut montrer [15] , que cette équation est équivalente à l'équation non linéaire proposée par Heisenberg [13] dans sa théorie des particules élémentaires. Du point de vue physique ce résultat témoigne que les propriétés non linéaires de la matière peuvent être remplacées par une description du mouvement dans l'espace munie de la métrique choisie de manière convenable.

BIBLIOGRAPHIE

[1] ANDREOTTI (A.) and ŁAWRYNOWICZ (J.).- On the generalized complex Monge-Ampère equation on complex manifolds and related questions, Bull. Acad. Polon. Sci. Sér. Sci. Math. Astronom. Phys. 25 (1977), 943-948.

[2] ANDREOTTI (A.) and ŁAWRYNOWICZ (J.).- The generalized complex Monge-Ampère equation and a variational capacity problem, ibid. 25 (1977), 949-955.

[3] BEDFORD (E.) and TAYLOR (B.A.).- The Dirichlet problem for a complex Monge-Ampère equation, Inventiones Math. 37 (1976), 1-44.

[4] BEDFORD (E.) and TAYLOR (B.A.).- Some potential theoretic properties of plurisubharmonic functions, à paraître.

[5] BEURLING (A.) and DENY (J.).- Dirichlet spaces, Proc. Nat.Acad. Sci. U.S.A. 45 (1959), 208-215.

[6] BRELOT (M.).- Eléments de la théorie classique du potentiel, Centre de Documentation, Paris 1959.

[7] BREMERMANN (H.J.).- On a generalized Dirichlet problem for plurisubharmonic functions and pseudoconvex domains. Characterization of Šilov boundaries, Trans. Amer. Math. Soc. 91 (1959), 246-276.

[8] CHERN (S.S.), LEVINE (H.I.) and NIRENBERG (L.).- Intrinsic norms on a complex manifolds, in : Global analysis, Papers in honor of K. Kodaira, ed. by D.C. Spencer and S. Iynaga, Univ. of Tokyo Press and Princeton Univ. Press, Tokyo 1969, pp. 119-139 ; reproduction dans : S.S. CHERN , Selected papers, Springer-Verlag, New-York-Heidelberg-Berlin 1978, pp. 371-391.

[9] FEFFERMAN (C.).- The Bergman kernel and biholomorphic mappings of pseudoconvex domains, Invent. Math. 26 (1974) , 1-65.

[10] FUKUSHIMA (M.).- Dirichlet forms and Markov processes, North-Holland Publ. Co. -Kodansha Ltd., Amsterdam-Oxford-New-York-Tokyo 1980.

[11] GAVEAU (B.).- Méthodes de contrôle optimal en analyse complexe. I. Résolution d'équations de Monge-Ampère, J.Functional Analysis 25 (1977), 391-411.

[12] GAVEAU (B.) et ŁAWRYNOWICZ (J.).- Espaces de Dirichlet invariant biholomorphes et capacités associées, Bull. Acad. Polon. Sci. Sér. Sci. Math., à paraître.

[13] HEISENBERG (W.).- Research on the non-linear spinor theory with indefinite metric in Hilbert space, in : 1958 Annual Internat. Conf. on High Energy Physics at CERN , Geneva 1958, CERN, Genève 1958, pp. 119-122 ; discussion, pp. 122-126.

[14] KALINA (J.), ŁAWRYNOWICZ (J.), LIGOCKA (E.) and SKWARCZYŃSKI (M.).- On some biholomorphic invariants in the analysis on manifolds, in : Analytic functions, Kozubnik 1979, Proceedings, ed. by J. Ławrynowicz (Lecture Notes in Math. 798), Springer-Verlag, Berlin-Heidelberg-New-York 1980, pp. 224-249.

[15] KRÓLIKOWSKI (W.).- On correspondence between the Dirac equation and the non-linear Heisenberg equation, à paraître.

[16] ŁAWRYNOWICZ (J.).- Condenser capacities and an extension of Schwarz's lemma, Bull. Acad. Polon. Sci. Sér. Sci. Math. Astronom. Phys. 23 (1975), 839-844.

[17] ŁAWRYNOWICZ (J.).- On quasiconformality of projections of biholomorphic mappings, ibid. 23 (1975), 845-851.

[18] ŁAWRYNOWICZ (J.).- Electromagnetic field and the theory of conformal and biholomorphic invariants, in : Complex analysis and its applications III, International Atomic Energy Agency, Vienna 1976, pp. 1-23.

[19] ŁAWRYNOWICZ (J.).- On a class of capacities on complex manifolds endowed with an hermitian structure and their relation to elliptic and hyperbolic quasiconformal mappings, (a) Ann. Polon. Math. 33 (1976), 178 (résumé), (b) Dissertationes Math 166 (1980), 48 pp. (in extenso).

[20] ŁAWRYNOWICZ (J.).- On biholomorphic continuability of regular quasiconformal mappings, in : Analytic functions, Kozubnik 1979, Proceedings, ed. by J. Ławrynowicz (Lecture Notes in Math. 798), Springer-Verlag, Berlin-Heidelberg-New-York 1980, pp. 326-349.

[21] ŁAWRYNOWICZ (J.) in cooperation with KRZYZ (J.).- Ouasiconformal mappings in the plane. Parametrical methods (Lecture Notes in Math.), Springer-Verlag, Berlin-Heidelberg-New-York, à paraître.

[22] ŁAWRYNOWICZ (J.) and SKWARCZYŃSKI (M.).- Conformal and biholomorphic invariants in the analysis on manifolds, Proc. of the First Finnish-Polish Summer School in Complex Analysis at Podlesice I, ed. by J. Ławrynowicz and O. Lehto, Uniwersytet Łódzki, Łódź 1977, pp. 35-113, 2ème ed. Łódź 1978, pp. 35-113.

[23] ŁAWRYNOWICZ (J.) and WOJTCZAK (L.).- A concept of explaining the properties of elementary particles in terms of manifolds, Z. Naturforsch. 29a (1974), 1407-1417.

[24] ŁAWRYNOWICZ (J.) and WOJTCZAK (L.).- On an almost complex manifold approach to elementary particles, ibid. 32a (1977), 1215-1221.

[25] NEUMANN (J. von).- Sur la théorie des jeux, C.R. Acad. Sci. Paris 186 (1928), 1689-1691 ; une version détaillée : Zur Theorie der Gesellschaftsspiele, Math. Ann. 100 (1928), 295-320 ; reproduction dans : Collected works, VI, Theory of games, astrophysics, hydrodynamics and meteorology, ed. by A.H. Taub, McMillan Co., New-York 1973, pp. 1-26.

[26] OKADA (M.).- A paraître. C.R.A.S., Paris, Janvier 1981.

[27] PÓLYA (G.) and SZEGÖ (G.).- Isoperimetric inequalities in mathematical physics, Princeton Univ. Press, Princeton 1951.

[28] SCHMUTZER (E.).- Relativistische Physik, 2ème ed., B.G. Teubnerverlagsgesellschaft, Leipzig 1968.

[29] SCHMUTZER (E.).- Symmetrien und Erhältungssätze der Physik, Akademie-Verlag, Berlin 1972.

[30] SICIAK (J.).- Extremal plurisubharmonic functions in $\mathbb{C}^n$, Proc. of the First Finnish-Polish Summer School in Complex Analysis at Podlesice I, ed. by J. Ławrynowicz and O. Lehto, Uniwersytet, Łódrki, Łódź 1977,pp.115-152 ; 2ème éd. Łódź 1978, pp. 115-152.

Bernard GAVEAU
Université Pierre
et Marie Curie
4, Place Jussieu
France-75230 Paris Cedex 05

Julian ŁAWRYNOWICZ
Institut Mathématique
Académie Polonaise des Sciences
Branche de Łódź, 86, rue Kilińskiego
PL-90-012 Łódź (Pologne)

CALCUL DU NOMBRE DENSITE $\nu(x,f)$ ET LEMME DE SCHWARZ POUR LES FONCTIONS PLURISOUSHARMONIQUES DANS UN ESPACE VECTORIEL TOPOLOGIQUE

par Pierre L E L O N G

1 - Dans l'article "Stabilité du nombre de Lelong par restriction à une sous-variété" (cf. ce volume, partie II), C. Kiselman pose le problème suivant. Soit f une fonction plurisousharmonique définie dans un espace de Banach E au voisinage d'un point x. Soit $r_o > 0$ le rayon de la plus grande boule $B(x,r)$ de centre x, de rayon r, dans laquelle il existe pour f une majoration finie. On note $M(x,r,f)$ cette majoration dans $B(x,r)$ et l'on pose

$$(1) \qquad \nu'(x,f) = \lim_{r=o} (\log r)^{-1} M(x,r,f) \quad , \quad 0 < r < \inf(1,r_o) .$$

La limite existe d'après la convexité du graphe de la fonction $\log r \to M(x,r,f)$. A-t-on alors l'égalité

$$(2) \qquad \nu'(x,f) = \nu(x,f)$$

où $\nu(x,f)$ est la densité de f en x, telle qu'elle a été définie dans [4,a] ?

Je rappellerai d'abord cette définition en complétant [4,a] ; je donnerai ensuite un énoncé qui est valable dans un espace topologique E . Il entraîne l'égalité (2) quand E est un espace de Banach. L'intérêt du problème est évident : (2) fournit en effet un lemme de Schwarz. L'étude montrera de plus que, en fait, $\nu(x,f)$ et $\nu'(x,f)$ sont déterminés en un point x par la restriction de f à un cône de droites issues de x, dès que ce cône est d'intérieur non vide. Si on suppose que E est un espace de Fréchet, il suffit même que ce cône soit non pluripolaire. Il apparaît ainsi que le nombre $\nu(x,f)$ qui est défini en dimension finie à partir d'une intégration est obtenu <u>en dimension infinie comme extremum</u> pour y variable sur un ensemble auquel on demande seulement de n'être pas "trop petit".

2 - Dans le cas $E = \mathbb{C}^n$, la définition de $\nu(x,f)$ est classique sous l'une des formes suivantes (équivalentes) : a) $\nu(x,f)$ est le nombre de Lelong du courant $\theta = \frac{1}{2\pi} dd^c f = \frac{i}{\pi} \partial\bar{\partial} f$ - b) plus directement : $\nu(x,f)$ est la densité en x de la mesure positive $\frac{1}{2\pi} \Delta f$ calculée en dimension réelle $(2n-2)$ sur le filtre des boules $B(x,r)$, $r \searrow 0$ - c) $\nu(x,f)$ est la valeur commune des limites, $(0 < r < 1)$:

(3) $$\nu(x,f) = \lim_{r=o} \frac{\partial}{\partial \log r} \lambda(x,r,f) = \lim_{r=o} (\log r)^{-1} \lambda(x,r,f)$$

où $\lambda(x,r,f)$ est la moyenne de f sur la sphère $S(x,r)$. Pour la suite on notera qu'on a :

(4) $$-\nu(x,f) = \lim_{r=o} \left(\log \frac{1}{r}\right)^{-1} \lambda(x,r,f)$$

ce qui montre : si A est une majoration de f, on a $\nu(x,y,f-A) = \nu(x,y,f)$ et $-\nu(x,y,f-A)$ s'obtient par (4) comme limite pour $r \searrow 0$ d'une suite croissante de fonctions plurisousharmoniques négatives obtenues en remplaçant f par $f - A$.

3 - Si $\operatorname{Dim} E$ n'est plus fini, aucune des 3 définitions précédentes ne s'applique directement. Ceci m'a conduit dans [4,a] à procéder à partir des nombres $\nu(x,y,f)$ définis par

(5) $$-\nu(x,y,f) = \lim_{r=o} \left(\log \frac{1}{r}\right)^{-1} \ell(x,y,r,f) \quad , \quad 0 < r < 1$$

où $$\ell(x,y,r,f) = \frac{1}{2\pi} \int_0^{2\pi} f(x + re^{i\theta} y)\, d\theta$$

pour $(x,y) \in E \times E$.

DÉFINITION 1 (cf. [4,a]). Soit $f(x)$ une fonction plurisousharmonique définie au voisinage de x dans un espace vectoriel topologique complexe E. On appelle nombre densité de f en x le nombre

(6) $$\nu(x,f) = \inf_{y \in E} \nu(x,y,f)$$

où $\nu(x,y,f)$ est défini par (5).

La définition est justifiée par les propriétés de $\nu(x,y,f)$ et l'énoncé suivant qui complète [4,a]. Dans la suite E est toujours un espace vectoriel topologique sur $\mathbb{C}$.

PROPOSITION 1 - Soit G un domaine dans E qu'on supposera de majoration finie $f(x) < A$ pour une fonction plurisousharmonique f. Soit $\xi \in G$ et U_1, U_2 deux voisinages disqués de 0 tels que $x+y$ demeure dans G pour $x \in \xi + U_1$, $y \in U_2$. Alors :

1°) Le nombre négatif $-\nu(x,y,f)$ est défini par (5) pour $x \in \xi + U_1$, $y \in E$. Il est localement la limite d'une suite de fonctions plurisousharmoniques négatives croissantes. On a $\nu(x,\lambda y,f) = \nu(x,y,f)$ pour tout $\lambda \neq 0$, $\lambda \in \mathbb{C}$.

2°) La régularisée inférieure $\nu_*(x,y,f) = \liminf_{y' \to y} \nu(x,y',f)$ est indépendante de y et a la valeur $\nu(x,f)$ définie par (6).

3°) L'ensemble (variable avec x) défini par :

$$(7) \qquad g_x = [\, y \in E \ ; \ \nu(x,y,f) > \nu(x,f) \,]$$

est un cône de sommet 0 (g_x est vide si et seulement si l'on a $f(x) \neq -\infty$) éventuellement réduit à 0, dans E. Ou bien on a $g_x = E$, ou bien g_x est un cône pluripolaire dans E. On a

$$(8) \qquad \nu(x,f) = \inf_{y \in \omega} \nu(x,y,f)$$

pour tout ouvert ω non vide dans E. Si E est un espace de Fréchet, ou si e^f est continu et E est un espace de Baire, (8) est vérifié dès que ω est non pluripolaire dans E ; g_x est alors pluripolaire.

4°) Pour tout $x \in G$, tout voisinage disqué W de l'origine tel qu'on ait $x + W \subset G$, on a la majoration pour tout $y \in W$, tout r, $0 \leq r < 1$:

$$(9) \qquad \ell(x,y,r,f) \leq \nu(x,f) \log r + A .$$

Démonstration. Soit D_o le disque unité compact $|u| \leq 1$ de $\mathbb{C}$. Le disque $D_{x,y} = x + D_o y$ est compact dans E et $\ell(x,y,r,f)$ est défini et majoré par A

pour $D_{x,y} \subset G$, $0 \leqslant r < 1$. Il est classique que $\ell(x,y,r,f)$ est une fonction plurisousharmonique de (x,y) pour $D_{x,y} \subset G$ et l'on a pour $0 < r < 1$:

(10) $-\nu(x,y,r,f) = -\nu(x,y,r,f-A) = \lim\limits_{r=o} (\log \frac{1}{r})^{-1} \ell(x,y,r,f-A)$

qui donne $-\nu(x,y,r,f)$ comme limite croissante pour $r \searrow 0$ d'une suite de fonctions plurisousharmoniques croissantes pour $x \in \xi + U_1$, $y \in U_2$. Soit $m > 1$ un entier; en remplaçant U_2 par $m\, U_2$ et faisant varier r dans $]0 , \frac{1}{m}]$, on voit que le disque $D_{x,ry}$ demeure dans G pour $x \in \xi + U_1$, $y \in m\, U_2$ et (5) définit encore $\nu(x,y,f)$ pour de tels (x, y) . **Finalement, comme** U_2 est absorbant, $\nu(x,y,f)$ est défini localement comme limite d'une suite croissante (10) de fonctions plurisousharmoniques de (x,y) négatives et ceci pour $x \in G$, $y \in E$. De plus, pour $\lambda \neq 0$, $\lambda \in \mathbb{C}$, on a $\ell(x, \lambda y, |\lambda|^{-1} r,f) = \ell(x,y,r,f)$ ce qui conduit dans (10) à

(11) $\nu(x,\lambda y,f) = \nu(x,y,f)$ pour tout $\lambda \neq 0$, $\lambda \in \mathbb{C}$

et établit 1°). Si l'on a $f(x) \not\equiv -\infty$, on a $\nu(x,y,f) = 0$ pour tout $y \in E$.

Si l'on a $f(x) = -\infty$, on a $\nu(x,0,f) = +\infty$. On a aussi $\nu(x,y,f) = +\infty$ si y appartient au cône polaire $g'_x = [\, y \in E \ ; \ f(D_{xy}) = -\infty]$; ces cas exceptés $-\nu(x,y,f)$ a une valeur finie, négative.

La propriété 2°) résulte d'une propriété classique : la régularisée supérieure en y (donc à x constant) de $-\nu(x,y,f)$ obtenue comme enveloppe supérieure négative d'une famille de fonctions plurisousharmoniques dans E est une fonction plurisousharmonique de $y \in E$. Comme elle est majorée sur E , elle est indépendante de y . Si on la note $-\nu(x,f)$, on a

(12) $-\nu(x,f) \geqslant -\nu(x,y,f)$ pour tout $y \in E$

(13) $-\nu(x,f) = \sup\limits_{y \in \omega} [-\nu(x,y,f)]$.

La dernière égalité est vraie pour tout ouvert ω non vide d'après $-\nu(x,f) = \limsup\limits_{y' \to y} [-\nu(x,y,f)]$. Le rapprochement de (12) et (13) établit 2°).

La valeur $-\nu(x,f)$ étant finie, si l'on a $f(x) = -\infty$, on a $0 \in g_x$ et g_x n'est pas vide. D'après $\nu(x,\lambda y) = \nu(x,y)$ pour $\lambda \neq 0$, g_x est un cône de sommet 0 .

Ceci posé, si il existe un point $y_o \in E$ où l'on a $-\nu(x,y_o,f) = -\nu(x,f)$, il existe une suite $r_n \searrow 0$ telle que les fonctions

$$v_n(y) = (\log \frac{1}{r_n})^{-1} \ell(x,y,r_n,f) + \nu(x,f) \leq 0$$

convergent vers zéro. Quitte à extraire une sous-suite, on peut supposer qu'on a $\Sigma|v_n(y_o)| < \infty$. Alors $S(y) = \Sigma v_n(y)$ définit pour $y \in U_2$ une fonction plurisousharmonique négative et l'on a

$$[\, g_x \cap U_2] \subset [\, y \in E \;,\; S(y) = -\infty\,]$$

où S est plurisousharmonique dans U_2 , g_x répond à la définition qu'on a donnée dans [4,b] pour un cône polaire ; (8) résulte de (13) . Enfin l'existence de $y_o \in E$ est acquise si E est de Fréchet d'après un énoncé de G.COEURE (cf. [2], p. 382) ; dans le second cas elle résulte du fait qu'on a $\nu(x,y,f) > \nu(x,f)$ sur un ensemble maigre pour y (cf. [4,b]) .

La propriété (9) est évidente si l'on a $f(D_{xy}) = -\infty$. Sinon la convexité du graphe $\log r \to \ell(x,y,r,f-A)$ entraîne

(14) $\quad \ell(x,y,r,f) - A \leq \nu(x,y,f)\log r \leq \nu(x,f)\log r$

pour $0 < r < 1$, et pour $y \in W$ où W est voisinage disqué de l'origine tel que l'on ait $f(x + W) < A$. L'énoncé est ainsi établi. Il entraîne :

COROLLAIRE 1 - Si E est un espace de Fréchet ou si f est continu et E est un espace de Baire, il existe en chaque point x et pour tout entier p un sous-espace L de dimension p (en particulier une droite complexe) pour lequel la restriction $f|_L$ vérifie :

$$\nu(x,f|_L) = \nu(x,f) .$$

On a alors :

THÉORÈME 1 - Soit G un domaine d'un espace vectoriel topologique complexe E et $f(x)$ une fonction plurisousharmonique dans G . Soit $x \in G$ et W un voisinage disqué de l'origine tel que $x + W$ appartienne à un ensemble de majoration finie A de f dans G . On pose

(15) $\quad M_W(x,r,f) = \sup f(x+x')$ pour $x' \in rW$, $0 < r < 1$

et

(16) $\quad \nu'_W(x,f) = \lim_{r=o} (\log r)^{-1} M_W(x,r,f)$.

Alors on a

(17) $\nu'_W(x,f) = \nu(x,f)$.

En particulier $\nu'_W(x,f)$ ne dépend pas du voisinage W de l'origine choisi.

Démonstration. On peut donner une démonstration simple de (17) en remarquant que si l'on a $f(D_{xy}) \neq -\infty$, la fonction $\varphi(u) = f(x+uy)$, $y \neq 0$, $y \in W$ est sous-harmonique de u dans le disque $|u| < 1$. Si l'on pose

$$m(x,y,r,f) = \sup f(x + uy) \quad \text{pour } |u| \leq r , \qquad 0 < r < 1$$

on a d'après un résultat établi par V.AVANISSIAN (cf. [1, théorème 1]) dans C^n :

(18) $\nu'(x,y,f) = \lim (\log r)^{-1} m(x,y,r,f) = \nu(x,y,f)$.

D'autre part si l'on a $f(D_{xy}) = -\infty$, on a dans (18) égalité avec valeur $+\infty$. On a donc $\nu'(x,y,f) = \nu(x,y,f)$ pour x fixé et tout y . On a aussi d'après la convexité du graphe $\log r \to m(x,y,r,f - A)$:

$$m(x,y,r,f-A) \leq \nu'(x,y,f)\log r \quad \text{pour } 0 < r < 1 , \ y \in W$$

$$M_W(x,r,f) = \sup_{y\in W} m(x,y,r,f) \leq [\inf_{y\in W} \nu(x,y,f)]\log r + A$$

et

(19) $M_W(x,r,f) \leq \nu(x,f) \log r + A$, $0 < r < 1$.

D'où, en utilisant (16) et remarquant qu'on a $\log r < 0$:

$$\nu'_W(x,f) \geq \nu(x,f) .$$

L'inégalité en sens contraire est évidente d'après $\ell(x,y,r,f) \leq m(x,y,r,f)$.

L'énoncé est ainsi établi. Il entraîne divers corollaires :

COROLLAIRE 2 - Pour une fonction plurisousharmonique f définie sur un ouvert d'un espace de Banach l'égalité (2) est vérifiée en tout point.

4 - De ce qui précède résulte un "lemme de Schwarz" qui est déjà donné par (19), mais on a un résultat plus précis en utilisant le 3°) de la proposition 1. On voit d'abord qu'on peut remplacer W par tout ensemble disqué et absorbant dans E , mais il suffit d'après (8) que W soit disqué et absorbe un ouvert non vide $\omega \subset E$.

COROLLAIRE 3 - Soit f(x) une fonction plurisousharmonique définie sur un espace vectoriel topologique complexe E au voisinage d'un point x , dans x + U , où U est un voisinage ouvert disqué de l'origine . Soit ω un ouvert non vide de E. Alors si $x + U$ est un domaine de majoration de f, on a en posant $\|f\|_r = \sup_{x'} f(x + x')$ pour $x' \in rU$:

(20) $$\|f\|_r \leq \|f\|_R - \nu(x,f) \log \frac{R}{r}, \quad 0 < r < R < 1$$

où l'on peut prendre

(21) $$\nu(x,f) = \inf_{y \in \omega} \nu(x,y,f) = \inf_{y \in \omega} \nu'(x,y,f)$$

ν et ν' étant définis par (5) et (18) respectivement, et on peut prendre pour ω tout ouvert non vide, et plus précisément, si E est de Fréchet, tout ensemble tel que le cône complexe de sommet 0 par ω soit non pluripolaire dans E. Il en est encore ainsi si e^f est continu et E est supposé seulement espace de Baire.

COROLLAIRE 4 - Si E est un espace de Banach, on a (20) si $B(x,R)$ est une boule de majoration pour la fonction plurisousharmonique f, en notant $\|f\|_r$ le sup de f dans $B(x,r)$; $\nu(x,f)$ est défini par (21), c'est-à-dire est le minimum de $\nu(x,y,f)$ pour y parcourant un cône quelconque de sommet 0 non pluripolaire dans E.

Ce type de lemme de Schwarz majore donc f soit à partir d'une majoration au voisinage de x sur un cône non pluripolaire de l'espace tangent T_x, soit, comme dans le lemme de Schwarz classique, à partir d'une majoration des masses du laplacien $\Delta_u f(x + uy)$, $u \in \mathbb{C}$, pour f restreint à des droites issues de x à condition qu'elles forment dans T_x un cône non pluripolaire.

5 - On va montrer que la différence $m(x,y,r,f) - \ell(x,y,r,f)$ demeure bornée pour $r \searrow 0$ hors d'un ensemble $y \in \omega$ pluripolaire dans E si E est un espace de Fréchet.

LEMME - Soit $\varphi(u)$ une fonction sousharmonique de u, pour $|u| < 1$, $u \in \mathbb{C}$. En posant $\ell(r) = \frac{1}{2\pi}\int_0^{2\pi} \varphi(re^{i\theta})d\theta$, $m(r) = \sup_\theta \varphi(re^{i\theta})$, on a pour $0 < r < R < 1$:

(22) $$0 \leq m(r) - \ell(r) \leq \mu(R) \log 2 + \varepsilon(r)$$

où $\varepsilon(r)$ tend vers 0 avec r et $\mu(R)$ est la masse $\frac{1}{2\pi}\Delta\varphi$ portée par le disque $|u| < R < 1$.

En effet à partir de l'expression de $\varphi(u)$ pour $|u| = r < R < 1$,

$$\varphi(u) = H_R(u) + \int d\mu(a) \log \left|\frac{R(u-a)}{R^2 - \bar{a}u}\right| = H'_R(u) + \int_{|a| \leq R} d\mu(a) \log |u - a|$$

où $H_R(u)$, $H'_R(u)$ sont harmoniques pour $|u| < R$, on a pour $|u| = r < R$, en posant $|a| = t$

$$\ell(r) = H'_R(0) + \int_o^r d\mu(t)\log r + \int_r^R d\mu(t)\log t$$

$$m(r) \leqslant H'_R(0) + \varepsilon(r) + \int_o^R d\mu(t)\log(r + t)$$

D'où $$0 \leqslant m(r) - \ell(r) \leqslant \int_o^r d\mu(t)\log \frac{r+t}{r} + \int_r^R d\mu(t)\log \frac{r+t}{t} + \varepsilon(r)$$

et (22) qui établit le lemme.

Revenons à la situation du théorème 1. Appliquons le lemme à $\varphi(u) = f(x+uy)$ pour $y \in W$. Soit comme plus haut $g'_x \subset g_x$ le cône pluripolaire des $y \in E$ pour lesquels on a $\varphi(u) = f(x+uy) \equiv -\infty$. Pour $y \notin g'_x$, $\varphi(u)$ est une fonction sousharmonique de u pour $|u|<1$ et le compact $|u| \leqslant R$, $(R<1)$, situé sur le disque D_{xy} porte une mesure positive $\frac{1}{2\pi}\Delta\varphi$ de masse $\mu(y,R)$ finie. Le lemme entraîne alors , pour $0<r<R<1$:

$$0 \leqslant m(x,y,r,f) - \ell(x,y,r,f) \leqslant C\mu(y,R) + \varepsilon(r)$$

où l'on peut prendre $C = \log 2$ et où $\lim\limits_{r=o} \varepsilon(r) = 0$.

On en déduit aisément en se ramenant au cas où l'on a $\mu(0) = 0$.

THÉORÈME 2 - Pour tout y n'appartenant pas au cône pluripolaire g'_x dans E formé des droites complexes issues de x sur lesquelles f est identique à $-\infty$, la différence $m(x,y,r,f) - \ell(x,y,r,f)$ est positive et tend vers zéro pour $r \to 0$.

6 - Le théorème 1 permet de montrer que $\nu(x,f)$ est fonction semi-continue de x dans certains espaces. En dimension finie la propriété résulte du fait que le moyenne $\lambda(x,r,f)$ est fonction continue de x ; (3) définit alors $\nu(x,f)$ comme limite décroissante de fonctions continues. On montrera :

THÉORÈME 3 - Dans un espace vectoriel complexe E muni d'une topologie localement convexe, le nombre densité $\nu(x,f)$ est une fonction semi-continue supérieurement de x .

Soit en effet $p_\alpha(x)$ une semi-norme continue sur E et B_α boule unité.

On pose

$$M_\alpha(x,r,f) = \sup_{u,y} f(x+uy) \quad \text{pour} \quad |u| < r \quad \text{et} \quad p_\alpha(y) \leqslant 1 .$$

La fonction $\log r \to M_\alpha(x,r,f)$ est continue pour $0 < r < R < 1$ si l'on choisit $R \in [0,1[$ et p_α tels que $W = x + RB_\alpha$ soit un ensemble de majoration de f. On a alors les inégalités évidentes pour $0 < \eta < r$, $p_\alpha(x-x') < \eta$, et $0 < r < R$:

(23) $\quad M_\alpha(x, r-\eta, f) \leqslant M_\alpha(x',r,f) \leqslant M_\alpha(x, r+\eta, f)$.

Utilisant la continuité de M_α par rapport à r pour x fixé, on choisit η tel que les deux extrêmes dans (23) diffèrent de moins de $\varepsilon > 0$ donné ; $M_\alpha(x,r,f)$ étant situé dans leur intervalle, on a alors :

$$|M_\alpha(x,r,f) - M_\alpha(x',r,f)| < \varepsilon \text{ pour } p_\alpha(x-x') < \eta$$

ce qui établit la continuité de $M_\alpha(x,r,f)$ en x. Si A est une majoration de f dans $W = x + RB_\alpha$, on a d'après le théorème 1

$$\nu(x,f) = \lim_{r=o} (\log r)^{-1} M_\alpha(x,r,f-A) \ ; \ 0 < r < R < 1 .$$

Ainsi $\nu(x,f)$ est encore localement la limite d'une suite décroissante de fonctions continues de x, ce qui établit l'énoncé. Remarquons d'autre part que la réunion pour $c > 0$ des ensembles $N(c,f) = [x \in G \ ; \ \nu(x,f) \geqslant c]$ est évidemment contenue dans l'ensemble pluripolaire des $-\infty$ de f.

On notera alors la conséquence suivante de l'énoncé précédent :

COROLLAIRE 5 - Dans un espace vectoriel topologique complexe E muni d'une topologie localement convexe, les ensembles de densité

$$N(c,f) = [x \in G \ ; \ \nu(x,f) \geqslant c \ , \ c > 0]$$

d'une fonction plurisousharmonique f définie dans un domaine $G \subset E$ sont des ensembles fermés localement pluripolaires.

Remarque. On conjecture que les ensembles $N(c,f)$ sont comme en dimension finie des sous-ensembles analytiques complexes, au moins dans les espaces de Fréchet à base avec propriété d'approximation. L'intérêt d'un tel énoncé sera toutefois limité si on ne possède aucune information sur la codimension de l'ensemble $N(c,f)$.

B I B L I O G R A P H I E

[1] AVANISSIAN (V.). - Fonctions plurisousharmoniques et fonctions doublement sousharmoniques. Ann. E.N.S., t. 78, 1961, p. 101-161.

[2] COEURÉ (G.). - Fonctions plurisousharmoniques sur les espaces vectoriels topologiques. Ann. Inst. Fourier, 1970, p. 361-432.

[3] KISELMAN (Ch.). - Stabilité du nombre de Lelong par restriction à une sous-variété - Ce volume.

[4] LELONG (P.). - a/ Plurisubharmonic functions in topological vector spaces. Polar sets and problems of measure. Lecture Notes, n° 364, 1973, p. 58-69.

- b/ Fonctions plurisousharmoniques et ensembles polaires sur une algèbre de fonctions holomorphes. Lecture Notes, n° 116, 1969, p. 1-20.

Boundary Regularity for the Cauchy - Riemann Complex

R. Michael RANGE

It is classical that on pseudoconvex domains $D\subset\mathbb{C}^n$ one can solve the $\bar{\partial}$ - equation, i.e., given $f\in C^{\infty}_{0,q}(D)$, $1\leq q\leq n$, with $\bar{\partial}f=0$, there is $u\in C^{\infty}_{0,q-1}(D)$ with

$$\bar{\partial}u = f.$$

It is of great interest to estimate a solution u in terms of f in suitable norms. Two methods have been studied to obtain solution operators which can be estimated:

A) The L^2 Theory. Here $u \equiv Kf$ is the Kohn solution, i.e., the unique solution which is perpendicular to ker $\bar{\partial}$ with respect to a given metric, for example the Euclidean metric on $\mathbb{C}^n$. The key problem is to establish subelliptic estimates; this leads to estimates of Kf in Sobolev norms. The fundamental results in this approach are due to Kohn ([9], and the references given there).

B) Explicit Integral Solution Operators. The classical example is the Cauchy transform in $\mathbb{C}$,

$$Tf(z) = -\frac{1}{2\pi i}\int_D \frac{f(\zeta)\,d\bar{\zeta}\wedge d\zeta}{\zeta - z}, \quad f\in L^1(D),$$

which satisfies $\partial/\partial\bar{z}\,(Tf)=f$ on D. This method was generalized to higher dimensions about 10 years ago; it leads to sup-norm and Hölder estimates.

In part I of this note we give a brief survey of some of the key results regarding B); in part II we discuss some recent new results in this area.

I. Hölder Estimates

We will always assume that D has a smooth boundary. To keep matters simple, we consider only the following type of estimate.

Definition. There is a Hölder estimate of order $\alpha<1$ for $\bar{\partial}$ on $D\subset\subset\mathbb{C}^n$, if, given $f\in C^{\infty}_{0,q}(D)\cap L^{\infty}_{0,q}(D)$ with $\bar{\partial}f=0$, there is a solution $u=Sf$ of $\bar{\partial}u=f$, such that

$$|u(z)-u(z')|\le \text{const.}\ \|f\|_{L^\infty}|z-z'|^{\alpha}, \text{ for } z,z'\in D.$$

A Hölder estimate of positive order expresses a smoothing property of the solution operator under consideration. This is analogous to a subelliptic estimate in the L^2 theory.

In $\mathbb{C}^1$ the situation is optimal; there is a Hölder estimate for any $\alpha<1$; a "good" solution is given by the

Cauchy transform Tf.

In $\mathbb{C}^n$, $n\geqq 2$, the situation is quite complicated; in general, one does not even have bounded solutions for $\bar{\partial}u=f$, f bounded, as is shown by a recent example of Sibony [19]. The complex geometry of the boundary of the domain plays a major role, as is illustrated by the following

Example (cf. [16]). For $m\in\mathbb{N}$, let

$$B^m=\{(z,w)\in\mathbb{C}^2:|z|^2+|w|^{2m}<1\}.$$

Then there is no Hölder estimate for $\bar{\partial}$ or B^m of order $\alpha>1/2m$.

In 1969, Henkin [4] and Ramirez [14] found generalizations of the Cauchy kernel for strictly pseudoconvex domains. Grauert/Lieb [2] and Henkin [5] then used these kernels to construct integral solution operators for $\bar{\partial}$ on such domains. Since then, these operators have been investigated by numerous authors and successfully applied to many problems in complex analysis (for example, see [17] for a bibliography until 1973). Regarding Hölder estimates, the following results have been obtained.

1) For D strictly pseudoconvex, there is a Hölder estimate of order α on D:

a) for q=1, with

$\alpha=0$ (i.e. sup-norm estimate), Grauert/Lieb [2] and Henkin [5]

$\alpha<1/2$, Kerzman [8]

$\alpha=1/2$, Henkin/Romanov [6];

b) for q>1, with

$\alpha<1/2$, Lieb [11]

$\alpha=1/2$, Range/Siu [17].

The order $\alpha=1/2$ is best possible in this case (see the example above, m=1).

2) for D Euclidean convex, with real analytic boundary, there is a Hölder estimate

a) of some order $\alpha>0$, if $D\subset\subset\mathbb{C}^2$; for B^m, one obtains the sharp estimate with $\alpha=1/2m$ (Range [15, 16]);

b) for $D=B^{(m_1,\ldots m_n)}=\{z\in\mathbb{C}^n : \Sigma|z_i|^{2m_i}<1\}$, $m_i\in\mathbb{N}$, with any order $\alpha<1/2M$, where $M=\max\{m_i\}$ (Range [16]).

2a) is false if one does not assume ∂D real analytic (see [16]).

The proofs involve detailed estimations of the integral solution operators. For convex domains there always exist integral solution operators, which are contructed in a canonical way from a defining function for D. However, the estimates are much more delicate if the Levi-form degenerates. This is the main difficulty in attempting to generalize 2a) to arbitrary dimensions. The proof of 2b) involves techniques which could be applied to other domains, but, so far, have not yet yielded the general convex case.

For arbitrary pseudoconvex domains with real analytic boundary, nothing seems to be known regarding Hölder estimates. No reasonable integral solution operator for $\bar{\partial}$ appears to be known in this generality; the main difficulty is intimately connected with the fact that such domains are not necessarily locally biholomorphic equivalent to a convex domain (cf. the Kohn/Nirenberg example [10]). However, Kohn [9] proved subelliptic estimates in this case, and thus it is reasonable to expect Hölder estimates as well.

II Estimates for Derivatives

Once one has a solution operator T_q for $\bar\partial$ which satisfies Hölder estimates, it is natural to consider the corresponding estimates for derivatives, i.e., if $f \in C^k_{0,q}(\bar D)$ for some k, $0 \le k \le \infty$, and $\bar\partial f = 0$, is then

(*) $|T_q f|_{C^{k+\alpha}} \le C_k |f|_{C^k}$?

This would be rather trivial if T_q would be defined by a convolution integral, as then differentiation commutes with T_q. However, the principal terms of the known integral solution operators for $\bar\partial$ are more complicated, and it is far from obvious how to exploit the regularity of f to prove regularity of $T_q f$.

Siu [20] proved (*) (with $\alpha = 1/2$) for the Henkin solution operator on (0,1) forms on strictly pseudoconvex domains. However, his proof makes essential use of q=1. Greiner/Stein [3] obtain (*) for the Kohn solution operator, but also for q=1 only.

Recently, the following result was proved by Lieb/Range [13].

Theorem 1. Let $D \subset\subset \mathbb{C}^n$ be strictly pseudoconvex. For $1 \le q \le n$, there exist linear integral operators

$$T_q^*: C^0_{0,q}(\bar{D}) \to C^0_{0,q-1}(D)$$

such that

i) if $\bar{\partial}f=0$, then $\bar{\partial}(T_q^* f)=f$;

ii) for $k=0,1,2,\ldots$, if $f \in C^k_{0,q}(D)$ with $\bar{\partial}f=0$, then

$$|T_q^* f|_{C^{k+1/2}} \leq C_k |f|_{C^k}$$

Moreover, T_q^* is (more than) pseudolocal:

If V is open, $f \in C^0_{0,q}(\bar{D}) \cap C^k_{0,q}(V \cap \bar{D})$ and $\bar{\partial}f=0$, then

$$T_q^* f \in C^{k+1/2}_{0,q-1}(V \cap \bar{D}).$$

Idea of proof. T_q^* is obtained by modifying the boundary integral of the Henkin operator T_q (as given in [17] for arbitrary q) as follows.

We fix a suitable neighborhood D^o of $\bar{D}$. By using the linear extension operator E of Seeley [18], the given form f is extended to $Ef \in C^k(D^o)$, such that Ef has compact support in D^o, and

$$|Ef|_{C^k(D^o)} \leq C_k |f|_{C^k(D)}.$$

Via Stokes' Theorem and other general formulas for the kernels, one decomposes $T_q f$ into integral operators defined over $D^o \setminus D$ and D^o:

$$T_q f = M_q(Ef) + S_q(\bar{\partial}Ef) + B_q(Ef),$$

where B_q is the Bochner-Martinelli transform. One then notes that $M_q(Ef)$ is $\bar{\partial}$-closed, hence $T_q^{*}f=T_qf-M_q(Ef)$ is a solution operator for $\bar{\partial}$.

For the estimations one uses an inductive procedure and integration by parts in order to express derivatives of $S_q(\bar{\partial}Ef)$ as a sum of integrals which can then be estimated by making use of the fact that one integrates over $D^{\circ}\setminus D$, and that $\bar{\partial}Ef$ vanishes to a certain order on ∂D.

By modifying the local kernels in the q-convex,resp. q-concave case (cf. [1],[7],[12]) as outlined above, and by passing to global via the bumping technique and interior elliptic estimates, one obtains

Theorem 2.([13]) Let X be a complex manifold and let $D\subset\subset X$ be a smoothly bounded domain which satisfies the Hörmander condition Z(q), i.e., at every $p\in\partial D$ the Levi form of ∂D has at least n-q positive or at least q+1 negative eigenvalues. Then there is a linear operator T_q^{*} for which the conclusions of Theorem 1 hold for $\bar{\partial}$-exact (0,q)-forms f on D.

REFERENCES

[1] W. Fischer/I. Lieb. Lokale Kerne und beschränkte Lösungen für den $\bar{\partial}$-Operator auf q-konvexen Gebieten. Math. Ann. 208, 249-265 (1974)

[2] H. Grauert/I. Lieb. Das Ramirezsche Integral und die Lösung der Gleichung $\bar{\partial}f=\alpha$ im Bereich der beschränkten Formen. Rice Univ. Studies 56, 26-50 (1970)

[3] P. Greiner/E. Stein. Estimates for the $\bar{\partial}$-Neumann problem. Princeton 1977

[4] G. M. Henkin. Integral representations of functions in strictly pseudoconvex domains and some applications. Mat. Sb. 78, 611-632 (1969). Maths USSR Sb. 7, 597-616 (1969)

[5] G. M. Henkin. Integral representations in strictly pseudoconvex domains and applications to the $\bar{\partial}$-problem. Mat. Sb. 82, 300-308 (1970). Maths. USSR Sb. 11, 273-281 (1970)

[6] G. M. Henkin/A. V. Romanov. Exact Hölder estimates for the solution of the $\bar{\partial}$-equation. Izv. Ak. Nauk USSR 35, 1171-1183 (1971). Maths. USSR Izv. 5, 1180-1192 (1971)

[7] M. Hortmann. Über die Lösbarkeit der $\bar{\partial}$-Gleichung mit Hilfe von L^p, C^k und D'-stetigen Integraloperatoren. Math. Ann. 223, 139-156 (1976)

[8] N. Kerzman. Hölder and L^p estimates for solutions of $\bar{\partial}u=f$ in strongly pseudoconvex domains. Comm. Pure Appl. Math 24, 301-379 (1971)

[9] J. J. Kohn. Subellipticity of the $\bar{\partial}$-Neumann problem on pseudoconvex domains: sufficient conditions. Acta Math. 142, 79-122 (1979)

[10] J. J. Kohn/L. Nirenberg. A pseudoconvex domain not admitting a holomorphic support function. Math. Ann. 201, 265-268 (1973)

[11] I. Lieb. Die Cauchy-Riemannschen Differentialgleichungen auf streng pseudokonvexen Gebieten. I. Math. Ann. 190, 6-44 (1970). II. Math. Ann. 199, 241-256 (1972)

[12] I. Lieb. Beschränkte Lösungen der Cauchy-Riemannschen Differentialgleichungen auf q-konkaven Gebieten. Man.math. 26, 387-409 (1979)

[13] I. Lieb/R. M. Range. Lösungsoperatoren für den Cauchy-Riemann-Komplex mit C^k-Abschätzungen. (1980 preprint)

[14] E. Ramirez de A. Ein Divisionsproblem und Randintegraldarstellungen in der komplexen Analysis. Math. Ann. 184, 172-187 (1970)

[15] R. M. Range. Hölder estimates for $\bar\partial$ on convex domains in $\mathbb{C}^2$ with real-analytic boundary. Proc. Symp. Pure Mathematics XXX, 2, 31-33 (1977)

[16] R. M. Range. On Hölder estimates for $\bar\partial u=f$ on weakly pseudoconvex domains. Proc. Conf. Complex Anal. Cortona 1976-77, Sc. Norm. Sup. Pisa 1978

[17] R. M. Range/Y. T. Siu. Uniform estimates for the $\bar\partial$-equation on domains with piecewise smooth strictly pseudoconvex boundaries. Math. Ann. 206, 325-354 (1973)

[18] R. T. Seeley. Extension of C^∞-functions defined in a half space. Proc. Amer. Math. Soc. 15, 625-626 (1964)

[19] N. Sibony. Un exemple de domaine pseudoconvexe régulier où l'équation $\bar\partial=f$ n'admet pas de solution bornée pour f bornée. (1980 preprint)

[20] Y. T. Siu. The $\bar\partial$-problem with uniform bounds on derivatives. Math. Ann. 207, 163-176 (1974)

R. Michael Range
Department of Mathematics
SUNY at Albany
Albany, N. Y. 12222

ALLOCUTION PRONONCÉE
PAR
MONSIEUR LE PROFESSEUR GÉRARD COEURÉ
DE L'UNIVERSITÉ DE LILLE I

EN L'HONNEUR
DU

PROFESSEUR PIERRE LELONG

COLLOQUE DE WIMEREUX
LE 12 MAI 1981

Monsieur le Professeur LELONG, c'est au nom de vos élèves et des mathématiciens qui ont eu l'honneur de travailler en étroite collaboration avec vous que j'ai le plaisir de présenter devant ce colloque votre oeuvre mathématique.

Dans un Mémoire désormais célèbre aux Annales de l'E.N.S., vous introduisiez en 1945 la notion de fonction p.s.h. et vous y développiez leurs principales propriétés à la suite de plusieurs notes à l'Académie. Il n'est guère, aujourd'hui, de problèmes sur les fonction analytiques de plusieurs variables complexes qui ne les utilisent. Les idées les plus fécondes sont souvent les plus simples ; c'est en remarquant que de nombreuses propriétés des fonctions analytiques intervenaient par celles du Log. de leur module que vous avez ouvert une voie de recherche dont la richesse est attestée par la place que prenaient ces fonctions p.s.h. dans les travaux qui sont exposés par les éminents mathématiciens de ce colloque. Vous avez déjà écrit plus de 70 Mémoires ou Notes sur les propriétés de ces fonctions et de nouveaux articles sont en cours d'élaboration. A partir des travaux d'Oka, connus tardivement en France du fait de la guerre, vous avez montré que les domaines d'holomorphie étaient convexes par rapport aux fonctions p.s.h. Vous avez achevé l'étude des propriétés géométriques de ces domaines en montrant leur équivalence.

C'est à cette même époque que sous l'influence de Henri Cartan, l'étude des ensembles analytiques s'est développée ; dès 1950, vous vous attachiez à l'étude de leurs propriétés métriques. Kodaira s'était interessé aux ensembles définis par une seule équation. En développant une théorie de l'intégration de formes différentielles convenables sur les ensembles analytiques vous réussissez à définir une "aire" de ces ensembles. Les très beaux résultats que vous obtenez dans cette voie continuent à inspirer nombre de travaux récents. Ainsi ceux de Thie et King montrent qu'un ensemble analytique est le support d'un courant positif fermé pour lequel une densité que vous aviez introduite dès 1956 et qui est souvent utilisée de nos jours, sous le nom de Nombre de Lelong, est un entier naturel. Ce nombre permet d'apporter sur les ensembles analytiques des renseignement quantitatifs qui échappaient aux méthodes héritées de la théorie des faisceaux. A la suite des travaux de Bombieri le nombre de Lelong a des applications importantes à la théorie des nombres.

Au cours de ces travaux, vous avez introduit et étudié d'autres objets mathématiques qui viennent encore récemment de bénéficier de développements importants. Je pense aux ensembles polaires et négligeables. Vous aviez remarqué depuis le début de vos recherches que la notion de capacité, si utile à l'étude des fonctions holomorphes d'une variable, était inadaptée pour plusieurs variables. L'utilisation des ensembles polaires est désormais indispensable pour "mesurer" les singularités analytiques. E. Borel avait remarqué que la série de Taylor avait, en général pour somme, une fonction admettant le cercle de convergence comme coupure. En utilisant les fonctions p.s.h. sur des espaces de dimension infinie, vous montrez que les fonctions holomorphes d'un domaine d'holomorphie, qui se prolongent en dehors, forment un ensemble polaire et négligeable. Tous vos travaux sur ces "petits" ensembles montrent l'intime relation entre la négligeabilité et la polarité. Les travaux récents de Bedford et Taylor ont confirmé la profondeur de vos vues. Je devrais parler aussi de vos recherches sur les fonctions entières qui ont beaucoup inspiré les travaux de Kiselman et de notre regretté collègue Martineau.

Les conférences qui vont se tenir durant ce colloque montreront, mieux que je ne saurais le faire, l'apport de votre oeuvre mathématique à l'Analyse complexe.

Permettez-moi, Monsieur le Professeur , **de conclure** ce trop bref aperçu de vos travaux en remerciant, en mon nom et en celui de mon collègue H. Skoda avec qui j'ai partagé l'élaboration de ce colloque, tous ceux qui ont contribué à son organisation. Je pense, tout d'abord, à Monsieur le Professeur Richard, Directeur du Centre de Biologie Maritime de Wimereux, qui, à travers l'Université de Lille I, a bien voulu mettre à notre disposition ses installations. Il m'est agréable de remercier Messieurs les Maires de Wimereux et de Boulogne qui, par leur présence à cette réception, ont montré leur intérêt pour le développement des activités scientifiques dans la région Nord - Pas de Calais.

Je ne voudrais pas terminer sans remercier les différents départements de Mathématiques présents à ce colloque qui par leur contribution financière nous ont permis d'inviter les éminents mathématiciens étrangers que nous avons le plaisir d'accueillir en cette région belle mais trop méconnue de notre pays.

Gérard COEURÉ

RÉPONSE DE Pierre LELONG

Chers Collègues et amis, mon cher Coeuré,

Je veux tout d'abord remercier ceux qui comme vous, mon cher Coeuré et vous aussi mon cher Skoda ont pris la peine d'organiser ce Colloque qui nous réunit dans cette belle station du bord de mer. Mes remerciements vont aussi à tous ceux qui ont rendu possible cette réunion de travail en mon honneur, et tout particulièrement aux Universités qui ont contribué à sa réalisation, telles Paris VI, Lille, Strasbourg. Enfin je remercie les nombreux mathématiciens qui sont venus participer à ce Colloque. A vrai dire rien ne pouvait me faire plus de plaisir que de voir certains de mes travaux inspirer un nombre croissant de recherches dont beaucoup sont profondes et dont la variété est remarquable.

Vous avez, mon cher Coeuré, souligné très justement l'intérêt que j'ai pris au développement de l'Analyse complexe. Le mot ne s'employait guère quand j'ai passé ma thèse en 1942 et l'on parlait plutôt de théorie des fonctions . Ce sont les propriétés -je dirais même les particularités- de la structure complexe qui ont été au centre de mes préoccupations au début. Ce que j'ai fait dans d'autres parties des mathématiques, par exemple l'étude de différents types de quasi-analyticité pour les classes de fonctions de n variables réelles, était souvent inspiré par le désir de vérifier que l'Analyse complexe que je cherchais à développer était bien l'instrument que d'autres branches des mathématiques utiliseraient. De même, ma thèse apparaît aujourd'hui comme le début d'une méthode, qui utilisant la théorie des potentiels plans, se limitait à $n = 2$, faute d'avoir encore à sa disposition la notion et les propriétés des fonctions plurisousharmoniques que je n'ai données que plus tard et rédigées à mon arrivée à votre Université de Lille.

Aujourd'hui du chemin a été parcouru : la géométrie analytique complexe, en particulier celle des ensembles analytiques que vous venez d'évoquer, mon cher Coeuré, participe à la fois de la géométrie algébrique et d'une certaine géométrie différentielle qu'on peut dire généralisée au point de vue des opérateurs. A l'heure où j'obtenais des résultats, c'est la première direction qui était chez nous en vogue, avec des énoncés qui assurent l'existence des solutions, ou, à défaut, fournissent les groupes d'obstruction. L'autre voie que j'ai suivie et peut-être, comme vous l'avez rappelé, quelque peu inaugurée, nous donne des méthodes plus constructives. De ce fait elle se trouve répondre aux besoins de certaines recherches actuelles en matière d'analyse plus fine et elle s'adapte bien aux problèmes à croissance devenus fondamentaux dans beaucoup d'applications. De ce fait aussi, je dois dire, son extension à la dimension infinie pose des problèmes tout à fait nouveaux que vous connaissez bien, mon cher Coeuré, puisque vous y avez apporté vous-même une contribution remarquée.

D'où est venu mon intérêt pour l'Analyse complexe ? Très tôt, je crois, et peut être déjà au lycée, le corps $\mathbb{C}$ m'avait paru très différent de R (ou de R^2 si vous préférez) : les êtres de la structure C^n bénéficient en particulier d'une orientation. Ce fait est très élémentaire, mais il est fondamental. Je crois que sa perception, plus ou moins consciente à l'époque, m'a guidé vers ces notions que vous avez rappelées et qui maintenant prennent place dans l'Analyse complexe . Je dois avouer que cette place ne leur a été reconnue que peu à peu, tellement grand était le prestige d'une "théorie des fonctions" axée plutôt sur la dimension un et les développements analytiques. Je dois dire, par contre, qu'aujourd'hui, après avoir écouté vos exposés, je me sens comme assiégé par ces notions où la positivité de certaines formes joue un si grand rôle.

De tout cela je me garderai de tirer vanité. Le mathématicien par force, est, plus que tout autre, un altruiste dans sa recherche. Au contraire de l'oeuvre littéraire qui, si elle est de qualité, s'individualise avec la durée, l'oeuvre du mathématicien s'y dissout. Si elle semble difficile et compliquée (ce qui parfois initialement peut lui donner de grands mérites), si, surtout, elle continue à le paraître, elle est rejetée et figurera peut être au rang des exercices. Quand, au contraire, elle demeure, c'est qu'assimilée à travers les séminaires et les cours que nous multiplions à cet effet, elle finit par paraître simple, puis classique, enfin triviale ; le progrès la saisit pour l'englober dans des vérités plus générales et de ce fait plus intéressantes. En mathématiques l'indispensable processus d'assimilation est un processus de trivialisation. Nous ne devons oublier ni cette évolution ni la modestie qu'elle conseille. J'oserai dire aussi qu'en présence d'une certaine diversification des notions mathématiques, le sentiment de cette évolution inexorable peut conduire le mathématicien à s'attacher aux notions les plus simples pour les approfondir. Ce sont parfois celles qui demeurent le plus longtemps utiles. En organisant ce Colloque, cher Collègues et amis, vous m'avez convaincu que je n'ai pas oeuvré inutilement et je vous en remercie.

Pierre L E L O N G

SUR LES FONCTIONS HARMONIQUES D'ORDRE QUELCONQUE ET LEUR PROLONGEMENT ANALYTIQUE DANS $\mathbb{C}^N$

V. A V A N I S S I A N (Strasbourg)

TABLE DES MATIERES

Chapitre 1 : Fonctions harmoniques d'ordre infini.

§ 1.1. - Analyticité liée à l'opérateur laplacien itéré 1

§ 1.2. - Harmonicité d'ordre infini 6

§ 1.3. - Distributions définissant des fonctions harmoniques d'ordre infini 14

§ 1.4. - Analyticité liée au signe de laplacien itéré 23

§ 1.5. - Applications 29

Chapitre 2 : Cellule d'harmonicité.

§ 2.1. - Propriétés générales - Chemins de P. Lelong 34

§ 2.2. - Exemples des cellules d'harmonicités 42

§ 2.3. - Boule de Lie et sa frontière de Bergman - Šilov . Le tube d'Elie Cartan 52

§ 2.4. - Prolongement analytique des fonctions harmoniques d'ordre quelconque 61

Chapitre 3 : Applications.

§ 3.1. - Fonctions harmoniques (d'ordre quelconque) arithmétiques 76

Bibliographie 84

INTRODUCTION

La complexification des fonctions harmoniques (resp. harmoniques d'ordre infini) dans un ouvert Ω de $\mathbb{R}^N$, a été abordé tout d'abord par N. ARONZAJN [3] et puis par P. LELONG notamment dans [19]. Son intérêt dépasse le cadre primitivement fixé, c'est-à-dire d'obtention des inégalités (pour ces classes de fonctions) en utilisant la technique de l'analyse complexe ou celle des fonctions plurisous harmoniques [17],[6] ; elle peut soulever des questions en analyse harmonique, tout particulièrement dans l'étude des domaines bornés symétriques homogènes, domaines de Siegel, etc [13] [16] [22] [23] .
Cette complexification liée au prolongement analytique permet aussi d'étudier les fonctions harmoniques d'ordre quelconque qui sont arithmétiques (i.e. $h(\mathbb{N}^N) \subset \mathbb{Z}$) et cela en utilisant les propriétés des fonctions entières arithmétiques dans $\mathbb{C}^N$ obtenues ces dernières années. Dans tout cela la notion de la cellule d'harmonicité introduite par N. ARONZAJN - P. LELONG [3] [18] [19] est fondamentale.

Ce travail comprend 3 chapitres.

Le chapitre 1 contient une étude des fonctions (ou distributions) harmoniques d'ordre infini dans un ouvert Ω de $\mathbb{R}^N$ et des propriétés d'analyticité liées à l'opérateur laplacien itéré Δ^m. Certains résultats de ce chapitre sont susceptibles d'être généralisés à des espaces Riemanniens symétriques de rang 1 ; ou bien dans $\mathbb{R}^N$, l'opérateur laplacien peut être remplacé par des opérateurs elliptiques et leurs itérés (voir par ex. [8] [9]).

Le chapitre 2 est consacré à la détermination de la cellule d'harmonicité des domaines tels que : cube, polyèdre convexe, demi-espace, boule et plus généralement des convexes de $\mathbb{R}^N$; on met en évidence que la cellule d'harmonicité de la boule unité de $\mathbb{R}^N$ (i.e. la boule de Lie dans $\mathbb{C}^N$), celle du demi-espace $\{x \in \mathbb{R}^N | x_N > o\}$ et le tube d'Elie Cartan $\{z = x + iy \in \mathbb{C}^N | y_N > \sqrt{y_1^2 + \ldots + y_{N-1}^2}\}$ sont analytiquement isomorphes. On retrouve par une démonstration directe et élémentaire la frontière de Bergman-Šilov de la boule de Lie. On montre que si u est une fonction harmonique d'ordre infini dans un ouvert étoilé Ω (resp. convexe) de $\mathbb{R}^N$, u se prolonge dans la cellule d'harmonicité de Ω.

Le chapitre 3 utilise le procédé de complexification et les résultats de [5],[1],[2] pour obtenir des énoncés concernant les fonctions harmoniques (d'ordre quelconque) arithmétiques. Par exemple : soit h harmonique arithmétique dans tout $\mathbb{R}^N$, vérifiant $|h(x)| \leq Ae^{B\|x\|}$: si $B < |\mathrm{Log}(\frac{3}{2} + i\frac{\sqrt{3}}{2})| = 0,7588\ldots$ alors h est un polynome. Ou bien, soit $f \in C^\infty(\mathbb{R}^N)$ vérifiant

$$|\Delta^m f(x)| \leq A(2m)!\, m^{-\alpha m} \exp(B\|x\|) \quad (A = \text{cte} \ ; \ B > o)$$

(f est harmonique d'ordre infini). Dans chaque cas suivant, f est un polynôme :

a - si $\alpha > 2$, $B < \text{Log } 2$ et $f(\mathbb{N}^N) \subset \mathbb{Z}$.

b - si $\alpha > 2$, $B < 1$ et $D^\nu f(o) \in \mathbb{N}$ pour tout $\nu \in \mathbb{N}^N$.

c - si $\alpha = 2$, $B < \text{Log}[\frac{1}{2}(3+\sqrt{5})] - \frac{2}{e} = 0{,}224\ldots$ et $f(\mathbb{Z}^N) \subset \mathbb{Z}$.

d - si $\alpha = 2$, $B + \frac{2}{e} < 1$ et $D^\nu f(o) \in \mathbb{N}$ pour tout $\nu \in \mathbb{N}^N$.

Ce travail a été exposé en Mai 1981, au colloque d'analyse complexe de Wimereux, en l'honneur du Professeur P. Lelong.

CHAPITRE 1

FONCTIONS HARMONIQUES D'ORDRE INFINI.

1.1. ANALYTICITÉ LIÉE A L'OPÉRATEUR LAPLACIEN ITÉRÉE.

Dans la classe $C^\infty_R(\Omega)$ des fonctions indéfiniment dérivables $f : \Omega \to R$ (Ω ouvert, non vide de R^N) les fonctions analytiques sont caractérisées habituellement par l'énoncé suivant :

1.1.1. Une fonction $f \in C^\infty_R(\Omega)$ est analytique dans Ω si et seulement si à tout compact $K \subset \Omega$ correspond un nombre $M_f(K) > 0$ tel que pour tout $\alpha \in \mathbb{N}^N$

$$(1) \qquad \max_{x \in K} |D^\alpha f(x)| \leq M_f(K)^{|\alpha|+1} \alpha!$$

Rappelons que $f : \Omega \to R$ est dite analytique si en tout point $a \in \Omega$ il existe une série multiple $\Sigma\, C_\alpha (x-a)^\alpha$ convergente dans un voisinage U_a de a et de somme égale à $f(x)$. D'après la théorie de familles sommables la série converge alors uniformément sur tout compact de U_a et on a $f \in C^\infty_R(\Omega)$, $C_\alpha = \frac{1}{\alpha!} D^\alpha f(a)$. Pour que $f \in C^\infty_R(\Omega)$ soit analytique, il faut et il suffit que dans le développement de Taylor de f en tout point $a \in \Omega$:

$$f(x) = \sum_{|\alpha| \leq k-1} \frac{1}{\alpha!} D^\alpha f(a)(x-a)^\alpha + \sum_{|\alpha| = k} \frac{1}{\alpha!} D^\alpha f(\xi)(x-a)^\alpha$$

$\xi \in [a,x]$, le reste

$$R_k = \sum_{|\alpha| = k} \frac{1}{\alpha!} D^\alpha f(\xi)(x-a)^\alpha$$

tende vers zéro dans un voisinage de a lorsque $k \to \infty$. C'est le cas notamment si (1) est vérifiée. En effet, dans un voisinage compact K de a, on a lors

$$|R_k| \leq k^N M(K)\,[M(K)|x-a|]^k \quad \text{où} \quad |x-a| = \operatorname{Max} |x_i - a_i|$$

Le second membre tend vers zéro dans l'ouvert $U_a = \{x \mid |x-a| < \frac{1}{M_f(K)}\} \subset K$. La réciproque se démontre en appliquant l'inégalité de Cauchy à la fonction $\hat{f}$ complexifiée de f définie ci-dessous dans la réunion $\hat{\Omega}$ des polycercles maximaux $U_a = \{z \in \mathbb{C}^N \mid |z-a| < r_a\}$, pour $a \in \Omega$.

1.1.2. LEMME. <u>L'espace $\mathbb{R}^N$ étant considéré comme un sous-espace fermé de $\mathbb{C}^N$, à tout couple (f,Ω) où f est analytique dans l'ouvert $\Omega \subset \mathbb{R}^N$, correspond un ouvert $\hat{\Omega} \subset \mathbb{C}^N$ dans lequel f se prolonge en $\hat{f}$ holomorphe. On a :</u>

$$\hat{\Omega} \cap \mathbb{R}^N = \Omega \qquad \hat{f}|_{\Omega} = f$$

<u>Démonstration</u>. Soit $S_a(x) = \Sigma\, C_\alpha (x-a)^\alpha$ la série convergente dans $\{x \mid |x-a| < r_a\}$ et de somme égale à $f(x)$. Si $U_a = \{z \in \mathbb{C}^N \mid |z-a| < r_a\}$, la série $S_a(z) = \sum_\alpha C_\alpha (z-a)^\alpha$ converge dans U_a et sa somme est une fonction holomorphe. Soient $\hat{U} = \bigcup_{a \in \Omega} U_a$ et $U_a \cap U_b = U_{a,b} \neq \emptyset$; on a $S_a(x) = S_b(x)$ dans $U_{a,b}$ puisque celui-ci est connexe. Si $U_{a,b} \neq \emptyset$ pour tout $c \in U_{a,b} \cap \mathbb{R}^N \neq \emptyset$, $D^\alpha S_a(c) = D^\alpha f(c) = D^\alpha S_b(c)$ et la fonction $\hat{f}(z)$ définie par $\hat{f}(z)|_{U_a} = S_a(z)$ répond à l'exigence du lemme grâce au principe du prolongement analytique. Cela étant, revenons à 1.1.1. Si K est un compact de Ω , soit $U \subset \mathbb{R}^N$ un voisinage de K d'adhérence $\overline{U}$ compacte dans Ω et $M_o = \sup_{\hat{U}} |\hat{f}(z)| < \infty$ (on peut supposer $\hat{U} \subset \hat{\Omega}$ compact) . D'après les inégalités de Cauchy, on a pour $z = x \in K$:

$$|D^\alpha \hat{f}(z)| = |D^\alpha f(x)| \leq M_o\, \alpha!\, \delta^{-|\alpha|} \qquad (\alpha \in \mathbb{N})$$

δ étant la distance de K au complémentaire de $\hat{U}$. D'où (1) avec $M(K)$ convenable.

Remarquer que si Ω est connexe, $\hat{\Omega}$ l'est aussi ; $\hat{\Omega}$ est la réunion des polycercles maximaux U_a de rayon égaux contenus dans le domaine d'holomorphie de f et centrés en des points $a \in \Omega$.

1.1.3. Dans l'énoncé 1.1.1. l'hypothèse de majoration concerne toutes les dérivées de f. On va raffiner cette condition en faisant intervenir l'opérateur laplacien itéré : soient

$$\Delta = \left(\frac{\partial}{\partial x_1}\right)^2 + \dots + \left(\frac{\partial}{\partial n_N}\right)^2$$

$$\Delta^{\circ} = I \ , \quad \Delta^m = \Delta(\Delta^{m-1}) \quad m = 1,2,\dots.$$

Si $f \in C^{\infty}_{\mathbb{R}}(\Omega)$, pour tout compact $K \subset \Omega$ on pose :

$$\text{(2)} \qquad M_m(f,K) = \sup_{x \in K} |\Delta^m f(x)| \qquad m = 0,1,2,\dots.$$

et on associe à (f,K) la série

$$\text{(3)} \qquad \sum_{m=o}^{\infty} \frac{M_m(f,K)}{(2m)!} z^m \qquad (z \in \mathbb{C})$$

Soit $R_f(K)$ le rayon de convergence de la série (3) :

$$R_f^{-1}(K) = \limsup_{m \to \infty} \left[\frac{M_m(f,K)}{(2m)!}\right]^{\frac{1}{m}}$$

1.1.4. THÉORÈME. <u>Une fonction</u> $f \in C^{\infty}_{\mathbb{R}}(\Omega)$ <u>est analytique dans</u> Ω <u>si et seulement si pour tout compact</u> $K \subset \Omega$, $R_f(K)$ <u>est non nul. Dans ce cas, à tout ouvert</u> Ω_o <u>d'adhérence compacte dans</u> Ω <u>correspond un intervalle ouvert</u> $I_{\Omega_o} \subset \mathbb{R}$ <u>centré en</u> 0 <u>et une fonction harmonique</u> $H_{\Omega_o}(x,t)$ <u>définie sur</u> $\Omega_o \times I_{\Omega_o}$ <u>tels que</u>

$$H_{\Omega_o}(x,0) = f(x)$$

$$H_{\Omega_o}(x,t) = M(x,-t) \ ; \quad (x,t) \in \Omega_o \times I_{\Omega_o} .$$

<u>Si</u> $R_f(K)$ <u>est indépendant de</u> K, <u>il existe un intervalle ouvert</u> $I \subset \mathbb{R}$ <u>centré en</u> 0 <u>et une fonction harmonique</u> $H(x,t)$ <u>vérifiant ces mêmes conditions dans</u> $\Omega \times I$.

Démonstration. Supposons f analytique. On a :

$$\Delta^m f(x) = \sum_{|\alpha|=m} \frac{m!}{\alpha!} D^{2\alpha} f(x)$$

Si $x \in K$, d'après (1), $|D^{2\alpha} f(x)| \leq M_1^{|2\alpha|+1}(K)(2\alpha)! \leq M_1^{2m+1}(K)|(2\alpha)!| = M_1^{2m+1}(K)(2m)!$

$M_1(K)$ étant une constante et $(2\alpha)! = (2\alpha_1)! \dots (2\alpha_N)!$

$$|(2\alpha)!| = (2\alpha_1 + \dots + 2\alpha_N)!$$

D'où $\sup_{x \in K} |\Delta^m f(x)| \leq M_1^{2m+1}(K)(2m)! \sum_{|\alpha|=m} \frac{m!}{\alpha!} = N^m M_1^{2m+1}(K)(2m)!$

(4) $\qquad M_m(f,K) \leq M^m(K)(2m)! \qquad$ ($M(K)$ convenable) $\quad m = 0,1,2,\dots$

et la série (3) a un rayon de convergence non nul.

La réciproque résulte du lemme suivant :

1.1.5. LEMME. Soient $f \in C^\infty_{\mathbb{R}}(\Omega)$ et $M_m = \sup_\Omega |\Delta^m f(x)| < \infty \quad m = 0,1,\dots$. Si la série $\sum_{m=0}^{\infty} \frac{M_m}{(2m)!} z^m$ a un rayon de convergence R non nul, la fonction f est analytique dans Ω. Précisément, il existe une fonction harmonique $H(x,t)$ dans $E = \Omega \times \{t \mid |t| < \sqrt{R}\}$ telle que

$$H(x,t) = H(x,-t); \quad f(x) = H(x,0) \quad ; \quad (x \in \Omega)$$

Démonstration. Introduisons les fonctions :

$$(5) \quad H(x,t) = f(x) + \sum_{n=1}^{\infty} (-1)^n \frac{t^{2n}}{(2n)!} \Delta^n f(x) \qquad (x,t) \in E$$

$$(6) \quad H_s(x,t) = f(x) + \sum_{n=1}^{s} (-1)^n \frac{t^{2n}}{(2n)!} \Delta^n f(x) \in C^\infty(\Omega \times \mathbb{R})$$

On a

$$|H(x,t) - H_s(x,t)| \leq \sum_{n=s+1}^{\infty} \frac{|t|^{2n}}{(2n)!} M_n \qquad |t| < \sqrt{R}$$

si $s \to \infty$, le second membre tend uniformément vers zéro pour $|t| \leq r < \sqrt{R}$

d'où la continuité de $H(x,t)$ dans E. Considérons dans $R^{N+1}(x,t)$ l'opérateur laplacien $\tilde{\Delta} = \frac{\partial^2}{\partial t^2} + \Delta_x$. La fonction $H(x,t)$ étant localement intégrable dans E elle définit une distribution $H \in \mathcal{D}'(E)$. Les calculs étant faits au sens des distributions, on a :

$$\tilde{\Delta} H(x,t) = \tilde{\Delta} f + \sum_{n=1}^{\infty} \tilde{\Delta} [(-1)^n \frac{t^{2n}}{(2n)!} \Delta_x^n f]$$

$$= \Delta f + \sum_{n=1}^{\infty} (-1)^n [\frac{t^{2n-2}}{(2n-2)!} \Delta_x^n f + \frac{t^{2n}}{(2n)!} \Delta_x^{n+1} f]$$

$$= \Delta f + \sum_{n=1}^{\infty} (-1)^n \frac{t^{2n-2}}{(2n-2)!} \Delta_x^n f + \sum_{n=2}^{\infty} (-1)^{n-1} \frac{t^{2n-2}}{(2n-2)!} \Delta_x^n f$$

$$= \sum_{n=2}^{\infty} (-1)^n \frac{t^{2n-2}}{(2n-2)!} \Delta_x^n f + \sum_{n=2}^{\infty} (-1)^{n-1} \frac{t^{2n-2}}{(2n-2)!} \Delta_x^n f = 0 .$$

Ainsi $H(x,t)$ étant continu sur E, représente en outre une distribution harmonique. Elle est donc une fonction harmonique au sens usuel dans E et on a $f(x) = H(x,t)|_{t=0}$ $(x \in \Omega)$. Le théorème 1.1.4. en résulte.

1.1.6. DÉFINITION. Une fonction $f : \Omega \to R$ est dite polyharmonique d'ordre p, ou p-harmonique, si $f \in C_R^{2p}(\Omega)$ et $\Delta^p f = 0$ dans Ω.

Soit f p-harmonique ; la série (3) correspondante converge pour tout z, elle est alors analytique dans Ω. Précisément, on a :

1.1.7. COROLLAIRE. Si f est p-harmonique dans Ω, il existe une fonction harmonique $H(x,t)$ dans $\Omega \times R$ telle que $f(x) = H(x,0)$. De plus, la fonction H est pour $x \in \Omega$ fixé, un polynôme en t^2 et de degré $\leq 2p-2$ en t. La fonction H est définie par (5).

§ 1.2. HARMONICITÉ D'ORDRE INFINI

1.2.1. Une classe intéressante des fonctions $f \in C^{\infty}_{\mathbb{R}}(\Omega)$ est constituée par les fonctions pour lesquelles la série (3) de 1.1.4. converge dans tout le plan complexe quel que soit le compact $K \subset \Omega$:

1.2.2. Exemple. Soit f une fonction analytique réelle qui est la restriction à $\mathbb{R}^N$ d'une fonction analytique complexe $\hat{f}$ dans tout $\mathbb{C}^N$. Pour une telle fonction, on a, si $a = (a_1,\dots,a_N) \in \mathbb{R}^N$ et $M(a,r) = \max\limits_{\substack{|z_j - a_j| \le r \\ 1 \le j \le N}} |\hat{f}(z_1,\dots,z_N)|$, l'inégalité de Cauchy :

$$|\Delta^m f(a)| = |\Delta_z^m \hat{f}(a)| \le (2m)!\ N^m M(a,r)\ r^{-2m}$$

Si a parcourt un compact $K \subset \mathbb{R}^N$, $\sup\limits_{a \in K} M(a,r) = M(K,r) < \infty$ et

$$\lim_{m \to \infty} \left[\sup_K \left| \frac{\Delta^m f(a)}{(2m)!} \right|\right]^{\frac{1}{m}} \le \frac{1}{r^2} \tag{7}$$

r étant arbitraire, le premier membre de (7) est nul, d'où la convergence de la série (3) quel que soit le compact $K \subset \mathbb{R}^N$.

Evidemment les fonctions harmoniques et les fonctions harmoniques d'ordre quelconque (1.1.6) dans Ω possèdent cette propriété.

1.2.3. DÉFINITION. Une fonction $f \in C^{\infty}_{\mathbb{R}}(\Omega)$ est dite harmonique d'ordre infini si la série (3) converge dans tout le plan complexe quel que soit le compact $K \subset \Omega$ (leur classe est notée $\mathcal{H}_{\infty}(\Omega)$).

Le théorème 1.1.4. implique :

1.2.4. COROLLAIRE. Une fonction harmonique d'ordre infini dans Ω est une fonction analytique. Précisément, il existe une fonction harmonique $H(x,t)$ dans $\Omega \times \mathbb{R}$ telle que $f(x) = H(x,t)|_{t=0}$ et $H(x,t) = H(x,-t)$.
La fonction H est donnée par la série (5).

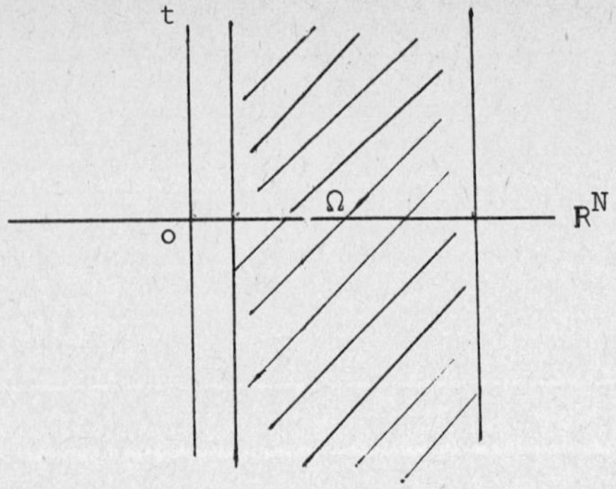

En effet, la série (5) de 1.1.5. converge uniformément sur tout compact $K_1(x) \times K_2(t) \subset \Omega \times \mathbb{R}$ et représente une fonction harmonique $H(x,t)$ vérifiant l'énoncé.

Remarque. Une fonction analytique réelle (même sur $\mathbb{R}^N$) n'appartient pas en général à $\mathcal{H}_\infty(\mathbb{R}^N)$.

Exemple 1 . $$f(x_1,\ldots,x_N) = \frac{1}{1+x_1^2} \qquad (x \in \mathbb{R})$$

$$\frac{\Delta^m f(x)}{(2m)!} = \frac{f^{(2m)}(x_1)}{(2m)!} = i\,\frac{(x_1-i)^{2m+1} - (x_1+i)^{2m+1}}{(x_1^2+1)^{2m+1}}$$

$$= \frac{\sum_{k=o}^{m} \binom{2m+1}{2k+1} (-1)^k x_1^{2(m-k)}}{(x_1^2+1)^{2m+1}}$$

$$\left| \frac{f^{(2m)}(0)}{(2m)!} \right| = 1$$ et la condition d'harmonicité d'ordre infini n'est pas vérifiée.

Exemple 2. La fonction $f(x) = \dfrac{1}{1-(x_1+\ldots x_N)}$ est analytique dans $\Omega = \mathbb{R}^N \setminus \{x | x_1+\ldots+x_N = 1\}$, mais elle n'est pas de la classe $\mathcal{H}_\infty(\Omega)$.

En effet :

$$\frac{\Delta^m f(x)}{(2m)!} = \frac{N^m}{1-(x_1+\ldots+x_N)^{2m+1}}$$

Si $K = \{0\}$ la série (3) ne converge pas dans tous le plan complexe :

1.2.5. PROPOSITION. La classe $\mathcal{H}_\infty(\Omega)$ des fonctions harmoniques d'ordre infini dans Ω est stable par dérivation.

La démonstration résulte immédiatement du lemme suivant :

1.2.6. LEMME. Soit $f \in C^\infty_{\mathbb{R}}(\Omega)$. A tout compact $K \subset \Omega$ et à tout voisinage ouvert $\Omega_o(K)$ de K d'adhérence compacte dans Ω correspondent deux constantes positives finies A et B ne dépendant pas de f (mais de K et de Ω_o) telles que :

$$(8) \qquad \operatorname*{Max}_K \left|\frac{\partial}{\partial x_i} \Delta^n f\right| \leq A \sup_{\Omega_o} |\Delta^n f| + B \sup_{\Omega_o} |\Delta^{n+1} f|$$

$n = 0,1,2,\ldots \quad 1 \leq i \leq N \quad (\Delta^0 f = f)$.

Démonstration. Il suffit d'établir (8) dans le cas $n = 0$.
Ecartons le cas $N = 1$ où (8) est une conséquence immédiate du développement de Taylor. Pour une grande analogie entre le cas du plan et de l'espace, plaçons-nous dans $\mathbb{R}^N$ $(N \geq 3)$.
Soit $K_o \subset \Omega_o$ un compact d'intérieur $\overset{\circ}{K}_o$ non vide égal à K si $\overset{\circ}{K} \neq \emptyset$ égal à un voisinage de K si $\overset{\circ}{K} = \emptyset$. Soit $\alpha(x) \in C^\infty_{\mathbb{R}}(\mathbb{R}^N)$ à support compact contenu dans Ω_o , $\alpha(x) = 1$ sur K_o et $0 \leq \alpha(x) \leq 1$ partout. Si la boule $B(x_o,\rho)$ de centre x_o et de rayon ρ a son adhérence dans K_o , de la formule de Green :

$$\int_D (u\,\Delta v - v\,\Delta u)\,d\xi = \int_{\partial D}\left(u\,\frac{\partial v}{\partial n_e} - v\,\frac{\partial U}{\partial n_e}\right) d\sigma$$

appliquée au domaine $D = \Omega_o \setminus \overline{B(x_o,\rho)}$ et aux fonctions

$$U = f \;, \quad V = V_x(\xi) = \alpha(x)\; h(x-\xi)$$

avec $$h(\xi) = \frac{1}{(N-2)\omega_N(1)} \frac{1}{\|\xi\|^{N-2}}$$

on obtient en faisant tendre ρ vers zéro :

$$f(x) = \int_{\Omega_o} f(\xi)\, \Delta_\xi[\alpha(\xi)h(x-\xi)]d\xi - \int_{\Omega_o} \alpha(\xi)h(x-\xi)\Delta f(\xi)d\xi \quad (x \in K) \,.$$

D'où

$$(9) \quad \frac{\partial f}{\partial x_i} = \int_{\Omega_o} f(\xi)\, \frac{\partial}{\partial x_i} \Delta_\xi[\alpha(\xi)h(x-\xi)]d\xi - \int_{\Omega_o} \alpha(\xi)\, \Delta_\xi f(\xi)\, \frac{\partial}{\partial x_i} h(x-\xi) \quad (x \in K) \,.$$

On remarquera que la première intégrale (9) se calcule sur

$$\mathrm{supp}\, \frac{\partial}{\partial x_i} \Delta_\xi[\alpha(\xi)h(x-\xi)] \subset \Omega_o \setminus K_o$$

donc est finie, et que la seconde intégrale est majorée par

$$\frac{1}{\omega_N(1)} \int_{\Omega_o} \frac{\Delta f(\xi)}{\|x-\xi\|^{N-1}}\, d\xi < \infty$$

Donc ,

$$\underset{K}{\mathrm{Max}} \left| \frac{\partial f}{\partial x_i} \right| \leq A(K,\Omega_o) \sup_{\Omega_o} |f(\xi)| + B(K,\Omega_o) \sup_{\Omega_o} |\Delta f(\xi)|$$

avec

$$A(K,\Omega_o) = \sup_{\substack{x \in K \\ 1 \leq i \leq N}} \int_{\Omega_o \setminus K_o} \left| \frac{\partial}{\partial x_i} \Delta_\xi[\alpha(\xi)h(x-\xi)] \right| d\xi$$

$$B(K,\Omega_o) = \sup_{x \in K} \frac{1}{\omega_N(1)} \int_{\Omega_o} \frac{d\xi}{\|x-\xi\|^{N-1}} \qquad (N \geq 3)$$

1.2.7. Particularité remarquable dans le cas $N = 1$.

La série de Mac-Laurin d'une fonction $f \in \mathcal{H}_\infty(]a,b[)$ en un point de l'intervalle $]a,b[\subset \mathbb{R}$ converge pour tout $x \in \mathbb{R}$.

<u>Précisément, un élément de</u> $\mathcal{H}_\infty(]a,b[)$ <u>est la restriction à</u> $\mathbb{R}$ <u>d'une fonction entière dans</u> $\mathbb{C}$.

<u>Démonstration</u>. Supposons f harmonique d'ordre infini dans un voisinage V de 0 . Considérons la série de Mac-Laurin de f :

$$\sum_{m=0}^{\infty} \frac{f^{(m)}(0)}{m!} x^m$$

Pour tout compact $K \subset V$ et tout $\varepsilon > 0$, il existe une constante $M(K,\varepsilon)$ telle que

$$\sup_K \left| \frac{f^{(2m)}(x)}{(2m)!} \right| \leq M(K,\varepsilon)\, \varepsilon^{2m} \qquad m = 0,1,2,\ldots$$

Or $f' = \frac{df}{dx}$ est aussi harmonique d'ordre infini dans V (1.2.5) donc

$$\sup_{x\in K} \left| \frac{d^{2m}}{dx^{2m}} \frac{f'(x)}{(2m)!} \right| = \sup_{x\in K} \left| \frac{f^{(2m+1)}(x)}{(2m+1)!}(2m+1) \right| \leq M_1(K,\varepsilon)\varepsilon^{2m}$$

il en résulte que :

$$\left| \frac{f^{(m)}(0)}{m!} \right| \leq M_2(\varepsilon)\, \varepsilon^{2E[\frac{m}{2}]} \quad (E[x] = \text{partie entière de } x\ ) \quad (M_2 = \text{cte}) .$$

D'où

$$\limsup_{m\to\infty} \frac{1}{m} \operatorname{Log} \left| \frac{f^{(m)}(0)}{m!} \right| = \operatorname{Log} \frac{1}{R} = \operatorname{Log} \varepsilon$$

$\varepsilon > 0$ étant arbitraire, $R = \infty$.

La fonction $\hat{f}(z) = \sum_{m=0}^{\infty} \frac{f^{(m)}(0)}{m!} z^m$ est entière avec $\hat{f}|_{\mathbb{R}} = f$.

1.2.8. THÉORÈME.

a) <u>Soit</u> $f \in \mathcal{H}_\infty(\Omega)$. <u>Pour toute boule</u> $B(x,R)$ <u>d'adhérence compacte dans</u> Ω , <u>la moyenne</u> $\lambda(f,x,R)$ <u>de</u> f <u>sur</u> $\partial B(x,R)$ <u>s'écrit</u> :

$$(10) \qquad \lambda(f,x,R) = f(x) + \sum_{m=1}^{\infty} a_m R^{2m} \Delta^m f(x)$$

avec

$$\lambda(f,x,R) = \frac{1}{\omega_N(1)} \int_{\|a\|=1} f(x+Ra)d\sigma(a) \qquad \text{si} \quad N \geq 2$$

$$\lambda(f,x,R) = \frac{f(x+R) + f(x-R)}{2} \qquad \text{si} \quad N = 1$$

$$a_m = \frac{\Gamma(\frac{N}{2})}{2^{2m}m!\Gamma(m+\frac{N}{2})} = \frac{1}{2^m m!\ N(N+2)\ldots(N+2m-2)}$$

Pour R fixé, la série figurant sans (10) converge absolument en tout point x, et uniformément en x sur tout compact $K \subset \Omega$ si on fixe $R < \delta(K)$ = distance de K au complémentaire de Ω.

b) Si f est analytique, on a la représentation (10) avec convergence absolue pour $R = R(x)$ suffisamment petit. De plus, pour tout compact $K \subset \Omega$ il existe $\rho_K > 0$ tel que la convergence soit uniforme en x sur K si on fixe $R < \rho_K$.

c) Pour f, p-harmonique, tous les termes de la série (10) à partir $m \geq p$ disparaissent.

Démonstration.

1.2.9. Développement de Pizetti : soit $f \in C^{2p}_{\mathbb{R}}(\Omega)$. On a pour toute boule $B(x,R)$ d'adhérence compacte dans Ω le développement :

$$(11) \qquad \lambda(f,x,R) = f(x) + \sum_{m=1}^{p-1} a_m R^{2m}\Delta^m f(x) + a_p R^{2p}\Delta^p f(\xi) \quad \text{où} \quad \xi \in \overline{B(x,R)}\ .$$

Démonstration.

1.2.10. LEMME. Si $f \in C^{2p}_{\mathbb{R}}[\overline{B(0,R)}]$ et $\lambda(f,0,R) = f(0) = \Delta f(0) = \ldots = \Delta^{p-1}f(0) = 0$ alors il existe $\xi \in B(0,R)$ tel que $\Delta^p f(\xi) = 0$.

En effet :

Soit $f \in C^2_{\mathbb{R}}[B(0,R)]$ et $\lambda(f,0,R) = f(0) = 0$.

Si $\lambda(t) = \lambda(f,0,t)$ $t \in [0,R]$, on a $\lambda(t) \in C^1[0,R]$ et $\lambda(0) = \lambda(R) = 0$ donc il existe $t_o \in]0,R[$ tel que $\lambda'(t_o) = 0$. D'autre part, d'après la formule de Green :

$$0 = \lambda'(t_o) = \frac{d}{dt}\left(\frac{1}{\omega_N(1)}\int_{\|a\|=1} f(ta)d\sigma(a)\right)_{t=t_o} = \left[\frac{1}{\omega_N(1)}\int_{\|a\|=1}\frac{\partial f}{\partial n_e}(ta)d\sigma(a)\right]_{t=t_o}$$

$$= \frac{1}{\omega_N(t_o)}\int_{\|a\|=t_o}\frac{\partial f}{\partial n_e}(a)d\sigma(a) = \frac{1}{\omega_N(t_o)}\int_{B(0,t_o)}\Delta f(x)dx$$

Δf étant continue, il existe nécessairement un point $\xi \in B(0,R)$ tel que $\Delta f(\xi) = 0$.

Soit $f \in C^4_{\mathbb{R}}[B(\overline{0,R})]$ avec $\lambda(f,0,R) = f(0) = \Delta f(0) = 0$. D'après le résultat précédent, il existe $t_1 \in]0,R[$ tel que $\int_{B(0,t_1)}\Delta f(x)dx = 0$. D'où

$$\int_0^{t_1}\left[\int_{\|a\|=t}\Delta f(x)d\omega(x)\right] t^{N-1}dt = 0$$

Donc, il existe $R_1 > 0$ tel que $\int_{\|a\|=R_1}\Delta f(x)d\omega(x) = 0$. Posons $\lambda_1(f) = \lambda(\Delta f,0,t)$ pour $t \in [0,R_1]$; on a $\lambda_1(0) = 0$, $\lambda_1(R_1) = 0$ puisque $\Delta f(0) = 0$ $\lambda_1(R_1) = 0$. La formule de Green montre que

$$\int_{B(0,t_1)}\Delta^2 f(x)dx = 0$$

D'où l'existence de $\xi \in B(0,R)$ tel que $\Delta^2 f(\xi) = 0$. On achève la démonstration par récurrence.

<u>Démonstration de</u> 1.2.9. Sans restreindre de la généralité, supposons $x = 0$.

Soient $f \in C^{2p}_{\mathbb{R}}[\overline{B(0,R)}]$ et h la fonction harmonique dans $B(0,R)$ égale à f sur $\partial B(0,R)$. Posons :

$$g(x) = f(x)-h(x) + \sum_{k=1}^{p-1} a_k(R^{2k} - \|x\|^{2k})\Delta^k f(0) + b(R^{2p} - \|x\|^{2p})$$

où $b = b(R)$ est une constante et

$$a_k^{-1} = \Delta^k\|x\|^{2k} = \Delta^k r^{2k} = 2^k.\ k!\ (N+2k-2)\ldots(N+2)\ N .$$

$k = 1,\ldots, p-1$.

On a $\qquad \Delta^m\|x\|^{2k} = \frac{a_{m+k}}{a_k}\|x\|^{2(k-m)} \qquad$ si $m \geq k$.

D'autre part

$$\lambda[g,0,R] = \lambda[f,o,R] - \lambda[h,0,R] + \sum_{k=1}^{p-1} a_k \lambda[R^{2k} - \|x\|^{2k},0,R]\ \Delta^k f(0)$$

$$+ b\ \lambda(R^{2p} - \|x\|^{2p},0,R) = 0$$

et pour $1 \leq m \leq p-1$:

$$\Delta^m g(0) = \Delta^m f(0) - a_m(\Delta^m\|x\|^{2k})(0)\ \Delta^m f(0) = 0$$

choisissons b de sorte que

$$g(0) = f(0) - \lambda(f,0,R) + \sum_{k=1}^{p-1} a_k R^{2k} \Delta^k f(0) + b\,R^{2p} = 0 .$$

(Remarquer que $\lambda(f,0,R) = h(0)$). Alors g vérifie les hypothèses du lemme 1.2.10 . Il existe $\xi \in B(0,R)$ tel que $\Delta^p g(\xi) = 0$ cela implique : $\Delta^p f(\xi) - b\ a_p^{-1} = 0$ d'où $b = a_p \Delta^p f(\xi)$ soit en remplaçant dans (13) :

$$\lambda(f,0,R) = f(0) + \sum_{k=1}^{p-1} a_k R^{2k} \Delta^k f(0) + a_p R^{2p} \Delta^p f(\xi)$$

D'où le développement (11) .

Démonstration de 1.2.8.

Pour établir (10) dans le cas a) considérons un R-voisinage compact de K . Pour tout $\epsilon > 0$ il existe une constante $M(K_1,\epsilon)$ telle que

$$|\Delta^m f(x)| \leq M(K_1,\epsilon)\ \epsilon^{2m}(2m)! \quad (x \in K_1) \qquad m = 0,1,2,\ldots$$

et

$$\sup_{x\in K} |a_m|\ |\Delta^m f(x)|R^{2m} \leq \frac{\Gamma(\frac{N}{a})(2m)!\,\epsilon^{2m}R^{2m}}{2^{2m}m!\ \Gamma(m+\frac{N}{2})}\ M(K_1,\epsilon)$$

Or, $\qquad m! \sim \sqrt{2\pi}\ m^{m+\frac{1}{2}}\ e^{-m} \quad (m \to \infty)$

$$\Gamma(m+\tfrac{N}{2}) \sim \Gamma(m)m^{\frac{m}{2}} \sim \sqrt{2\pi}(m-1)^{m-\frac{1}{2}}\ m^{\frac{N}{2}} .\ e^{-(m-1)}$$

$$\frac{(2m)!}{2^{2m}m!\Gamma(m+\frac{N}{2})} = \frac{1}{\sqrt{\pi}\, m^{\frac{N-1}{2}}} + \theta_N(m) \qquad (\lim_{m\to\infty} \theta_n(m) = 0 \ , \ \theta_1(m) = 0 \ )$$

Dans (11) le reste $a_p R^{2p}\Delta^p f(\xi)$ est majoré pour toute boule $\overline{B(x,R)} \subset \Omega$, $x \in K$ par

(12) $$\Gamma(\tfrac{N}{2})\, M(K_1,\epsilon) \frac{1}{\sqrt{\pi}\, p^{\frac{N-1}{2}}} + \theta_N(p))(\epsilon^2 R^2)^p$$

l'expression qui tende vers zéro, quand $p \to \infty$ si on choisit $\epsilon < \frac{1}{R}$.

Dans le cas b) on a $|\Delta^m f(x)| \leq M(K,\epsilon)(\rho_K+\epsilon)^m(2m)!$ $(x \in K)$ $\rho_K > 0$ et, dans l'expression analogue à (12), $(\epsilon^2 R^2)^p$ est remplacée par

$$[(\rho_K+\epsilon)^2 R^2]^p \quad \text{d'où l'obligation du choix} \quad R < \frac{1}{\rho+\epsilon}$$

dans les boules $B(x,R)$ $(x \in K)$ pour obtenir la convergence uniforme sur K.

Remarque. L'exemple qui suit le corollaire 1.2.4. montre que pour les fonctions analytiques la convergence de la série figurant dans (10), n'est pas assurée pour toute boule fermée contenue dans le domaine de définition contrairement aux fonctions harmoniques d'ordre infini.

§ 1.3. DISTRIBUTIONS DEFINISSANT FONCTIONS HARMONIQUES D'ORDRE INFINI [7]

1.3.1. Dans ce paragraphe, nous étudions les distributions qui vérifient une condition rappelant celle figurant dans 1.2.3. On montrera que ces distributions sont en réalité fonctions harmoniques d'ordre infini.

Nous rappelons tout d'abord quelques énoncés bien connus de la théorie des distributions. Pour tout ouvert Ω de R^N, on note $\mathcal{D}(\Omega)$ l'espace des fonctions indéfiniment dérivables à valeurs réelles ou complexes et à support compact dans Ω. Pour toute partie K de R^N, $\mathcal{D}_K$ désigne la partie de $\mathcal{D} = \mathcal{D}(\mathbb{R}^N)$ constituée par les fonctions dont le support est contenu dans K. La topologie de $\mathcal{D}_K$ étant définie par un système fondamental (N_p) indexée par $\mathbb{N}^N$ des semi-normes définies par

$$N_p(\varphi) = \sup_{x \in K} \left| \frac{\partial^p \varphi(x)}{\partial x^p} \right| \qquad \varphi \in \mathcal{D}_K$$

$\mathcal{D}(\Omega)$ sera muni de la topologie de la limite inductive des $\mathcal{D}_K$ où K parcourt une suite croissante de parties compactes de Ω dont les intérieurs $\overset{o}{K}$ recouvrent Ω. L'espace des distributions sur Ω muni de sa topologie usuelle est noté $\mathcal{D}'(\Omega)$ ou $\mathcal{D}'$ s'il n'y a pas de confusion possible. Une distribution est dite fonction indéfiniment dérivable (resp. analytique) dans Ω s'il existe une fonction f au sens usuel, indéfiniment dérivable (resp. analytique) dans Ω telle que la distribution définie par f soit égale à T ($< T,\varphi > = \int f\,\varphi$ pour tout $\varphi \in \mathcal{D}(\Omega)$). Une telle fonction est évidemment unique. On identifie alors T et f. Dans la suite, il sera commode d'utiliser une autre base de voisinages pour la topologie de $\mathcal{D}_K$ celle-ci sera définie en effet, par la famille de semi-normes $\varphi \longmapsto \|\Delta^m \varphi\|_{L^2}$ $(m \in \mathbb{N})$. On fera usage des énoncés suivants fort connus en théorie de distribution.

- Une distribution dont toutes les dérivées successives sont des mesures est une fonction indéfiniment dérivable au sens usuel.

- Les distributions T vérifiant $\Delta^p T = 0$ sont des fonctions indéfiniment dérivables (précisément sont p-harmoniques).

Précisons que les fonctions définies sur $\Omega \subset \mathbb{R}^N$ à valeurs complexes sont dites indéfiniment dérivables (resp. analytiques) si leurs parties réelles et complexes le sont.

1.3.2. DÉFINITION. Soient Ω un ouvert non vide de $\mathbb{R}^N$ et $\alpha = (\alpha_m)_{m \in \mathbb{N}}$ une suite de nombres réels strictement positifs. Une distribution $T \in \mathcal{D}'(\Omega)$ est dite de la classe $A(\alpha,\rho,\Omega)$ si la série

(13) $$\sum_{m=0}^{\infty} \frac{<\Delta^m T,\varphi>}{\alpha_m} z^m \quad (z \in \mathbb{C})$$

<u>a pour tout</u> $\varphi \in \mathcal{D}(\Omega)$ <u>un rayon de convergence</u> $R(\varphi)$ <u>avec</u>

$$\rho = \inf_{\varphi} R(\varphi) > 0$$

<u>On remarquera que</u> $A(\alpha,\rho,\Omega)$ <u>est stable par dérivation</u>

1.3.3. DÉFINITION. <u>Une distribution</u> $T \in \mathcal{D}'(\Omega)$ <u>est dite harmonique d'ordre infini, si la série</u>

(14) $$\sum_{m=0}^{\infty} \frac{<\Delta^m T,\varphi>}{(2m)!} z^m \quad (z \in \mathbb{C})$$

converge dans tout le plan complexe quel que soit $\varphi \in \mathcal{D}(\Omega)$

1.3.4. THÉORÈME.

1) <u>Une distribution</u> $T \in A(\alpha,\rho,\Omega)$ <u>est une fonction indéfiniment dérivable dans</u> Ω . <u>De plus</u> :

2) <u>Si</u> $\alpha = ((2m)!)_{m\in \mathbb{N}}$, T <u>supposée réelle, est une fonction analytique dans</u> Ω .

3) <u>Si</u> $\alpha = ((2m)!)_{m\in \mathbb{N}}$ <u>et</u> $\rho = \infty$, T <u>supposée réelle, est une fonction harmonique d'ordre infini</u>.

La démonstration de 1.3.4. sera la conséquence des énoncés qui vont suivre.

1.3.5. Pour tout élément $\varphi \in \mathcal{D}$ on pose :

$$\|\varphi\|_\infty = \sup_x |\varphi(x)|$$

$$\|\varphi\|_{L^2} = [\int |\varphi(x)|^2 dx]^{\frac{1}{2}}$$

$$\Lambda^m\varphi = \frac{\partial^{mN}}{\partial x_1^m \cdots \partial x_N^m} \quad (\Lambda^0\varphi = \varphi) \quad m \in \mathbb{N} .$$

Pour $m \in \mathbb{N}$, Q_m note la forme sesquilinéaire sur $\mathcal{D} \times \mathcal{D}$ définie par

$$(\varphi_1, \varphi_2) \longmapsto \int \Lambda^m \varphi_1 . \overline{\Lambda^m \varphi_2} dx$$

Remarquons que la famille

$$\| \Lambda^m(\varphi)\|_{L^2} = [\int |\Lambda^m \varphi|^2 dx]^{\frac{1}{2}} \quad (m \in \mathbb{N})$$

est une famille de semi-normes continue sur $\mathcal{D}$ et que la famille de leurs carrés $(Q_m(\varphi))_{m \in \mathbb{N}}$ est une famille de formes quadratiques positives continues sur $\mathcal{D}$.

1.3.6. LEMME. <u>Si dans</u> $\mathbb{R}^N$ <u>le cube</u>

$$\{ x \in \mathbb{R}^N | \ |x_j| < a, \ j = 1,\dots,N \}$$

<u>contient le compact</u> K <u>alors pour tout</u> $\varphi \in \mathcal{D}_K$ <u>on a</u> :

1) $$\|\varphi\|_{L^2} \leq a^{\frac{N}{2}} \|\varphi\|_\infty$$

2) $$\|\varphi\|_\infty \leq a \| \frac{\partial}{\partial x_j} \varphi\|_\infty \qquad (1 \leq j \leq N)$$

3) $$\|\varphi\|_\infty \leq a^{\frac{N}{2}} \|\Lambda^1 \varphi\|_{L^2}$$

<u>et pour tout</u> $m \in \mathbb{N}$:

4) $$\|\Lambda^m \varphi\|_{L^2} \leq a^N \|\Lambda^{m+1} \varphi\|_{L^2}$$

5) $$\|\Lambda^{2m} \varphi\|_{L^2} \leq (2\pi)^{2m(N-1)} \|\Delta^m \varphi\|_{L^2}$$

<u>Démonstration</u>. Les inégalités 1) et 2) résultent de la formule de la moyenne. Pour établir 3) soit m un entier ≥ 0 , considérons une famille orthonormale d'éléments de $\mathcal{D}_K$ pour Q_{m+1} , et pour $t \in \mathbb{R}^N$ fixé, la fonction $x \longmapsto h_t(x)$ défini sur $\mathbb{R}^N$ telle que

$$h_t(x) = \begin{cases} 1 & \text{si } x_1 \geq t_1, \dots, x_N \geq t_N \\ 0 & \text{ailleurs} \end{cases}$$

alors pour tout $\varphi \in \mathcal{D}_K$ on a

$$\varphi(t) = \int \Lambda^1\varphi(x)\, h_t(x)dx$$

on en déduit l'inégalité 3) par application de l'inégalité de Schwartz.

L'inégalité 4) résulte de 1) et 3) (il suffit de la vérifier pour $m = 0$) .

Pour établir 5) désignons par $\mathfrak{F}$ la transformation de Fourier. On a :

$$\|\Lambda^{2m}\varphi\|^2_{L^2} = \int \Big| \frac{\partial^{2mN}\varphi}{\partial x_1^{2m}\dots\partial x_N^{2m}}\Big|^2 dx = \int \Big| \mathfrak{F}\, \frac{\partial^{2mN}\varphi}{\partial x_1^{2n}\dots\partial x_N^{2m}}\Big|^2 dy$$

$$= \int \prod_{j=1}^{N} |2\pi y_j|^{4m} |\mathfrak{F}\varphi|^2 dy \le (2\pi)^{4m(N-1)} \int \Big[\sum_{j=1}^{N} (2\pi y_j)^2\Big]^{2m} |\mathfrak{F}\varphi|^2 dy$$

$$= (2\pi)^{4m(N-1)} \int |\mathfrak{F}\Delta^m\varphi|^2 dy = (2\pi)^{4m(N-1)} \int |\Delta^m\varphi|^2 dy$$

$$= (2\pi)^{4m(N-1)} \, \| \Delta^m\varphi \|^2_{L^2}$$

1.3.7. <u>Δ - boules.</u>

Les inégalités 2) et 3) montrent que toute semi-norme de la famille $(N_p)_{n\in \mathbb{N}^N}$ est majorée par une semi-norme de la forme $\varphi \longmapsto \|\Lambda^m\varphi\|_{L^2}$.
La famille de ces semi-normes engendre donc la topologie usuelle de $\mathcal{D}_K$.
L'inégalité 4) montre que pour toute famille finie (m_i, a_i) $(m_i \in \mathbb{N}, a_i \in \mathbb{R}_+^*)$ il existe un couple $(m,a) \in \mathbb{N} \times \mathbb{R}_+^*$ tel que

$$\{\varphi \in \mathcal{D}_K \mid \|\Lambda^m\varphi\|_{L^2} \le a\} \subset \bigcap_i \{\varphi \in \mathcal{D}_K \mid \|\Lambda^{m_i}\varphi\|_{L^2} \le a_i\}$$

<u>Cela montre que l'ensemble des Λ - boules</u>

$$\{\varphi \in \mathcal{D}_K \mid \|\Lambda^m\varphi\|_{L^2} \le \alpha\} \quad \alpha > 0 \text{ , } m \in \mathbb{N}$$

constitue une base de voisinage de l'origine dans $\mathcal{D}_K$. Les inégalités 4) et 5) permettent alors d'utiliser comme base les Λ - boules d'indice pair, puis les Δ-boules

$$\{\varphi \in \mathcal{D}_K \mid \|\Delta^m\varphi\|_{L^2} \le \alpha\} \quad m \in \mathbb{N}, \alpha > 0 .$$

D'où

1.3.8. PROPOSITION. <u>Pour tout voisinage</u> V <u>de l'origine dans</u> $\mathcal{D}_K$ <u>il existe un couple</u> (p_o,c) , $p_o \in \mathbb{N}$, $c \in \mathbb{R}_+^*$ <u>tel que</u>

$$\{\varphi \in \mathcal{D}_K | \, \|\Delta^{p_o}\varphi\|_{L^2} < c\} \subset V .$$

1.3.9. LEMME. <u>Soit</u> T <u>une distribution de la classe</u> $A(\alpha,\rho,\Omega)$ <u>définition</u> 1.3.) <u>alors pour tout</u> $\sigma > \frac{1}{\rho}$, <u>la suite</u> $((\alpha_m\sigma^m)^{-1}\Delta^m T)_{m \in \mathbb{N}}$ <u>est bornée dans</u> $\mathcal{D}'(\Omega)$. <u>Pour tout ouvert</u> Ω_o <u>d'adhérence compacte dans</u> Ω , <u>il existe un couple</u> $(p_o,a) \in \mathbb{N} \times \mathbb{R}_+^*$ <u>tel que</u> :

$$(15) \qquad | < \Delta^m T,\varphi > | \le a\alpha_m\sigma^m \|\Delta^{p_o}\varphi\|_{L^2}$$

<u>quels que soient</u> $\varphi \in \mathcal{D}(\bar{\Omega}_o)$ et $m \in \mathbb{N}$.

<u>Démonstration</u>. Pour tout $m \in \mathbb{N}$, considérons l'application $R_{\sigma,m} : \mathcal{D}(\Omega) \to \mathbb{R}_+$

$$R_{\sigma,m}(\varphi) = \sum_{p=o}^{m} (\alpha_p\sigma^p)^{-1} | < \Delta^p T,\varphi > |$$

La suite $R_{\sigma,m}(\varphi))_{m \in \mathbb{N}}$ converge en croissant vers une limite $R_\sigma(\varphi)$ en vertu de (13)

La suite $((\alpha_m\sigma^m)^{-1} < \Delta^m T,\varphi >)_{m \in \mathbb{N}}$ a alors tous ses termes majorés par $R_\sigma(\varphi)$ cela prouve que la suite forme un ensemble borné dans $\mathcal{D}'(\Omega)$ et qu'il existe un nombre M et un voisinage V de 0 dans $\mathcal{D}(\Omega)$ tels que

$$| < \Delta^m T,\varphi | \le \alpha_m\sigma^m M$$

pour tout $\varphi \in V$, $m \in \mathbb{N}$. Appliquons la proposition 1.3.8. à la trace de V sur $\mathcal{D}(\bar{\Omega}_o)$: il existe un couple $(p_o,a') \in \mathbb{N} \times \mathbb{R}$ tel que

$$(\varphi \in \mathcal{D}_{\bar{\Omega}_o} \text{ et } \|\Delta^{p_o}\varphi\|_{L^2} \le a') \Rightarrow$$

$$(|<\Delta^m T,\varphi > | \le \alpha_m\sigma^m M \text{ pour tout } m \in \mathbb{N})$$

Par homogéneité, on obtient (15) avec $a = \alpha M$.

1.3.10. COROLLAIRE. Soient $T \in A(\alpha,\rho,\Omega)$, $\sigma > \frac{1}{\rho}$ et K un compact de Ω . Il existe un voisinage V de l'origine dans $\mathbb{R}^N$ et un couple $(p_o,a) \in \mathbb{N} \times \mathbb{R}_+^*$ tels que

$$|(\Delta^m T * \varphi)(x)| \leq a\alpha_m \sigma^m \|\Delta^{p_o}\varphi\|_{L^2}$$

pour tout $\varphi \in \mathcal{D}(V)$, $x \in K$, $m \in \mathbb{N}$.

Démonstration. On choisit un voisinage V d'adhérence compacte de l'origine dans $\mathbb{R}^N$ tel que $\Omega_o = K \setminus V$ soit un ouvert d'adhérence compacte dans Ω . Soient $x \in K$ et $\varphi \in \mathcal{D}$ supp $\varphi \subset V$; la fonction $t \longmapsto \varphi(x-t)$ a son support dans Ω_o . L'inégalité (15) s'écrit :

$$(16) \qquad |(\Delta^m T * \varphi)(x)| \leq |<\Delta^m T_t,\varphi(x-t)>| \leq a\alpha_m \sigma^m \|\Delta^{p_o}\varphi(x-t)\|_{L^2}$$
$$= a\alpha_m \sigma^m \|\Delta^{p_o}\varphi\|_{L^2}$$

D'où le résultat.

1.3.11. LEMME. Toute distribution $T \in A(\alpha,\rho,\Omega)$ est une fonction (au sens usuel) indéfiniment dérivable.

Démonstration. Soient $\sigma > \frac{1}{\rho}$, Ω_o un ouvert d'adhérence compacte dans Ω . Soit $\mathcal{H}$ le sous-espace de $L^2(\mathbb{R}^N)$ engendré algébriquement par les $\Delta^{p_o}\varphi$ où φ parcourt $\mathcal{D}(\Omega_o)$ et p_o étant défini dans la proposition 1.3.8. . L'inégalité (15) appliquée à $n = p_o$ montre que T est une forme linéaire continue sur $\mathcal{H}$ pour la topologie induite par celle de $L^2(\mathbb{R}^N)$. Elle peut donc se prolonger en une forme linéaire $\hat{T}$ continue sur $L^2(\mathbb{R}^N)$. D'après les propriétés de cet espace, $\hat{T}$ est défini par une fonction $f \in L^2(\mathbb{R}^N)$; pour tout $\varphi \in \mathcal{D}(\Omega_o)$ on a alors (chaque terme en indice précise la dualité définissant la forme sesquilinéaire utilisée) :

$$< \Delta^{p_o} T, \varphi >_{\mathcal{D}',\mathcal{D}} < T, \Delta^{p_o}\varphi >_{\mathcal{D}',\mathcal{D}} = < T, \Delta^{p_o}\varphi >_{\mathcal{H}',\mathcal{H}} = < \hat{T}, \Delta^{p_o}\varphi >_{L^2(\mathbb{R}^N),(L^2(\mathbb{R}^N)}$$

$$= \int f(x)\Delta^{p_o}\varphi(x)dx = < f, \Delta^{p_o} >_{\mathcal{D}',\mathcal{D}} = < \Delta^{p_o} f, \varphi >_{\mathcal{D}',\mathcal{D}}$$

d'où $\Delta^{p_o}(T-f) = 0$ dans Ω_o au sens des distributions. Cela implique que $T-f$ est une fonction et que T est une fonction localement intégrable dans Ω_o. La stabilité de $A(\alpha,\rho,\Omega)$ par dérivation montre qu'il en est de même pour toutes les dérivées de T. On en déduit que T est dans Ω_o une fonction indéfiniment dérivable au sens usuel (1.3.1.) ; d'où le résultat en considérant une suite croissante d'ouverts d'adhérence compacte et de réunion Ω.

1.3.12. PROPOSITION. <u>Soit</u> $T \in A(\alpha,\rho,\Omega)$ <u>pour toute partie compacte</u> K <u>de</u> Ω <u>et tout</u> $\sigma > \frac{1}{\rho}$, <u>il existe un nombre</u> $c = c(K,\sigma)$ <u>et un entier</u> $k = k(K,\sigma)$ positifs tels que :

$$(17) \qquad |(\Delta^m T)(x)| \leq c\, \sigma^m(\alpha_m + \alpha_{m+k}\sigma^k)$$

pour tout $x \in K$, $m \in \mathbb{N}$.

<u>Démonstration</u>. Soient V un voisinage de l'origine d'adhérence compacte dans $\mathbb{R}^N$ et (p_o,a) $p_o \in \mathbb{N}$, $a \in \mathbb{R}_+^*$ tels qu'on puisse appliquer aux données de l'énoncé l'inégalité (16). Soient de plus k un entier $> p_o$, $\gamma \in \mathcal{D}(V)$ égale à 1 sur un voisinage de l'origine de Ω et E_k la solution élémentaire de l'opérateur Δ^k, un argument élémentaire de prolongement montre que γE_k qui a son support dans V est assez régulière pour qu'on puisse la substituer à φ dans l'inégalité (16) ; comme par ailleurs la différence $(\Delta^k(\gamma E_k) - \delta)$ est une fonction ζ indéfiniment dérivable à support compact contenu dans V à laquelle on peut appliquer l'inégalité (16) on obtient

$$|(\Delta^m T)(x)| = |(\Delta^m T * \delta)(x)| = |(\Delta^m T * \Delta^k(\gamma\, E_k))(x) - (\Delta^m T * \zeta)(x)$$

$$\leq |(\Delta^{m+k} T * \gamma\, E_k)(x) + |(\Delta^m T * \zeta)(x)| \leq$$

$$\leq a[\alpha_{m+k}\sigma^{m+k}\|\Delta^{P_o}(\gamma\, E_k)\|_{L^2} + \alpha_m\sigma^m\|\Delta^{P_o}\zeta\|_{L^2}]$$

$$\leq \sup(\|\Delta^{P_o}(\gamma\, E_k)\|_{L^2}, \|\Delta^{P_o}\zeta\|_{L^2})\; a(\alpha_m\sigma^m + \alpha_{m+k}\sigma^{m+k}$$

$$\leq c\, \sigma^m(\alpha_m + \alpha_{m+k}\sigma^k)$$

Démonstration du théorème 1.3.4. La partie 1 , c'est le lemme 1.3.8.

Si $\alpha = ((2m)!_{m\in\mathbb{N}}$, pour entier k et $\sigma > \frac{1}{\rho}$ fixés, on a

$$\limsup_{m\to\infty} \left[1 + \frac{\alpha_{m+k}}{\alpha_m} \sigma^k \right]^{\frac{1}{m}} < \infty$$

Donc, d'après la proposition 1.3.12. , pour tout compact $K \subset \Omega$ et $\sigma > \frac{1}{\rho}$ fixé, on peut trouver une constante $M(K,\sigma) > 0$ telle que

$$(18) \qquad |(\Delta^m T)(x)| \leq M^m(K,\sigma)(2m)! \qquad (m \in \mathbb{N}) \qquad (x \in K)$$

d'où l'analycité de T (Th. 1.1.4.) .

La propriété 3 résulte du fait que $\sigma = \epsilon$ peut être choisi arbitrairement petit, auquel cas, la constante $M(K,\sigma)$ de (18) sera de la forme $M_1(K,\sigma)\epsilon$ ce qui implique l'harmonicité d'ordre infini de T (1.2.3.) .

1.3.13. COROLLAIRE. Une fonction $f \in C^\infty_{\mathbb{R}}(\Omega)$ est harmonique, d'ordre infini, si et seulement si pour tout compact $K \subset \Omega$ la série

$$(19) \qquad \sum_{m=o}^{\infty} \frac{\int_K |\Delta^m f|\,dx}{(2m)!} z^m$$

converge dans tout le plan complexe.

Démonstration. Si $f \in \mathcal{H}_\infty(\Omega)$, la condition est évidente; réciproquement si la série (19) converge pour tout $z \in \mathbb{C}$, quel que soit $\epsilon > 0$, il existe $M(K,\epsilon)$ constante > 0 telle que

$$\int_K |\Delta^m f(x)| dx \leq M(K,\epsilon)\, \epsilon^m (2m)! \quad m = 0,1,\ldots$$

alors la distribution $T = f$ vérifie :

$$|<\Delta^m f,\varphi>| \leq \int_{K = \operatorname{supp}\varphi} |\Delta^m f(x)|\, |\varphi(x)| dx \leq M_1(K,\epsilon)\epsilon^m (2m)!$$

$M_1(K,\epsilon)$ constante > 0 .

Cela implique que $f \in A(\alpha,\infty,\Omega)$ $(\alpha = ((2m)!)_{m \in \mathbb{N}})$, et par conséquent f est une fonction harmonique d'ordre infini.

§ 1.4. ANALYTICITÉ LIÉE AU SIGNE DE LAPLACIEN ITÉRÉ.

Dans $C_R^\infty(\Omega)$ certains sous-ensembles caractérisés par la condition que leur laplacien itéré, conservent un signe constant dans leur domaine de définition , forment des classes particulières de fonctions analytiques.

Dans le cas $N = 1$ les premiers résultats dans cette voie sont dus à J. Bernstein on connaît par exemple :

Soit $f \in C_R^\infty(]a,b[)$

a) si pour tout $n \in \mathbb{N}$: $f^{(2n)}(x) \geq 0$ sur $]a,b[$ alors f est analytique.

b) si pour tout $n \in \mathbb{N}$: $(-1)^n f^{(2n)}(x) \geq 0$ $x \in]a,b[$ f est analytique.

Ces énoncés se généralisent à plusieurs variables avec une différence essentielle dans les démonstrations suivant qu'on considère le cas a) ou le cas b) .

1.4.1. THÉORÈME [4]

Si $f \in C^{\infty}_{\mathbb{R}}(\Omega)$ vérifie

$$\Delta^m f(x) \geq 0 \quad m = 0,1,2,\ldots \quad (x \in \Omega)$$

alors f est analytique dans Ω.

1.4.2. THÉORÈME. Soit $f \in C^{\infty}_{\mathbb{R}}(\Omega)$ vérifiant

$$(-1)^m \Delta^m f(x) \geq 0 \quad m = 0,1,3,\ldots \quad (x \in \Omega)$$

a - La fonction f est une fonction harmonique d'ordre infini dans Ω (donc analytique).

b - Si $\Omega = \mathbb{R}^N$, f est une constante.

1.4.3. THÉORÈME [25]

Soit dans $\mathbb{R}^N$ $(N \geq 3)$ $f \in C_{\mathbb{R}}^{2(E[\frac{N}{2}]+1)}(\mathbb{R}^N)$

($E(x)$ = partie entière de x).

Si $\quad (-1)^p \Delta^p f(x) \geq 0 \quad p = 0,1,\ldots, E[\frac{N}{2}]+1$

alors f est une constante.

Le résultat est inexact en général si la fonction dépend au plus de deux variables ou bien si le domaine de définition de f n'est pas $\mathbb{R}^N$ tout entier.

Démonstration de 1.4.1. Considérons le développement (11). Tous les termes sont positifs, donc :

$$a_m \Delta^m f(x) R^{2m} \leq \lambda\,[f,x,R]\ , \ \overline{B(x,R)} \subset \Omega$$

Si x décrit un compact $K \subset \Omega$ et si $R \leq R_o <$ distance de K au complémentaire de Ω est fixé, la fonction $x \longmapsto \lambda\,[f,x,R]$ étant continue, on a :

$$\sup_{x \in K} \Delta^m f(x) \leq \sup_{x \in K} \lambda\,[f,x,R] \times \frac{1}{a_m R^{2m}} = \frac{\theta(K,R)}{a_m R^{2m}}$$

d'où

$$(20) \qquad \left[\sup_{x \in K} \frac{\Delta^m f(x)}{(2m)!} \right]^{\frac{1}{m}} \leq \frac{\theta^{\frac{1}{m}}(K,R)}{R^2} \; \frac{1}{[(2m)!\,a_m]^{\frac{1}{m}}}$$

Or $$a_m(2m)! \sim \Gamma\left(\frac{N}{2}\right) \times \frac{1}{\sqrt{\pi}\, m^{\frac{N-1}{2}}} \qquad (m \to \infty)$$

le premier membre de (20) est alors majoré par une constante $M(K,R)$ quel que soit m. D'où l'analyticité de f (Th. 1.1.4.).

Démonstration de 1.4.2. La première affirmation est un résultat de P. Lelong [20]. Si $N \geq 2$ soit Ω^* un domaine de Green d'adhérence compacte dans Ω (si $N = 1$, $\Omega^* =]a,b[\subset \mathbb{R}$).

Considérons la formule de Green :

$$(21) \quad f(x) = \frac{1}{C_N}\int_{\partial\Omega^*} f(y)\,\frac{\partial G}{\partial n_i}(x,y)\,d\sigma(y) - \frac{1}{C_N}\int_{\Omega^*} \Delta f(y)\,G(x,y)\,d\tau(y)$$

$$C_N = \begin{cases} (N-2)\,\omega_N(1) & \text{si } N \geq 3 \\ 2\pi & \text{si } N = 2\,. \end{cases}$$

Si $N = 1$,

$$(21') \qquad f(x) = Ax + B - \tfrac{1}{2}\int_a^b f''(t)\,G\,(x,t)\,dt$$

Dans (21), la première intégrale notée $H_o(x)$ est la fonction harmonique qui coïncide avec f sur $\partial\Omega^*$; dans (21') la fonction affine $Ax + B$ est égale à $f(a)$ pour $x = a$ et à $f(b)$ si $x = b$. La fonction de Green G de $]a,b[$ est représentée dans la figure ci-dessous :

$G(x,t) = -|x-t| + \ell_a^b(x,t)$, $\ell_a^b(x,t)$ est fonction affine en t telle que :

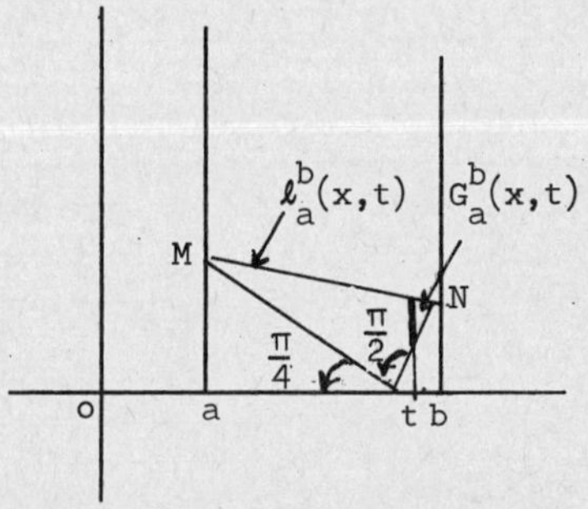

$$\ell_a^b(x,a) = \ell_a^b(x,b) = 0$$

On a

$$G_a^b(x,t) = \begin{cases} \dfrac{2(b-x)(t-a)}{b-a} & a \leq t \leq x \leq b \\[2ex] \dfrac{2(b-t)(x-a)}{b-a} & a \leq x \leq t \leq b \end{cases}$$

Avec ces considérations, la démonstration ci-dessous est valable pour $N = 1$.

Plaçons-nous dans le cas $N > 1$.

Posons :

$$G_n(x,y) = \int_{\Omega^*} G_{n-1}(x,t)\, G(t,y)d\tau(t) \quad (n \geq 2)$$

$$G_2(x,y) = \int_{\Omega^*} G(x,t)\, G(t,y)\, d\tau(y)$$

Si K est un compact de Ω^* d'intérieur non vide, on a, d'après les propriétés de la fonction de Green :

$$(22) \qquad \inf_{x,y \in K} G(x,y) = \gamma(K) > 0 \,.$$

donc

$$G_n(x,y) \geq \gamma^n(K)\, [\text{Mes } K]^{n-1} \qquad (x,y \in K)$$

(Mes K = mesure de Lebesgue de K) .

Soient $H_o, H_1, \dots, H_n, \dots$ respectivement les fonctions harmoniques dans Ω^* qui coïncident avec f , $\Delta f, \dots, \Delta^n f, \dots$, sur $\partial\, \Omega^*$.

Appliquons (21) à $\Delta f (c = c_N^{-1})$:

$$(23) \qquad \Delta f(x) = H_1(x) - c \int_{\Omega^*} \Delta^2 f(y)\, G(x,y)\, d\tau(y)$$

$$f(x) = H_o(x) - c \int_{\Omega^*} H_1(y) G(x,y) d\tau(y) + (-1)^2 c^2 \int_{\Omega^*}\int_{\Omega^*} \Delta^2 f(y) G(y,t) G(x,y) d\tau(t) d\tau(y)$$

Donc, après n-itérations :

$$(24) \qquad f(x) = H_o(x) + \sum_{p=2}^{n-1} (-1)^{p-1} c^{p-1} \int_{\Omega^*} H_{p-1}(y) G_{p-1}(x,y) d\tau(y)$$

$$+ (-1)^n c^n \int_{\Omega^*} \Delta^n f(y) G_n(x,y) d\tau(y) \qquad (x \in \Omega^*)$$

L'hypothèse $(-1)^n \Delta^n f(x) \geq 0$ implique que tous les termes de (24) sont ≥ 0 .

D'où

$$|(-1)^n c^n \int_{\Omega^*} \Delta^n f(y) G_n(x,y) d\tau(y)| \leq f(x) \quad (x \in \Omega^*)$$

$$\int_{\Omega^*} |\Delta^n f(y)| G_n(x,y) d\tau(y) \leq c_N^n f(x) \quad (x \in \Omega^*)$$

D'après (22) on a :

$$\int_K |\Delta^n f(y)|\, d\tau(y) \leq c_N^n \operatorname{Max}_K |f(x)|\ \gamma^{-n} [\operatorname{Mes} K]^{-(n-1)}$$

Finalement sur tout compact $K \subset \Omega$:

$$\lim_{n \to o} \left[\frac{\int_K |\Delta^n f(y)| d\tau(y)}{(2n)!} \right]^{\frac{1}{n}} = 0$$

c'est-à-dire $f \in \mathcal{H}_\infty(\Omega)$ (1.2.3.). Si $\Omega = R^N$, la fonction $H(x,t)$ associée à f (1.2.4) est harmonique ≥ 0 dans tout R^{N+1} donc constante. Par suite $f = H(x,o)$ cte.

<u>Démonstration de 1.4.3.</u>

Reprenons la représentation (24) où Ω^* est la boule $B(0,R)$ de centre 0 et de rayon R. Si $x \in B(0,R)$,

$$(25) \quad f(x) = H_o(x) + \sum_{p=2}^{E[\frac{N}{2}]} (-1)^{p-1} c^{p-1} \int_{B(0,R)} H_{p-1}(y) G_{p-1}(x,y) d\tau(y)$$

$$+ (-1)^{E[\frac{N}{2}]+1} c^{E[\frac{N}{2}]+1} \int_{B(0,R)} \Delta^{E[\frac{N}{2}]+1} f(y)\, G_{E[\frac{N}{2}]+1}(x,y) d\tau(y)$$

1.4.4. LEMME. <u>Si</u> $G(x,y) = G^R(x,y)$ <u>est la fonction de Green de la boule</u> $B(0,R)$ <u>on a</u> :

$$\inf_{\substack{\|y\| \leq \frac{R}{2} \\ \|y\| \leq \frac{R}{2}}} G^R(x,y) \geq \frac{A}{R^{N-2}} \qquad (A = \text{cte})$$

$$N \geq 3$$

Démonstration. En effet, $G^R(x,y)$ pour x fixé dans $B(0,\frac{R}{2})$ est surharmonique dans $B(0,R)$, donc, pour x fixé dans $B(0,\frac{R}{2})$,

$$\inf_{\|y\|\leq \frac{R}{2}} G^R(x,y) = \inf_{\|y\|=\frac{R}{2}} G^R(x,y)$$

Or
$$G(x,y) = [\|x\|^2 - 2\|x\|\|y\| \cos\theta + \|y\|^2]^{1-\frac{N}{2}} -$$
$$- [R^2 - 2\|x\|\|y\|\cos\theta + \frac{\|x\|^2\|y\|^2}{R^2}]^{1-\frac{N}{2}}$$

un calcul élémentaire montre l'existence d'une constante $A > 0$ telle que

$$\left[\inf_{\|x\|\leq \frac{R}{2}} \inf_{\|y\|=\frac{R}{2}} G^R(x,y)\right] \geq \frac{A}{R^{N-2}}$$

Dans le développement (25) , tous les termes sont ≥ 0 quel que soit R , donc :

$$(26) \qquad (-1)^{p-1}c^{p-1}\int_{B(0,R)} H_{p-1}(y)G^R_{p-1}(x,y)d\tau(y) \leq f(x)$$

$p = 2,\ldots, E[\frac{N}{2}]$, $x \in B(0,R)$

Or, d'après le lemme 1.4.4. et (22) :

$$\inf_{x,y\in B(0,R)} G^R_{p-1}(x,y) \geq \left(\frac{A}{R^{N-2}}\right)^{p-1} [\text{Mes } B(0,\tfrac{R}{2})]^{p-2} \geq B R^{2p-2-N} \qquad (B = \text{cte})$$

En particulier

$$(27) \qquad \inf_{x,y\in B(0,\frac{R}{2})} G^R_{E[\frac{N}{2}]+1}(x,y) \geq B\, R^{2(E[\frac{N}{2}]+1)-N}$$

D'après (26)

$$\alpha\, R^{2p-2-N}\int_{B(0,R)} (-1)^{p-1}H_{p-1}(y)d\tau(y) \leq f(x) \qquad (\|x\| \leq \tfrac{R}{2})$$

α étant une constante numérique > 0 .

Comme $(-1)^{p-1} H_{p-1}(y)$ est harmonique ≥ 0 et que l'intégrale ci-dessus est à un facteur numérique près, sa moyenne spatiale, il résulte que

$$\beta\, R^{2p-2}(-1)^{p-1}H_{p-1}(0) \leq f(x) \quad (\beta = \text{cte} > 0\,,\ x \in B(0, \tfrac{R}{2})).$$

Cela est impossible si $H_{p-1}(0) \neq 0$. D'où $H_{p-1}(0) = 0$; donc $(-1)^{p-1} H_{p-1}(x)$ qui est harmonique > 0 dans $B(0,\frac{R}{2})$, nulle à l'origine, est identiquement nulle. D'où

$$H_{p-1} = 0 \quad p = 2,\ldots, E[\tfrac{N}{2}]$$

De même d'après (25) et (27) :

$$(28)\quad B_1 R^{2(E[\frac{N}{2}]+1)-N} \int_{B(0,\frac{R}{2})} (-1)^{E[\frac{N}{2}]+1} \Delta^{E[\frac{N}{2}]+1} f(y)\, d\tau(y) \leq f(x)\,(\|x\| < R),$$

($B_1 > 0$ constante numérique) .

Comme $2(E[\frac{N}{2}]+1) - N \geq 0$ l'intégrale (28) est nécessairement nulle, sinon pour R assez grand, l'inégalité (28) est en défaut. Finalement $f(x) = H_o(x)$ dans (25) ; $H_o(x)$ étant harmonique > 0 dans $B(0,R)$ est nécessairement la restriction à $B(0,R)$ d'une fonction harmonique > 0 dans tout $\mathbb{R}^N$, donc est une constante .

1.4.5. Contre-exemple. La fonction $f(x) = \sin x_1 + \ldots + \sin x_N$ vérifie les hypothèses du théorème 1.4.3. dans $\Omega = \{x \mid 0 < x_i < \frac{\pi}{2}\,,\ i = 1,\ldots,N\}$ sans être constante. La fonction vérifie la condition du théorème sans être constante.

§ 1.5. APPLICATIONS.

1.5.1. Soient $\alpha \in \mathbb{R}^*$ et $f \in C^2_{\mathbb{R}}(\Omega)$ telle que

$$(29)\qquad (\Delta + \alpha)f = 0$$

La fonction f définit une distribution sur Ω et on a $\Delta^m f = (-\alpha)^m f \quad m = 0,1,\ldots$ on en déduit que f est une distribution harmonique d'ordre infini et finalement une fonction harmonique d'ordre infini (Th. 1.3.4.) . D'après 1.2.8. la moyenne $\lambda(f,x,R)$ s'écrit :

(30) $$\lambda(f,x,R) = \left(\sum_{m=0}^{\infty} \frac{\Gamma(\frac{N}{2})(-1)^m \alpha^m R^{2m}}{4^m m!\ \Gamma(m+\frac{N}{2})}\right) f(x)$$

$$= \mathring{J}_\alpha(R)\ f(x) \qquad (\overline{B(x,R)} \subset \Omega)$$

Réciproquement, si f continue vérifie (30) pour toute $B(x,R) \subset \Omega$, alors elle est de la classe $C^\infty_{\mathbb{R}}(\Omega)$ et vérifie (29).

En effet, soit $\theta_r(x) \geq 0$, $\theta_r \in C^\infty_{\mathbb{R}}(\Omega)$, $\theta_r(x) = \theta_r(\|x\|)$, $\operatorname{supp} \theta_r = \overline{B(0,r)}$ $\int \theta_r = 1$, on a :

(31) $$(f*\theta_r)(x) = N\omega_N(1)\int_0^r \lambda(f,x,t)\theta_r(t)t^{N-1}dt = N\omega_N(1)\left[\int_0^r \mathring{J}_\alpha(t)\,\theta_r(t)t^{N-1}dt\right]f(x)$$

Le premier membre de (31) est de la classe $C^\infty_{\mathbb{R}}(\Omega)$ donc $f \in C^\infty_{\mathbb{R}}(\Omega)$.

D'autre part

$$\frac{\lambda(f,x,R)-f(x)}{R^2} = \frac{\mathring{J}_\alpha(R)-1}{R^2} f(x)$$

et

$$\Delta f(x) = \lim_{R\to 0} 2N\frac{\lambda(f,x,R)-f(x)}{R^2} = \lim_{R\to 0}\left(2N\frac{\mathring{J}_\alpha(R)-1}{R^2}\right) f(x) = -\alpha f(x).$$

d'où

$$(\Delta+\alpha)f = 0.$$

Remarque. Il existe une relation intime entre $\mathring{J}_\alpha(R)$ et la fonction de Bessel $J_{\frac{N}{2}-1}$.

1.5.2. Soit $u \in C^{2k}(\Omega)$, $u \not\equiv 0$, vérifiant la relation :

(32) $$\Delta^k u = \alpha_{k-1}\Delta^{k-1}u + \alpha_{k-2}\Delta^{k-2}u + \dots + \alpha_0 u \qquad (\alpha_0 \neq 0,\ k \geq 1)$$

où les α_k sont des constantes. Considérant u comme une distribution, la relation (32) implique

$$\Delta^{m+k}u = \alpha_{k-1}\Delta^{m+k-1}u + \alpha_{k-2}\Delta u^{m+k-2} + \dots + \alpha_o \Delta^m u$$

pour tout $m \in \mathbb{N}$. La méthode de résolution des suites récurrentes appliquée à la suite $f_m = \Delta^m u$ qui vérifie :

$$(33) \qquad f_{m+k} = \alpha_{k-1} f_{m+k-1} + \alpha_{k-2} f_{m+k-2} + \dots + \alpha_o f_m$$

permet d'exprimer $\Delta^m u$ pour tout $m \in \mathbb{N}$ comme une combinaison linéaire des $f_o,\dots,f_{k-1}$. Soient $\chi_1,\dots,\chi_p$ les valeurs propres de la matrice :

$$\begin{pmatrix} 0 & 1 & 0 & \cdots & 0 \\ \vdots & 0 & 1 & & \\ & & \ddots & \ddots & \\ 0 & & & & 1 \\ \alpha_o & \alpha_1 & & & \alpha_{k-1} \end{pmatrix}$$

et $r_1,\dots,r_p$ leur multiplicité $(r_1 + \dots + r_p = k)$. L'ensemble des suites f_m qui vérifient (33) étant un espace vectoriel (sur $\mathbb{C}$) de dimension k, f_m s'exprime comme une combinaison linéaire de suites du type $(\chi_i^m m^q)$ $\begin{matrix} 1 \le i \le p \\ 0 \le q \le r_i - 1 \end{matrix}$:

$$f_m = \sum_{\substack{1 \le i \le p \\ 0 \le q \le r_i - 1}} a_{i,q} \chi_i^{m_i} m^q$$ où les k constantes $a_{i,q}$ sont déterminées par la connaissance de $f_o,\dots,f_{k-1}$ et sont fonctions linéaires de celles-ci : il en résulte que

$$\Delta^{m+k}u = P_{k-1}(m)\Delta^{k-1}u + P_{k-2}(m)\Delta^{k-2}u + \dots + P_o(m)u$$

où $P_{k-j}(m)$ pour tout $1 \le j \le k$ est une combinaison linéaire des $\chi_i^m m^q$. D'après le théorème 1.3.4. , u est une fonction harmonique d'ordre infini dans Ω si elle est à valeurs réelles. D'où

1.5.3. PROPOSITION. <u>Une fonction</u> $u \in C^{2k}_{\mathbb{R}}(\Omega)$, <u>vérifiant une relation de la forme</u> (32) <u>est une fonction harmonique d'ordre infini dans</u> Ω .

1.5.4. Considérons le cas particulier $u \in C^4_{\mathbb{R}}(\Omega), u = u(x,y,z), \Omega \subset \mathbb{R}^3$.

$$(34) \qquad \Delta^2 u + \Delta u + u = 0 \; .$$

En tant que distribution (34) implique pour tout $p \in \mathbb{N}$:

$$\Delta^{3p} u = u$$

$$\Delta^{3p+1} u = \Delta u$$

$$\Delta^{3p+2} u = -\Delta u - u$$

ces relations montrent que u est une fonction harmonique d'ordre infini dans Ω (Th. 1.3.4.) .

Nous allons montrer que $u(x,y,z)$ change de signe à l'extérieur de toute sphère de $\mathbb{R}^3$.

D'après 1.2.8. ,

$$(35) \qquad \lambda(u,x,R) = \sum_{m=o}^{\infty} \frac{R^{2m}}{(2m+1)!} \Delta^m u(x_o) \; (\overline{B(x,R)} \subset \Omega)$$

le Σ figurant dans (35) s'écrit :

$$\sum_{m=o}^{\infty} = (\sum_{p=o}^{\infty} \frac{R^{6p}}{(6p+1)!})u + (\sum_{p=o}^{\infty} \frac{R^{6p+2}}{(6p+3)!}) \Delta u$$

$$+ (\sum_{p=o}^{\infty} \frac{R^{6p+4}}{(6p+5)!}) (-\Delta u - u)$$

ou

$$(36) \qquad \lambda(u,x,R) = \sum_{p=o}^{\infty} [\frac{R^{6p}}{(6p+1)!} - \frac{R^{6p+4}}{(6p+5)!}] u(x) +$$

$$(\sum_{p=o}^{\infty} [\frac{R^{6p+2}}{(6p+3)!} - \frac{R^{6p+4}}{(6p+5)!}]) \Delta u(x)$$

Ecrivons (36) sous la forme :

$$(37) \qquad R\lambda(u,x,R) = A(R)u(x) + B(R)\,\Delta u(x)$$

on vérifie que les séries $A(R)$ et $B(R)$ sont solutions de l'équation différentielle $y^6 - y = 0$ et on déduit alors leur expression :

$$A(R) = \frac{2\sqrt{3}}{3}\,\mathrm{ch}\,\frac{R}{2}\,\sin\frac{\sqrt{3}}{2}R$$

$$B(R) = \frac{\sqrt{3}}{3}\,\mathrm{ch}\,\frac{R}{2}\,\sin\frac{\sqrt{3}}{2}R - \mathrm{sh}\,\frac{R}{2}\,\cos\frac{\sqrt{3}}{2}R\,.$$

Supposons $\Omega = \mathbb{R}^3$. Ecrivons (36) pour $R > 0$ et au point $x = x_o$

$$(38) \qquad R\lambda(u,x_o,R) = e^{\frac{R}{2}}\left[\alpha(x_o)\sin\frac{\sqrt{3}}{2}R - \tfrac{1}{2}\cos\frac{\sqrt{3}}{2}R\right] + h(R)$$

avec
$$\alpha(x_o) = \frac{\sqrt{3}}{3}u(x_o) + \frac{\sqrt{3}}{6}\Delta u(x_o)$$

$$\lim_{R\to\infty} h(R) = 0$$

Ecartons le cas $\alpha(x_o) = 0$ et supposons $u(x_o) \neq 0$. La relation (38) s'écrit :

$$(39) \qquad R\lambda(u,x_o,R) = e^{\frac{R}{2}}\,k(x_o)\sin\left(\frac{\sqrt{3}}{2}R + \theta\right) + h(R)$$

considérons les suites suivantes :

$$R_n = \frac{2n\pi - 2\theta}{\sqrt{3}}$$

$$R'_n = \frac{(4n+1)\pi - 2\theta}{\sqrt{3}}$$

$$R''_n = \frac{(4n-1)\pi - 2\theta}{\sqrt{3}}$$

$n = 1,2,\dots$ Pour $(R_n)_{n\geq 1}$ la fonction $k(x_o)e^{\frac{R}{2}}\sin\left(\frac{\sqrt{3}}{2}R + \theta\right)$ change de signe. Pour les deux autres, le second membre de (39) tend respectivement vers $+\infty$ et $-\infty$ si $n \to \infty$.

Donc, quel que soit R_o, la fonction $\lambda(u,x_o,R)$, pour $R > R_o$ et par conséquent $u(x)$, $\|x\| > R_o$, ne peuvent garder un signe constant. (même raisonnement dans le cas $\Delta u(x_o) \neq 0$.

CHAPITRE 2

CELLULE D'HARMONICITE

§ 2.1. DÉFINITIONS ET PROPRIETES GENERALES - CHEMINS DE P. LELONG

2.1.1. L'espace euclidien $\mathbb{R}^N$ est confondu avec le sous-espace $\{z = (z_1,\ldots,z_N) = x + iy \in \mathbb{C}^N | y = o\}$. Nous écrivons aussi $\mathbb{C}^N(z) = \mathbb{R}^N(x)+i\mathbb{R}^N(y)$. Dans tout le chapitre le cas $N = 1$ de dimension réelle est sans intérêt. On supposera donc $N \geq 2$. On note :

$$Q : z \mapsto Q(z) = z^2 + \ldots + z_N^2$$

$$<z,a> = z_1\,\bar{a}_1 + \ldots + z_N\,\bar{a}_N \quad (z,a \in \mathbb{C}^N)$$

$$\cos(\widehat{x,a}) = \frac{<x,a>}{\|x\|\,\|a\|} \quad (x,a \in \mathbb{R}^N, \|x\|\,\|a\| \neq 0) .$$

Les normes $\|z\|$ ou $\|x\|$ sont déduites du produit $<z,a>$.

Si $\mathcal{P}(\mathbb{R}^N)$ est l'ensemble des parties de $\mathbb{R}^N$, l'application $\begin{array}{l} T : \mathbb{C}^N \to \mathcal{P}(\mathbb{R}^N) \\ z \mapsto T(z) \end{array}$ est définie par :

$$T(z) = \{t \in \mathbb{R}^N | Q(z-t) = 0\} .$$

Le cône isotrope (dans $\mathbb{C}^N$) de sommet $t \in \mathbb{R}^N$ est

$$\Gamma(t) = \{z \in \mathbb{C}^N | Q(z-t) = 0\} .$$

On a :

(1) $\quad T(x+iy) = T(z) = \{t \in \mathbb{R}^N | \|t-x\| = \|y\| \text{ et } <t-x,y> = 0\} .$

Pour $N \geq 3$, l'image d'un point $z \in \mathbb{C}^N$ par T , est la sphère $(N-2)$ dimentionnelle S_z^{N-2} de centre $x = \text{Rez}$ et de rayon $\|y\|$ $(y = \text{Im} z)$ contenue dans l'hyperplan Π d'équation $<t-x,y> = 0$ (t étant la variable). Si $N = 2$, la sphère S_z^o est réduite à deux points de $\mathbb{R}^2$ (identifié à $\mathbb{C}$) $z_1 + iz_2$, $\bar{z}_1 + i\bar{z}_2$ ces points sont symétriques par rapport au point $z = x$ et sont situés sur la droite d'équation

$$y_1t_1 + y_2t_2 - x_1y_1 - x_2y_2 = <t-x,y> = 0 .$$

(Fig. 1)

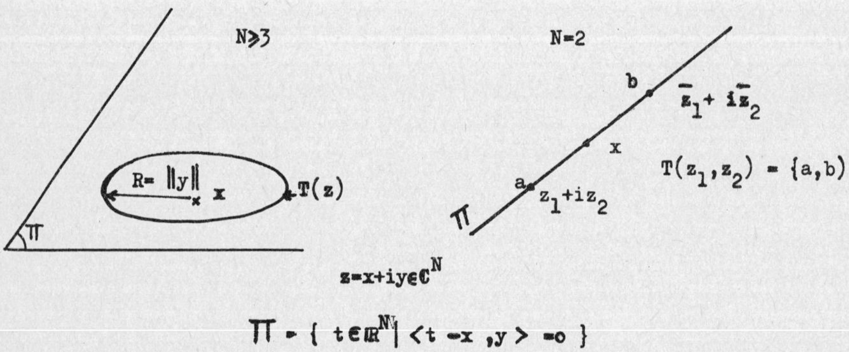

On remarque les propriétés suivantes :

2) $T(z) = T(\bar{z})$ $(z \in \mathbb{C}^N)$.

Si $x \in \mathbb{R}^N$, $T(x) = \{x\} = S_x^{N-2}$ = sphère de centre x et de rayon nul.

3) Si $N \geq 3$, l'ensemble $T(z)$ est connexe.

4) Pour $t \in \mathbb{R}^N$, $\Gamma(t) \cap \mathbb{R}^N = \{t\}$ $(\Gamma(t) = \{t\}$ si $N = 1)$

5) Pour $z \in \mathbb{C}^N$ et $t \in \mathbb{R}^N$ les deux propriétés $z \in \Gamma(t)$ et $t \in T(z)$ sont équivalentes.

6) Pour tout α réel non nul, on a : $T(x+iy) = x + \alpha T(i\,\frac{y}{\alpha})$.

2.1.2. <u>CHEMINS DE P. LELONG</u>

Soient A une partie non vide de $\mathbb{R}^N$, $a \in \mathbb{C}^N$, $z \in \mathbb{C}^N$.

Un chemin de P. Lelong ou brièvement un L_A-chemin d'origine z et d'extrêmité a , est une application $\gamma : [0,1] \to \mathbb{C}^N$ continue telle que pour tout $\tau \in [0,1]$, la sphère $T[\gamma(\tau)]$ est contenue dans A.

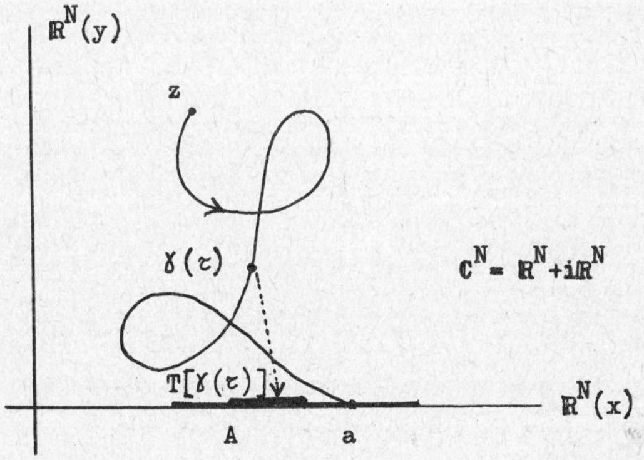

Autrement dit un L_A-chemin joignant le point z au point $a \in \mathbb{C}^N$ est un chemin continu dont l'image par T de ses points est une famille continue de sphères $S^{N-2} \subset A$. Il est évident que si on peut joindre le point z , au point b , et puis le point b au point a respectivement par des L_A-chemins γ , et γ' , alors il existe un L_A-chemin joignant z au point a .

De même, d'après la propriété 2) de 2.1.1. tout chemin continu γ dans A est un L_A-chemin, la famille S^{N-2} de sphères correspondante étant constituée par les points de l'image de γ .

2.1.3. PROPOSITION.

Soit A un ouvert connexe de $\mathbb{R}^N (N \geq 2)$. L'image par T d'un L_A-chemin γ d'origine $z \in \mathbb{C}^N$ et d'extrémité $a \in A$ est un compact connexe de A.

2.1.4. PROPOSITION.

Soit A un sous-ensemble de $\mathbb{R}^N$ tel que $\partial A \neq \emptyset$. L'ensemble

$$\Lambda(A) = \bigcup_{t \in \partial A} \Gamma(t)$$

est fermé dans $\mathbb{C}^N$ et on a :

$$\Lambda(A) \cap \mathbb{R}^N = \partial A$$

(Si $N = 1$, $\Lambda(A) = \partial A$).

Démonstration de 2.1.3. : Soient

$$\gamma : I = [0,1] \to \mathbb{C}^N, \quad T(\gamma) = \bigcup_{\tau \in I} T[\gamma(\tau)] \subset A .$$

1. L'ensemble $T(\gamma)$ est fermé dans A.

En effet, soit $t_o = \lim_{k \to \infty} t_k$, $t_k \in T[\gamma(\tau_k)]$. Il existe une suite extraite $(\tau_{k_m})_{m \in \mathbb{N}}$ et un point $\tau_o \in I$ tels que $\tau_o = \lim_{m \to \infty} \tau_{k_m}$. Les applications Q et γ étant continues on a

$$0 = \lim_{m \to \infty} Q[t_{k_m} - \gamma(\tau_{k_m})] = Q[t_o - \gamma(\tau_o)]$$

d'où $t_o \in T[\gamma(\tau_o)]$ et la conclusion.

2. L'ensemble $T(\gamma)$ est borné.

Soient t' , $t'' \in T(\gamma)$. Il existe τ' , τ'' tels que $t' \in T[\gamma(\tau')]$, $t'' \in T[\gamma(\tau'')]$. On a :

$$\|t'-t''\| \leq \|t'-z'\| + \|z'-z''\| + \|z''-t''\| \ , \ z' = \gamma(\tau'), z'' = \gamma(\tau'')$$

or $$\|z'-z''\| \leq 2 \operatorname*{Max}_{z \in \gamma(I)} \|z\| = 2M \quad \text{et}$$

$$\|t'-z'\|^2 = \|t'-(x'+iy')\|^2 = \|t'-x'\|^2 + \|y'\|^2 = 2\|y'\|^2 \leq 2M^2$$

de même $\|z''-t''\|^2 \leq 2M^2$.

D'où $\|t'-t''\| \leq 2(1+\sqrt{2})M$.

3. L'ensemble $T(\gamma)$ est connexe.

Supposons tout d'abord $N \geq 3$. Si $T(\gamma) = F_1 \cup F_2$ avec $F_1 \cap F_2 = \emptyset$, $F_j \neq \emptyset$, F_j fermé $j = 1,2$, posons $I_j = \{\tau \in I \mid T[\gamma(\tau)] \cap F_j \neq \emptyset\}$, $j = 1,2$. On a $I_j \neq \emptyset$ et $I = I_1 \cup I_2$. Or pour $\tau_o \in I$ l'ensemble $T[\gamma(\tau_o)]$ est connexe d'où $T[\gamma(\tau_o)] \subset F_1$ ou $T[\gamma(\tau_o)] \subset F_2$. Donc, $I_1 \cap I_2 = \emptyset$ et comme d'autre part I_1, I_2 sont fermés cela est en contradiction avec la connexité de I . Si $N = 2$, soient $\gamma_j : I \to \mathbb{C}$ $(j = 1,2)$ $\gamma = (\gamma_1, \gamma_2)$, $F_1 = (\gamma_1 + i\gamma_2)(I)$ $F_2 = (\bar{\gamma}_1 + i\bar{\gamma}_2)(I)$ (voir 2.1.1). Les ensembles F_1 et F_2 sont connexes, avec une intersection contenant le point a (donc non vide), il en résulte la connexité de leur réunion.

Démonstration de 2.1.4 : La deuxième égalité résulte de la propriété 4) . Montrons l'égalité $\Lambda(A) = \overline{\Lambda(A)}$. Soit $z_{(k)}, k = 1,2,\ldots$, une suite de points de $\Lambda(A)$ de limite z_o , avec $z_{(k)} \in \Gamma(t_{(k)})$, $t_{(k)} \in \partial A$. Soit $M = \operatorname{Sup}\|z_{(k)}\| < \infty$. On a quels que soient $k, \ell \in \mathbb{N}$:

$$\|t_{(k)}-t_{(\ell)}\| \leq \|t_{(k)}-z_{(k)}\| + \|z_{(k)}-z_{(\ell)}\| + \|z_{(\ell)}-t_{(\ell)}\| \leq \|t_{(k)}-z_{(k)}\| + \|t_{(\ell)}-z_{(\ell)}\| + 2M$$

$$\|t_{(k)}-z_{(k)}\|^2=\|t_{(k)}-(x_{(k)}+iy_{(k)})\|^2=\|t_{(k)}-x_{(k)}\|^2+\|y_{(k)}\|^2=2\|y_{(k)}\|^2\leq 2M^2 .$$

D'où $\|t_{(k)}-t_{(\ell)}\|^2 \leq 2(1+\sqrt{2})M$. L'ensemble ∂A étant fermé et la suite $(t_{(k)})$ bornée, il existe une suite extraite $(t_{(k_m)})$ convergente. Soit t_o sa limite. Alors $Q(z_o-t_o) = \lim_{m\to\infty} Q[z_{(k_m)}-t_{(k_m)}] = 0$ c'est-à-dire $z_o \in \Gamma(t_o)$. D'où le résultat.

2.1.5. Soient D un ouvert connexe non vide de $\mathbb{R}^N$, $\partial D \neq \emptyset$, $\mathcal{H}(D)$ = composante connexe contenant D de

$$\mathbb{C}^N \setminus \Lambda(D) = \mathbb{C}^N \setminus \bigcup_{t\in\partial D} \Gamma(t)$$

(on pose $\mathcal{H}(\mathbb{R}^N) = \mathbb{C}^N$) et

$W(D) = \{z \in \mathbb{C}^N |$ il existe $a \in D$ et un L_D-chemin joignant z au point $a\}$.

D'après 2.1.4., $\mathcal{H}(D)$ est un domaine ainsi que $W(D)$.

2.1.6. PROPOSITION.

<u>Soit</u> D <u>un ouvert connexe non vide de</u> $\mathbb{R}^N$, $\partial D \neq \emptyset$.

a) <u>On a</u>

$$\mathcal{H}(D) = W(D) \supset D \text{ , } \mathcal{H}(D) \cap \mathbb{R}^N = D \text{ , } \partial D \subset \partial\mathcal{H}(D) \subset \Lambda(D)$$

b) <u>Si</u> D <u>est étoilé au point</u> $a_o \in D$, <u>alors</u> $\mathcal{H}(D)$ <u>est étoilé au même point et</u>

$$\mathcal{H}(D) = \{z \in \mathbb{C}^N | T(z) \subset D\}$$

<u>en particulier si</u> D <u>est convexe</u>, $\mathcal{H}(D)$ <u>l'est aussi</u>.

c) Si $N = 2$, $\mathcal{H}(D) = \{z = (z_1,z_2) \in \mathbb{C}^2 | T(z) = \{z_1+iz_2, \overline{z}_1+i\overline{z}_2\} \subset D\}$.

2.1.7. DÉFINITION.

Le domaine $\mathcal{H}(D)$ défini ci-dessus est appelé la cellule d'harmonicité de D .

Démonstration de 2.1.6.

a) L'ensemble $W(D)$ est ouvert, connexe par arcs (donc connexe), c'est donc un domaine. On a $W(D) \cap \Lambda(D) = \emptyset$. En effet, si $z \in W(D) \cap \Lambda(D)$, il existe $t \in \partial D$ tel que $t \in T(z)$ et $T(z) \cap \partial D = \emptyset$ en contradiction avec $T(z) \subset D$; d'où $W(D) \subset \mathcal{H}(D)$. Pour établir l'inclusion inverse, il suffit de montrer que pour tout $z \in \mathcal{H}(D)$, $a \in D$, il existe un L_D-chemin joignant z au point a . Soit $\gamma : [0,1] \to \mathcal{H}(D)$ un chemin continu dans $\mathcal{H}(D)$, $\gamma(0) = z$, $\gamma(1) = a$. (γ existe puisque $\mathcal{H}(D) \supset D$). On va établir que $T[\gamma] \subset D$. D'après 2.1.3., $T(\gamma)$ est un compact connexe de $\mathcal{H}(D)$ et $T[\gamma] \cap \partial D = \emptyset$, $T[\gamma] \cap D = \emptyset$. Si $T[\gamma] \not\subset D$, écrivons

$$T[\gamma] = [T[\gamma] \cap D] \cup [T[\gamma] \cap (\mathbb{R}^N \setminus \overline{D})]$$

où $T[\gamma] \cap (\mathbb{R}^N \setminus \overline{D}) \neq \emptyset$. Cela est en contradiction avec la connexité de $T[\gamma]$. D'où $\mathcal{H}(D) \subset W(D)$ et finalement $\mathcal{H}(D) = W(D)$. Les inclusions $\partial D \subset \partial\mathcal{H}(D) \subset \Lambda(D)$ se vérifient aisément.

b) L'ensemble D étant étoilé au point a_o , il est connexe par arcs. Il suffit alors de montrer que la propriété $(z \in \mathbb{C}^N, T(z) \subset D)$ implique : il existe un L_D-chemin joignant z à un point a de D . Soit $\gamma : I = [0,1] \to \mathbb{C}^N$ $\gamma(\tau) = \tau a_o + (1-\tau)z$; γ est continu $\gamma(0) = z$, $\gamma(1) = a_o$ et d'après la propriété 6) de 2.1.1. :

$$T[\gamma(\tau)] = \tau a_o + (1-\tau)T(z) \subset D \quad (\tau \in I) .$$

Ainsi γ est un L_D-chemin joignant z au point a_o. On vérifie que $W(D)$ est étoilé au point a_o.

c) Soit $z = (z_1, z_2)$ tel que $T(z) = \{z_1+z_2, \bar{z}_1+i\bar{z}_2\} \subset D$

Si $a \in D$ est un point arbitrairement fixé, considérons les deux chemins continus γ_1 et γ_2 de $I \to D$ tels que $\gamma_1(0) = z_1+iz_2, \gamma_1(1) = a$; $\gamma_2(0) = \bar{z}_1+i\bar{z}_2, \gamma_2(1) = a$. L'application $\sigma : I \to \mathbb{C}^2$ définie par

$$\sigma(t) = \{\frac{\gamma_1(t) + \overline{\gamma_2(t)}}{2}, \frac{\gamma_1(t) - \overline{\gamma_2(t)}}{2i}\}$$

est un chemin continu dans $\mathbb{C}^2$ avec $\sigma(0) = z$, $\sigma(1) = a$, et pour tout $\tau \in I$ on a $T[\sigma(\tau)] = \{\gamma_1(\tau), \gamma_2(\tau)\} \subset D$ donc $\{z \in \mathbb{C}^2 | T(z) \subset D\} \subset W(D)$, d'où le résultat.

2.1.8. COROLLAIRE.

<u>Soient</u> D_1 <u>et</u> D_2 <u>deux domaines de</u> $\mathbb{R}^N$, $\partial D_1 \neq \emptyset$, $\partial D_2 \neq \emptyset$.

a) <u>Si</u> $D_1 \cap D_2 = \emptyset$ <u>alors</u> $\mathcal{H}(D_1) \cap \mathcal{H}(D_2) = \emptyset$

b) <u>Si</u> $D_1 \subset D_2$ <u>alors</u> $\mathcal{H}(D_1) \subset \mathcal{H}(D_2)$

2.1.9. PROPOSITION.

<u>Soient</u> (D_ν) <u>une famille filtrante croissante de domaines de</u> $\mathbb{R}^N$ <u>et</u> $D = \bigcup_{\nu \in J} D_\nu$. <u>Alors la famille des cellules</u> $(\mathcal{H}(D_\nu))_{\nu \in J}$ <u>est filtrante croissante et on a</u> :

$$(2) \qquad \mathcal{H}(D) = \bigcup_{\nu \in J} \mathcal{H}(D_\nu)$$

<u>Démonstration</u> : La première affirmation résulte de 2.1.8. Pour établir

l'égalité (2), soient $z \in \mathcal{H}(D)$, $a \in D$. Il existe un L_D-chemin $\gamma : [0,1] \to \mathbb{C}^N$ joignant z au point a . L'ensemble

$$K = \bigcup_{0 \leq \tau \leq 1} T[\gamma(\tau)] \subset D$$

est compact (2.1.3.) et il existe $\nu_o \in J$ tel que $K \subset D_{\nu_o}$ c'est-à-dire $z \in \mathcal{H}(D_{\nu_o})$. D'où $\mathcal{H}(D) \subset \bigcup_{\nu \in J} \mathcal{H}(D_\nu)$. L'inclusion inverse étant évidente.

2.1.10. Si U est une matrice réelle, orthogonale d'ordre N et D un domaine de $\mathbb{R}^N$, $\partial D \neq \phi$, on a $\mathcal{H}(UD) = U\mathcal{H}(D)$.

En effet, pour tout $z \in \mathbb{C}^N$, $T(Uz) = UT(z)$ et l'image par U d'un L_D-chemin d'origine z et d'extrémité $a \in D$ est un L_{UD}-chemin d'origine Uz et d'extrémité Ua .

§ 2.2. EXEMPLES DE CELLULES D'HARMONICITÉ.

2.2.1. Cellule d'harmonicité d'un domaine convexe quelconque.

Soit $\delta_D(a)$ la fonction support d'un domaine convexe D de $\mathbb{R}^N$:

$$\delta_D(a) = \sup\{\langle a,b\rangle \mid b \in D\}$$

on a :

$$D = \{x \in \mathbb{R}^N \mid \langle x,a\rangle < \delta_D(a) \text{ pour tout } a \neq 0\}$$

et d'après 2.1.6. b) :

$$\mathcal{H}(D) = \{z = x + iy \mid T(z) \subset D\} = \{z \mid x+T(iy) \subset D\} .$$

La condition $x + \xi \in D$ $(\xi \in T(iy))$ équivaut à

$$(3) \qquad \langle x+\xi,a\rangle < \delta_D(a) \text{ pour tout } a \neq 0 .$$

Mais $\delta_D(a)$ étant positivement homogène, il suffit que (3) soit vérifiée pour tout $a, \|a\| = 1$. Donc,

(4) $$\mathcal{H}(D) = \{z = x+iy \mid \sup_{\xi \in T(iy)} [\, \max_{\|a\|=1} (\langle x+\xi, a\rangle - \delta_D(a))] < 0\}\ .$$

2.2.2. Cellule d'harmonicité d'une boule

Soit $B = B(0,R) \subset \mathbb{R}^N$ la boule de centre 0 et de rayon $R > 0$.

La fonction support de B est $\delta_D(a) = R\|a\|$. On a :

$$\max_{\|a\|=1} (\langle x + \xi, a\rangle - \delta_B(a)) = \|x + \xi\| - R$$

et $$\mathcal{H}[B(0,R)] = \{z = x+iy \in \mathbb{C}^N \mid \sup_{\xi \in T(iy)} \|x+\xi\| - R < o\}$$

Calculons $\sup_{\xi \in T(iy)} \|x+\xi\|^2 = \|x\|^2 + \|y\|^2 + 2 \max_{\xi \in T(iy)} \langle x, \xi\rangle$

2.2.3. LEMME

Pour tout $z = x + iy \in \mathbb{C}^N$:

$$\begin{matrix} M(z) \\ m(z) \end{matrix} = \begin{matrix} \mathrm{Max} \\ \mathrm{Min} \end{matrix} [\langle \xi, x\rangle \ ,\ \xi \in T(iy)] = \pm[\|x\|^2 . \|y\|^2 - \langle x, y\rangle^2]^{\frac{1}{2}} =$$

$$\pm \|x\| \ \|y\| \ |\sin(\widehat{x,y})| = \pm \tfrac{1}{2} \sqrt{\|z\|^4 - \|\sum_{j=1}^{N} z_j^2\|^2}\ .$$

Démonstration : Rappelons (2.1.1.) que $\xi \in T(iy)$ équivaut à $\|\xi\|^2 = \|y\|^2$ et $\langle \xi, y\rangle = 0$. Cherchons les extrema de la fonction

$$L_o(\xi) = \xi_1 x_1 + \ldots + \xi_N x_N$$

sous les conditions

$$(5)\qquad \begin{aligned} L_1(\xi) &= \|\xi\|^2 - \|y\|^2 = 0 \\ L_2(\xi) &= \langle \xi, y\rangle = 0 \end{aligned}$$

utilisons la méthode des multiplicateurs de Lagrange : soit

$$L(\xi) = (L_o + \alpha L_1 + \rho L_2)(\xi) \qquad (\alpha, \beta \ \text{constantes}).$$

Les conditions extrema libres sont :

$$\frac{\partial L}{\partial \xi_j} = x_j + 2\alpha\xi_j + \beta y_j = 0 \ (j = 1, \dots, N) .$$

Supposons tout d'abord x et y linéairement indépendants sur $\mathbb{R}$; les points ξ donnant les extrema sont de la forme : $\xi = ax + by$ $(a, b \in \mathbb{R})$, les constantes a et b étant déterminées par les conditions (5) . On trouve

$$(a,b) = \left(\epsilon \frac{\|y\|^2}{M(z)}, - \epsilon \frac{\langle x,y\rangle}{M(z)}\right) \ (\epsilon = \pm 1) .$$

D'où

$$\langle ax + by, x\rangle = \epsilon \frac{\|y\|^2\|x\|^2}{M(z)} - \epsilon \frac{\langle x,y\rangle^2}{M(z)}$$

$$\mathop{\mathrm{Max}}_{\mathrm{Min}}[\langle \xi, x\rangle, \xi \in T(iy)] = \epsilon[\|x\|^2 . \|y\|^2 - \langle x,y\rangle^2]^{\frac{1}{2}} .$$

Si x et y sont linéairement dépendants, pour tout $\xi \in T(iy)$ on a $\langle \xi, x\rangle = 0$. D'où le lemme.

Il en résulte que

$$\mathcal{H}[B(0,R)] = \{z = x+iy \in \mathbb{C}^N | (\|x\|^2 + \|y\|^2 + 2\sqrt{\|x\|^2\|y\|^2 - \langle x,y\rangle^2})^{\frac{1}{2}} < R\}$$

$$= \{z \in \mathbb{C}^N | [\|z\|^2 + \sqrt{\|z\|^4 - \|\sum_{j=1}^{N} z_j^2\|^2}\,]^{\frac{1}{2}} < R\}$$

On notera que $\mathcal{H}[B(0,R)]$ est un domaine équilibré (i.e. si $\alpha \in \mathbb{C}$, $|\alpha| < 1$, on a $\alpha\mathcal{H}[B(0,R)] \subset \mathcal{H}[B(0,R)]$).

Dans le paragraphe suivant nous reviendrons sur cette cellule qui a fait l'objet des études en analyse harmonique [13] [16] [22] [23] .

<u>Remarque</u>. Dans le cas $N = 2$ on peut retrouver directement la cellule d'harmonicité du disque $B(R) = \{(x_1,x_2) \in \mathbb{R}^2 | x_1^2 + x_2^2 < R^2\}$. En effet, d'après 2.1.6. c)

$$\mathcal{H}(B(R)) = \{z = (z_1,z_2) \in \mathbb{C}^2 | \ |z_1 + iz_2| < R, |\bar{z}_1 + i\bar{z}_2| < R\} .$$

<u>Remarque</u>. Si D est borné, $\mathcal{H}(D)$ l'est aussi. En effet, D est contenu dans une boule, le corollaire 2.1.8. s'applique.

2.2.4. <u>Cellule d'harmonicité d'un cube, d'un demi-espace, d'un tube</u>.

Soit

$$D = \{x = (x_1,\dots,x_N) | a_j < x_j < b_j \quad j = 1,\dots,N\} .$$

Le cube D étant étoilé par rapport à son centre, d'après 2.1.6. :

$$\mathcal{H}(D) = \{z \in \mathbb{C}^N | T(z) \subset D\} .$$

Soit $z = x + iy$. On a $T(z) = x + T(iy)$. La condition $T(z) \subset D$ équivaut à : pour tout $\xi = (\xi_1,\dots,\xi_N) \in T(iy)$, $a_j < x_j + \xi_j < b_j$ $j = 1,\dots,N$.
Il en résulte que $T(z) \subset D$ équivaut à :

$$(6) \qquad \begin{aligned} &x_j + \operatorname{Max}(\xi_j | \xi \in T(iy)) < b_j \\ &x_j + \min(\xi_j | \xi \in T(iy)) > a_j \end{aligned} \qquad j = 1,\dots,N$$

Or la méthode des multiplicateurs de Lagrange montre que

$$\underset{\text{Min}}{\text{Max}}[\xi_j | \xi \in T(iy)] = \underset{\text{Min}}{\text{Max}}[\xi_j | \xi_1^2+\dots+\xi_N^2 = \|y\|^2, \xi_1 y_1 +\dots+ \xi_N y_N = 0]$$

$$= \pm\left(\sum_{k\neq j} y_k^2\right)^{\frac{1}{2}}$$

des inégalités (6) on déduit :

$$\mathcal{H}(D) = \{z = x+iy \in \mathbb{C}^N | a_j < x_j - \left(\sum_{k\neq j} y_k^2\right)^{\frac{1}{2}} \leq x_j + \left(\sum_{k\neq j} y_k^2\right)^{\frac{1}{2}} < b_j, j=1,\dots,N\}$$

2.2.5. En particulier si $D = \{x = (x_1,\dots,x_N) | |x_j| < a_j \ j = 1,\dots,N\}$, on a

$$\mathcal{H}(D) = \{z = x+iy \in \mathbb{C}^N | |x_j| + \left(\sum_{k\neq j} y_k^2\right)^{\frac{1}{2}} < a_j \quad j = 1,\dots,N\} .$$

On retrouve un résultat figurant dans [15] .

2.2.6. Pour le demi-espace $\Pi_N = \{x \in \mathbb{R}^N | x_N > 0\}$ on a :

$$\mathcal{H}(\Pi_N) = \{z = x+iy \in \mathbb{C}^N | x_N - \left(\sum_{k\neq N} y_k^2\right)^{\frac{1}{2}} > 0\} .$$

et pour le tube

$$D =]a_1,b_1[\times\dots\times]a_{N-1},b_{N-1}[\times \mathbb{R}(x_N) ,$$

$$\mathcal{H}(D) = \{z = x + iy \in \mathbb{C}^N | a_j < x_j - \left(\sum_{k\neq j} y_k^2\right)^{\frac{1}{2}} \leq x_j + \left(\sum_{k\neq j} y_k^2\right)^{\frac{1}{2}} < b_j, j = 1,\dots,N-1\}$$

Si $D =]a,b[\times \mathbb{R}^{N-1}(x_2,\dots,x_N)$, on a :

$$\mathcal{H}(D) = \{z = x + iy \in \mathbb{C}^N | a < x_1 - \left(\sum_{k\neq 1} y_k^2\right)^{\frac{1}{2}} \leq x_1 + \left(\sum_{k\neq 1} y_k^2\right)^{\frac{1}{2}} < b\} .$$

2.2.7. Cellule d'harmonicité d'un polyèdre convexe.

$$D = \{x \in \mathbb{R}^N | a_1^i x_1 + \ldots + a_N^i x_N = \langle a^i, x \rangle < \alpha^i \quad 1 \le i \le p\} .$$

Le domaine D étant étoilé par rapport à un de ces points on a :

$$\mathcal{H}(D) = \{z \in \mathbb{C}^N | T(z) \subset D\} = \{z = x+iy | x+T(iy) \in D\}$$

soit ξ un point de $T(iy)$.

On a

$$(*) \quad (\xi \in T(iy)) \Leftrightarrow (\xi_1^2 + \ldots + \xi_N^2 = \|y\|^2 \ , \ \langle \xi, y \rangle = 0)$$

et

$$(x+\xi \in D_y) \Leftrightarrow (\langle a^i, x+\xi \rangle < a^i \quad 1 \le i \le p)$$

$$(*) \quad \Leftrightarrow (\langle a^i, x \rangle + \underset{\xi \in T(iy)}{\text{Max}} \langle a^i, \xi \rangle < a^i \quad 1 \le i \le p \ )$$

or

$$\underset{\xi \in T(iy)}{\text{Max}} \ \langle a^i, \xi \rangle = \sqrt{\|a^i\|^2 \|y\|^2 - \langle a^i, y \rangle^2} .$$

Finalement

$$\mathcal{H}(D) = \{z = x+iy \in \mathbb{C}^N | \langle a^i, x \rangle + \sqrt{\|a^i\|^2 \|y\|^2 - \langle a, y \rangle^2} < \alpha^i\} \quad 1 \le i \le p .$$

2.2.8. Le domaine constitué par la différence de deux boules concentriques n'étant pas étoilé par rapport à un de ses points, on ne peut utiliser pour la détermination de sa cellule, la partie b) de la proposition 2.1.6. Nous procédons comme dans [14] .

2.2.9. Cellule d'harmonicité de la différence de deux boules centrées en 0 [14].

Soit

$$B(R_1, R_2) = \{x \in \mathbb{R}^N | R_1 < \|x\| < R_2\} .$$

Si $z = x + iy \in \mathbb{C}^N$, posons

$$L_\pm(z) = L_\pm(x,y) = [\|z\|^2 \pm \sqrt{\|z\|^4 - \|\sum_{j=1}^{N} z_j^2\|^2}]^{\frac{1}{2}}$$

$$= [\|x\|^2 + \|y\|^2 \pm 2\sqrt{\|x\|^2\|y\|^2 - <x,y>^2}]^{\frac{1}{2}}$$

on remarquera que $L_+(z) \geq \|z\|$, $L_-(z) \geq |\|x\|-\|y\||$ (voir aussi 2.3.1., 2.3.9) .

Remarquer que $L_\pm(e^{i\theta}z) = L_\pm(z)$ $(\theta \in \mathbb{R})$.

2.2.10. LEMME.

<u>Pour tout</u> $z \in \mathbb{C}^n$, <u>la condition</u> $T(z) \subset B(R_1,R_2)$ <u>équivaut à</u>

$$R_1 < L_-(z) \leq L_+(z) < R_2 .$$

<u>Démonstration.</u> Comme dans 2.2.2, la condition $T(z) \subset B(R_1,R_2)$ équivaut à

$$R_1^2 < \|x\|^2 + \|y\|^2 + 2\min_{\xi \in T(iy)} <\xi,x> < \|x\|^2 + \|y\|^2 + 2\mathrm{Max}<\xi,x> < R_2^2 .$$

D'où le lemme d'après 2.2.3.

2.2.11. LEMME.

<u>Si</u> $T(z) \subset B(R_1,R_2)$, <u>il existe un point</u> $a \in B(R_1,R_2)$ <u>et un</u> $L_{B(R_1,R_2)}$<u>-chemin</u> γ <u>joignant</u> z <u>à</u> a .

<u>Démonstration</u>.

Soient $z = x + iy \in \mathbb{C}^N$ et $T(z) \subset B(R_1,R_2)$. Distinguons plusieurs cas :

1) Si $y = 0$, on a $T(z) = \{x\} \subset B(R_1,R_2)$; on choisit $a = x$ et γ l'application constante : $\gamma(\tau) = x$, $\tau \in [0,1]$.

2) Si $y \neq 0$, $x = 0$, l'inclusion $T(iy) \subset B(R_1,R_2)$ implique

$R_1 < \|y\| < R_2$ d'où $y \in B(R_1)$. On choisit $a = y$ et $\gamma : \tau \mapsto \tau y + i\sqrt{1-\tau^2}\, y$ $(\tau \in [0,1])$.

3) Si x, y sont différents de zéro, et $x = \alpha y$ $(\alpha \in \mathbb{R})$, le lemme 2.2.3. montre que $R_1^2 < (1+\alpha^2)\|y\|^2 < R_2^2$. Le $L_{B(R_1,R_2)}$-chemin :

$$\gamma_1 : \tau \to (1-\tau)\alpha y + i\sqrt{[1-(1-\tau)^2]\alpha^2+1}\; y$$

permet de joindre le point $z = \gamma_1(0)$ au point $b = \gamma_1(1) = i\sqrt{1+\alpha^2}\, y \neq 0$.
(On remarquera que $L_\pm[\gamma_1(\tau)] = (1+\alpha^2)\|y\|^2)$.
D'autre part, il existe (d'après le cas 2)) un point $a \in (R_1, R_2)$ et un $L_{B(R_1,R_2)}$-chemin γ_2 joignant le point b au point a. D'où le résultat.

4) Soit $U(z) = \bigcup_{\alpha \in [0,1]} T(x+i\alpha y)$.

Si $U(z) \subset B(R_1,R_2)$, on a $T(x) = \{x\} \subset B(R_1,R_2)$ et on choisit $a = x$, $\gamma : \tau \mapsto x + i\tau y$ $(\tau \in [0,1])$.

5) Supposons x et y linéairement indépendants et $U(z) \not\subset B(R_1,R_2)$.
Considérons le chemin continu

$$\gamma : \tau \to x+\tau(\mu y - x)+iy = \gamma(\tau) \quad (\tau \in [0,1]) \quad (\mu \in \mathbb{R}).$$

On a $\gamma(0) = x + iy = z$; $\gamma(1) = \mu y + iy$.

D'après le cas 3, il existe un point $a \in B(R_1,R_2)$ et un $L_{B(R_1,R_2)}$-chemin joignant le point $\mu y + iy$ au point a. Il suffit donc de montrer que pour une certaine valeur de μ, γ est un L_{B_1,B_2}-chemin joignant z au point $\mu y + iy$.
D'après le lemme 2.2.10., cela revient à vérifier que pour tout $\tau \in [0,1]$:

$$R_1 < L_-[\gamma(\tau)] \le L_+[\gamma(\tau)] < R_2 .$$

On a :

$$L_{\pm}^2[\gamma(\tau)] = \|x+\tau(\mu y-x)\|^2+\|y\|^2 \pm 2\sqrt{\|x+\tau(\mu y-x)\|^2\|y\|^2-\langle x+\tau(\mu y-x),y\rangle^2}$$

$$\|x+\tau(\mu y-x)\|^2 = \|x\|^2+\tau^2\|\mu y-x\|^2 + 2\tau \langle x,\mu y-x\rangle$$

$$= \|x\|^2 + \tau^2[\mu^2\|y\|^2 + \|x\|^2 - 2\mu \langle x,y\rangle] + 2\tau\mu \langle x,y\rangle - 2\tau\|x\|^2$$

$$= (1-\tau)^2\|x\|^2+ \tau^2\mu^2\|y\|^2 - 2\mu\tau(\tau-1) \langle x,y\rangle$$

$$\langle x+\tau(\mu y-x),y\rangle = \langle x,y\rangle + \tau \langle \mu y-x,y\rangle .$$

Choisissons μ tel que $\langle \mu y-x,y\rangle = 0$ soit $\mu = \dfrac{\langle x,y\rangle}{\|y\|^2}$. Alors

$$f_{\pm}(t) = L_{\pm}^2[\gamma(\tau)] = t^2 \frac{M(x,y)}{\|y\|^2} \pm 2tM(x,y) + \frac{\langle x,y\rangle^2}{\|y\|^2} + \|y\|^2$$

avec

$$M^2 = \|x\|^2\|y\|^2 - \langle x,y\rangle^2 , \quad t = 1-\tau \in [0,1] .$$

Remarquons que $f_+(t)$ est une fonction croissante de $t \in [0,1]$ et par conséquent $f_+(t) \le f_+(1) = L_+^2(z) < R_2^2$ pour tout $t \in [0,1]$ (Lemme 2.2.3).
Reste à vérifier l'inégalité $f_-(t) > R_1^2, t \in [0,1]$. Considérons la fonction

$$\varphi(\alpha) = L_-^2(x+i\alpha y) - R_1^2 = \alpha^2\|y\|^2 - 2\alpha M + \|x\|^2 - R_1^2 \quad (\alpha \in [0,1]) .$$

L'hypothèse $U(z) \not\subset B(R_1,R_2)$ implique l'existence d'un $\alpha_o \in [0,1]$ tel que $\varphi(\alpha_o) \le 0$ (Lemme 2.2.10) et comme d'autre part $\varphi(1) > 0$, on en déduit que le trinôme

$$\alpha^2\|y\|^2 - 2\alpha M + \|x\|^2 - R_1^2 \quad (\alpha \in \mathbb{R})$$

a deux racines réelles et que la plus grande des racines est < 1. Donc,

$$\delta = M^2 - \|y\|^2(\|x\|^2 - R_1^2) = R_1^2\|y\|^2 - \langle x,y\rangle^2 \ge 0$$

et

$$\frac{M+\sqrt{\delta}}{\|y\|^2} < 1 .$$

D'où

$$(7) \qquad \frac{\langle x,y\rangle^2}{\|y\|^2} \geq R_1^2 \quad \text{et} \quad \frac{M}{\|y\|^2} < 1 \ .$$

Or, le trinôme

$$t^2 \frac{M}{\|y\|^2} - 2tM + \frac{\langle x,y\rangle^2}{\|x\|^2} + \|y\|^2 \quad (t \in \mathbb{R})$$

est minimum pour $t_o = \frac{\|y\|^2}{M} > 1$ avec la valeur minimale égale à $\frac{\langle x,y\rangle^2}{\|y\|^2} \geq R_1^2$ d'après (7)). Il en résulte que sur $[0,1]$, on a $f_-(t) > R_1^2$. Ce qui achève la démonstration du lemme 2.2.11.

D'où :

<u>La cellule d'harmonicité de</u> $B(R_1,R_2) = \{x \in \mathbb{R}^N | R_1 < \|x\| < R_2\}$ <u>est le domaine circulaire</u> :

$$\mathcal{H}(B_{R_1,R_2}) = \{z \in \mathbb{C}^N | R_1 < L_-(z) \leq L_+(z) < R_2\} \ .$$

<u>En particulier</u>

<u>La cellule d'harmonicité de</u> $E = \{z \in \mathbb{C}^N | \|z\| > R\}$ <u>est</u>

$$\mathcal{H}(E) = \{z = x+iy \in \mathbb{C}^N | \|x\|^2 + \|y\|^2 - 2\|x\|\|y\| |\operatorname{Sin}(x,y)| > R^2\}$$

ou

$$\mathcal{H}(E) = \{z \in \mathbb{C}^N | [\|z\|^2 - \sqrt{\|z\|^4 - \|\Sigma\, z^2\|^2}]^{\frac{1}{2}} > R\} \ .$$

2.2.12. <u>Remarque</u>. La formule (3) de 2.2.1. et la propriété

$$\delta_{D_1+D_2}(a) = \delta_{D_1}(a) + \delta_{D_2}(a)$$

de la fonction support d'un ouvert convexe de $\mathbb{R}^N$ peuvent dans certains cas nous fournir la cellule d'harmonicité de la somme d'un ouvert convexe et d'une boule.

2.2.13. Remarque. Si ω_1 et ω_2 sont deux domaines de $\mathbb{R}^N$, en général $\mathcal{H}(\omega_1 \times \omega_2) \neq \mathcal{H}(\omega_1) \times \mathcal{H}(\omega_2)$ (exemple:le tube 2.2.6.) mais si ω est un ouvert étoilé de $\mathbb{R}^N$ on a

$$\mathcal{H}(\omega \times \mathbb{R}) \supset \mathcal{H}(\omega) \times \{0\}$$

En effet, on vérifie que si $(z,o) \in \mathcal{H}(\omega) \times \{0\}$ donc si $T(z) \subset \omega$, alors $T[(z,o)] \subset \omega \times \mathbb{R}$.

§ 2.3. BOULE DE LIE ET SA FRONTIÈRE DE BERGMAN - ŠILOV. LE TUBE D'ELIE CARTAN.

2.3.1. Dans [11], Elie Cartan avait déterminé tous les domaines bornés homogènes symétriques de $\mathbb{C}^N$. Un domaine $D \subset \mathbb{C}^N$ est dit homogène si le groupe de l'automorphisme $Au(D)$ agit transitivement sur D (i.e. pour tout $a,b \in D$, il existe $\sigma \in Au(D)$ tel que $\sigma a = b$). Le domaine D est dit symétrique par rapport au point $a \in D$, s'il existe $\sigma \in Au(D)$ involutif (i.e. $\sigma^2 = Id$) laissant le point a invariant. Le domaine D est dit symétrique s'il l'est en tous ses points (pour qu'un domaine homogène soit symétrique, il suffit qu'il le soit en un de ses points). Elie Cartan a montré qu'il existe six types de domaines bornés homogènes symétriques (irréductibles). A part les deux types particuliers correspondant aux cas $N = 16$, $N = 27$, il reste 4 types dont les trois premiers sont des domaines bien connus (à un homéomorphisme analytique complexe près) : boules, polydisques, etc... Le quatrième est un domaine de type nouveau : c'est l'ensemble des points $\zeta = (\zeta_1,\dots,\zeta_N) \in \mathbb{C}^N$ vérifiant les deux inégalités suivantes :

$$(8) \quad \begin{cases} |\zeta_1^2 + \dots + \zeta_N^2| < R^2 \\ |\zeta_1^2 + \dots + \zeta_N^2|^2 - 2R^2(\zeta_1 \bar{\zeta}_1 + \dots + \zeta_N \bar{\zeta}_N) + R^4 > 0 \end{cases}$$

il est aisé de voir [13] que l'ensemble des inégalités (8) équivaut à l'unique inégalité

$$(9)\qquad \zeta_1\bar{\zeta}_1+\dots+\zeta_N\bar{\zeta}_N+\sqrt{(\zeta_1\bar{\zeta}_1+\dots+\zeta_N\bar{\zeta}_N)^2-|\zeta_1^2+\dots+\zeta_N^2|^2}<R^2$$

on reconnaît la cellule d'harmonicité de la boule $B(0,R)\subset\mathbb{R}^N$ (2.2.2.).
L'application $\zeta\to L(\zeta)=(\|\zeta\|^2+\sqrt{\|\zeta\|^4-|\zeta_1^2+\dots+\zeta_N^2|^2})^{\frac{1}{2}}$ définie une norme sur $\mathbb{C}^N$ appelée <u>la norme de Lie</u>. Pour cette raison le domaine (9) i.e. $\mathcal{H}[B(0,R)]$, est appelé la boule de Lie(de centre 0 et de rayon R notée $BL(0,R)$) dans $\mathbb{C}^N$ (disque de Lie si $N=1$).
Si nous identifions le vecteur $\zeta=(\zeta_1,\dots,\zeta_N)\in\mathbb{C}^N$ avec la matrice colonne

$\zeta=\begin{pmatrix}\zeta_1\\ \vdots\\ \zeta_N\end{pmatrix}$ et si on note $t_\zeta=(\zeta_1\ \dots\ \zeta_N)$, $\bar{\zeta}=(\bar{\zeta}_1\ \dots\ \bar{\zeta}_N)$

$\zeta^*=t_{\bar{\zeta}}$, la boule unité de Lie dans $\mathbb{C}^N$ est définie par

$$BL(0,1)=\{\zeta\in\mathbb{C}^N|\zeta^*\zeta+\sqrt{(\zeta^*\zeta)^2-|t_\zeta\cdot\zeta|^2}<1\}$$

2.3.2. Dans [11] Elie Cartan montre que la boule unité de Lie est analytiquement homéomorphe au tube de $\mathbb{C}^N$:

$$(10)\qquad \{z=x+iy\in\mathbb{C}^N|y_N>\sqrt{y_1^2+\dots+y_{N-1}^2}\}$$

Nous appelons le domaine (10) <u>le tube d'Elie Cartan</u>.

2.3.3. THÉORÈME.

<u>La boule unité de Lie, le tube d'Elie Cartan et la cellule d'harmonicité de demi espace</u> $\{x\in\mathbb{R}^N|x_N>0\}=\Pi_N$ <u>sont identiques (à un homéomorphisme analytique complexe près)</u>.

En effet, la transformation Cayley généralisé de $\mathbb{C}^N(\zeta) \underset{z}{\overset{\zeta}{\rightleftarrows}} \mathbb{C}^N(z)$

$$\begin{cases} \zeta_k = - \dfrac{2i\ z_k}{Z} & k = 1,\ldots,N-1 \\[2ex] \zeta_N = i - \dfrac{2\ (z_N+i)}{Z} \end{cases}$$

$$\begin{cases} z_k = \dfrac{-\ 2i\ \zeta_k}{W} & k = 1,\ldots,N-1 \\[2ex] z_N = -\ i + \dfrac{2(\zeta_N - i)}{W} \end{cases}$$

$$Z = z_1^2 + \ldots + z_{N-1}^2 - (z_N+i)^2 = -\ 4\ W^{-1}$$

$$W = \zeta_1^2 + \ldots + \zeta_{N-1}^2 + (\zeta_N - i)^2 \ , \quad \frac{D(\zeta_1\ldots\zeta_N)}{D(z_1\ldots z_N)} = -2^N\ (-i)^{N+1}\ Z^{-N}$$

définit un homéomorphisme analytique complexe entre la boule unité de Lie et le tube d'Elie Cartan (voir par exemple [22]). La cellule d'harmonicité de demi-espace Π_N est d'après 2.2.6. le domaine :

$$\mathcal{H}(\Pi_N) = \{z = x + iy \in \mathbb{C}^N \mid x_N > \sqrt{y_1^2+\ldots+y_{N-1}^2}\}$$

qui est homéomorphe (analytique complexe) au tube d'Elie Cartan par l'homéomorphisme

$$f : (z_1,\ldots,z_N) \mapsto (z_1,\ldots,z_{N-1}, iz_N)$$

D'où le théorème 2.3.3.

$$\begin{array}{ccc} \mathbb{C}^N \supset \text{Boule unité de Lie} & \xrightarrow[\tilde{\zeta}]{\text{Transformation de Cayley}} & \text{Tube d'Elie Cartan} \subset \mathbb{C}^N \\ \big\downarrow p_r & & \big\downarrow f \\ & & \mathcal{H}(\Pi_N) \subset \mathbb{C}^N \\ & & \big\downarrow p_r \\ \mathbb{R}^N \supset \text{boule } B(0,1) & \xleftarrow[\text{inversion}]{} & \Pi_N \subset \mathbb{R}^N \end{array}$$

$p_r : \mathbb{C}^N \to \mathbb{R}^N \quad z = x + iy \mapsto x$

$f \ : \mathbb{C}^N \to \mathbb{C}^N \quad (z_1,\ldots,z_N) \mapsto (z_1,\ldots,z_{N-1},iz_N)$

2.3.4. Le résultat 2.3.3. et le fait qu'une inversion convenable transforme le demi espace Π_N à $B(0,1)$ nous suggèrent une question et une conjecture :

A. Quelles relations doivent exister entre deux domaines bornés de $\mathbb{R}^N$ pour que leurs cellules d'harmonicités dans $\mathbb{C}^N$ soient analytiquement homéomorphes ?

B. Soient D_1 et D_2 deux domaines de $\mathbb{C}^N$ identifié à $\mathbb{R}^{2N}$.

Si D_1 et $D_2 \neq \mathbb{C}^N$ sont analytiquement homéomorphes leur cellule d'harmonicité sont analytiquement homéomorphes dans $\mathbb{C}^{2N}$ (?)

La conjecture B a une démonstration dans $\mathbb{C}$. On a en effet :

2.3.5. THÉORÈME [14]

Soient D_1, $D_2 \neq \mathbb{C}$ deux domaines du plan complexe, $f = u + iV$ un homéomorphisme analytique de $D_1 \to D_2$. Alors $\mathcal{H}(D_1)$ et $\mathcal{H}(D_2)$ sont analytiquement homéomorphes. L'application holomorphe $\tilde{f} : \mathcal{H}(D_1) \to \mathbb{C}^2$ **définie par :**

$$\tilde{f} = \begin{pmatrix} \tilde{u} \\ \tilde{v} \end{pmatrix} = \begin{pmatrix} \frac{1}{2}[f(z_1+iz_2) + \overline{f(\bar{z}_1 + i\bar{z}_2)}] \\ \frac{1}{2i}[f(z_1+iz_2) - \overline{f(\bar{z}_1 + i\bar{z}_2)}] \end{pmatrix}$$

réalise cet homéomorphisme.

2.3.6. COROLLAIRE.

<u>La cellule d'harmonicité de tout domaine</u> $D \subset \mathbb{C}$, <u>simplement connexe,</u> ∂D <u>ayant au moins deux points distincts, est analytiquement homéomorphe au disque unité de Lie de</u> $\mathbb{C}$.

<u>Démonstration de</u> 2.3.5. D'après 2.1.6. c) les fonctions $\tilde{u}$, $\tilde{v}$ sont bien définies dans $\mathcal{H}(D_1)$, y sont holomorphes, et $T[\tilde{f}(z)] \subset f[T(z)]$ si $z \in \mathcal{H}(D_1)$. Donc, $\tilde{f}(\mathcal{H}(D_1)) \subset \mathcal{H}(D_2)$.

Soit $w = (w_1, w_2) \in \mathcal{H}(D_2)$ alors $\{w_1 + iw_2, \bar{w}_1 + i\bar{w}_2\} \subset D_2$ et il existe ξ_1 , $\xi_2 \in D_1$ tels que $w_1 + iw_2 = f(\xi_1)$, $\bar{w}_1 + i\bar{w}_2 = f(\xi_2)$. Soient $z_1 = \frac{1}{2}(\xi_1 + \bar{\xi}_2)$, $z_2 = \frac{1}{2i}(\xi_1 - \bar{\xi}_2)$, $z = (z_1, z_2) \in \mathbb{C}^2$. On a $T(z) = \{\xi_1, \xi_2\} \subset D_1$, donc $z \in \mathcal{H}(D_1)$

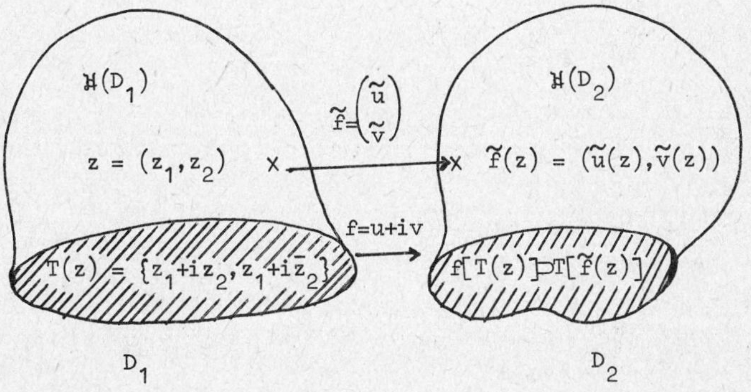

et $W = \tilde{f}(z) \subset \tilde{f}(\mathcal{H}(D_1))$ c'est-à-dire $\mathcal{H}(D_2) \subset \tilde{f}(\mathcal{H}(D_1))$. D'où

$$\tilde{f}(\mathcal{H}(D_1)) = \mathcal{H}(D_2)$$

et $\tilde{f}$ est une application de $\mathcal{H}(D_1)$ sur $\mathcal{H}(D_2)$. Si $g = f^{-1} : D_2 \to D_1$, on définit $\tilde{g}$ d'une manière analogue. Les applications $\tilde{f}$, $\tilde{g}$ sont holomorphes avec $\tilde{f} \circ \tilde{g} = id_{D_2}$, $\tilde{g} \circ \tilde{f} = id_{D_1}$. Il en résulte que $\tilde{f} \circ \tilde{g} = id_{\mathcal{H}(D_1)}$, $\tilde{f} \circ \tilde{g} = id_{\mathcal{H}(D_2)}$. Finalement $(\tilde{f})^{-1} = \tilde{g}$ et $\tilde{f}$ est un homéomorphisme analytique de $\mathcal{H}(D_1) \to \mathcal{H}(D_2)$.

Le corollaire 2.3.6. résulte du fait que D est homéomorphe au disque unité.

2.3.7. Frontière de Bergman - Šilov de la boule de Lie $\mathcal{H}(R) = BL(0,R)$.

THÉORÈME.

La plus petite partie fermée de $\partial \mathcal{H}(R)$ telle que toute fonction holomorphe dans $\mathcal{H}(R)$ et continue sur $\overline{\mathcal{H}(R)}$ atteint son maximum en module (Frontière de Bergman-Šilov de $\mathcal{H}(R)$) est l'ensemble

$$\check{\mathcal{B}}_R = \{\zeta = x\, e^{i\theta} \in \mathbb{C}^N | x \in \mathbb{R}^N ,\ \|x\| = R ,\ \theta \in \mathbb{R}\} \subset \partial\mathcal{H}(R)$$

$$(\partial\mathcal{H} = \{\zeta \in \mathbb{C}^N | \zeta^*\zeta + \sqrt{(\zeta^*\zeta)^2 - |{}^t\zeta\zeta|^2} = R^2\})$$

Démonstration :

Cet énoncé qui figure déjà dans [13] peut être établi en utilisant les résultats de [16] (th. 3.2). La démonstration qui va suivre est directe et n'utilise que des considérations élémentaires. Il suffit en effet de montrer :

a) Quel que soit $b \in \check{\mathcal{B}}_R$ il existe une fonction holomorphe dans $\mathcal{H}(R)$ et continue sur $\overline{\mathcal{H}(R)}$ qui atteint son maximum en module dans $\mathcal{H}(R)$, au point unique b .

b) Si $a \in \partial\mathcal{H}(R) \setminus \check{\mathcal{B}}_R$, alors a n'appartient pas à la frontière de Bergman-Šilov.

Sans restreindre de la généralité on supposera $R = 1$, $\check{\mathcal{B}} = \check{\mathcal{B}}_1$, $\mathcal{H} = \mathcal{H}(1)$.

Pour établir a) remarquons que le groupe orthogonal $SO(N)$ agit transitivement sur la sphère $\|x\| = 1$ de $\mathbb{R}^N$, et comme un point de $\check{\mathcal{B}}$ est de la forme $e^{i\theta}x(\|x\|=1)$, le groupe des matrices de la forme $e^{i\theta}k, \theta \in \mathbb{R}$, $k \in SO(N)$ agit transitivement sur $\check{\mathcal{B}}$. Donc, tout point $b \in \check{\mathcal{B}}$ est de la forme : $b = e^{i\theta} k a_o$ avec $a_o = (1,0,\ldots,0) \in \check{\mathcal{B}}$, $k \in SO(N)$, $\theta \in \mathbb{R}$. La fonction $f(\zeta_1,\ldots,\zeta_N)=\zeta_1 + 1$ prend sa valeur maximale en module au seul point a_o et la fonction $\zeta \mapsto f(e^{-i\theta}k^{-1}\zeta)$ $(\zeta \in \mathcal{H})$ au seul point b.

Démonstration de b).

LEMME 1.

<u>Soient</u> $U = \{z \in \mathbb{C} \mid |z| < 1\}$ et $h : \mathbb{C} \to \mathbb{C}^N$ <u>définie par</u>

$z \mapsto (\frac{1-z}{2i}\ \frac{1+z}{2}\ 0 \ldots 0) = \zeta$. <u>Alors</u>

$$\Omega = h(U) \subset \partial\mathcal{H} \setminus \check{\mathcal{B}}$$
$$\partial\Omega = h(\partial U) \subset \check{\mathcal{B}}$$

En effet,
$$\zeta^*\zeta = \frac{1-z}{2i}\left(\overline{\frac{1-z}{2i}}\right) + \frac{1+z}{2} - \left(\overline{\frac{1+z}{2}}\right)$$
$$= \tfrac{1}{2}(1 + |z|^2)$$
$$t_\zeta \cdot \zeta = -\left(\frac{1-z}{4}\right)^2 + \left(\frac{1+z}{4}\right)^2 = z$$

et $|z| < 1$ implique $\zeta^*\zeta < 1$, $|t_\zeta \cdot \zeta| = |z| < 1$; comme

$$\zeta^*\zeta + \sqrt{(\zeta^*\zeta)^2 - |t_\zeta \cdot \zeta|^2} = 1$$

alors $\zeta \in \partial\mathcal{H}$. Mais $\zeta \notin \check{\mathcal{B}}$, puisque $\|\zeta\| = (\zeta^*\zeta)^{\frac{1}{2}} < 1$ si $|z| < 1$. Donc

$$\Omega = h(U) \subset \partial\mathcal{H} \setminus \check{\mathcal{B}}$$

De même,

$$\partial\Omega = h(\partial U) = \{h(z) \mid |z|=1\} = \{\zeta \mid \zeta^*\zeta = 1 \ , \ |t_\zeta \cdot \zeta|=1\} \subset \check{\mathcal{B}}$$

LEMME 2.

<u>Si</u> $a \in \partial\mathcal{H}\setminus\check{\mathcal{B}}$ <u>il existe une matrice de la forme</u> $e^{i\theta}k$, $k \in SO(N)$, $\theta \in \mathbb{R}$ <u>telle que</u> $e^{i\theta}ka \in \Omega$.

En effet, posons $u = t_a \cdot a$. On a $|u|<1$ et

$$\eta_a = \begin{pmatrix} \frac{1-u}{2i} \\ \frac{1-u}{2} \\ 0 \\ \vdots \\ 0 \end{pmatrix} = b' + ib'' \in \Omega \quad (b', b'' \in \mathbb{R}^N)$$

$$2a^*a = 1 + |u|^2 \ , \ |t_{\eta_a} \cdot \eta_a| = |t_a \cdot a| = |u|$$

$$b' = \tfrac{1}{2}(\eta_a + \bar{\eta}_a) = \tfrac{1}{4} \begin{pmatrix} i(u-\bar{u}) \\ 2+u+\bar{u} \\ 0 \\ \vdots \\ 0 \end{pmatrix}$$

$$b'' = \frac{1}{2i}(\eta_a - \bar{\eta}_a) = \tfrac{1}{4} \begin{pmatrix} u+\bar{u}-2 \\ i(\bar{u}-u) \\ 0 \\ \vdots \\ 0 \end{pmatrix}$$

D'où

$$(11) \quad \begin{cases} 4\|b'\|^2 = 1 + 2\,\mathrm{Re}\, u + |u|^2 \\ 4\|b''\|^2 = 1 - 2\,\mathrm{Re}\, u + |u|^2 \\ \langle b', b''\rangle = \tfrac{1}{2}\,\mathrm{Im}\, u \end{cases}$$

De même, si $a = a' + ia''$ $(a', a'' \in \mathbb{R}^N)$ on trouve :

$$\|a'\|^2 = t_{a'} \cdot a' = \tfrac{1}{4}(t_a + t_{\bar{a}})(a+\bar{a}) = \tfrac{1}{4}(u+\bar{u}+|u|^2+1)$$

$$(12)\quad \begin{cases} 4\|a'\|^2 = 1 + 2\,\mathrm{Re}\,u + |u|^2 \\ 4\|a''\|^2 = 1 - 2\,\mathrm{Re}\,u + |u|^2 \\ \langle a',a''\rangle = t_{a'}\cdot a'' = \frac{1}{2}\,\mathrm{Im}\,u \end{cases}$$

Finalement

$$(13)\quad \begin{cases} \|a'\| = \|b'\| \\ \|a''\| = \|b''\| \\ \langle a',a''\rangle = \langle b',b''\rangle = \frac{1}{2}\,\mathrm{Im}\,u = \frac{1}{2}\,\mathrm{Im}(t_a\cdot a) \end{cases}$$

En considérant dans (13), $e^{i\theta}a$ au lieu de a (pour un θ tel que $t_{e^{i\theta}a}\cdot e^{i\theta}a$ soit réel) on peut supposer que a',a'' d'une part, b',b'' d'autre part sont orthogonaux. Il existe alors $k \in SO(N)$ telle que $ka' = b'$, $ka''=b''$.

Alors $e^{i\theta}k\,a = \eta_{e^{i\theta}ka} \in \Omega$. D'où le lemme 2.

Pour achever la démonstration de b), supposons qu'il existe une fonction non constante f, holomorphe dans $\mathcal{H}$, continue sur $\mathcal{H}$ et qui atteint son maximum en module au point $a \in \partial\mathcal{H}\setminus\check{\mathcal{B}}$. Soient θ,k tels que $\eta_0 = e^{i\theta}k\,a \in \Omega$ (Lemme 2) considérons la fonction $\zeta \mapsto F(\zeta) = f(e^{-i\theta}k^{-1}\zeta)$ qui est holomorphe dans $\mathcal{H}$ et atteint son maximum en module au point $e^{-i\theta}k^{-1}\zeta = a$ c'est à dire au point $\zeta = \eta_0 \in \Omega$. La fonction composée $F \circ h$ est holomorphe dans U continue sur $\bar{U}$ et atteint son maximum en module au point $u_0 \in U$ tel que $h(u_0) = \eta_0$ cela est en contradiction avec le principe de maximum. D'où le théorème 2.3.6.

2.3.8. <u>Remarque</u>. La démonstration ci-dessus montre que $\check{\mathcal{B}}_R$ est aussi la frontière de Bergman-Šilov de $\mathcal{H}(R)$ relative à la classe constituée par les fonctions F holomorphes dans $\mathcal{H}(R)$ continue sur $\mathcal{H}(R)$ avec $F|_{R^N}$ harmonique. En effet, les fonctions $\zeta_1 + 1$ et $F(e^{-i\theta}k^{-1}\zeta)$ figurant dans la démonstration ci-dessus sont de cette classe.

2.3.9. Remarque. On notera que si $L(\zeta)$ est la norme de Lie dans C^N (2.3.1) i.e.

$$L^2(\zeta) = \|\zeta\|^2 + \sqrt{\|\zeta\|^4 - |(\Sigma\zeta_j^2)|^2}$$

on a

$$(\|\zeta\|^4 - |\Sigma\zeta_j^2|^2) \leq 4 \sum_{j<k} |\zeta_j|^2 |\zeta_k|^2$$

d'où

$$L^2(\zeta) \leq \|\zeta\|^2 + 2\sqrt{\sum_{j<k} |\zeta_j|^2 |\zeta_k|^2}$$

et

$$G_R = \{\zeta \in C^N \mid \|\zeta\|^2 + 2\sqrt{\sum_{j<k} |\zeta_j|^2 |\zeta_k|^2} < R^2\} \subset BL(0,R)$$

$$H_R = \{\zeta \in C^N \mid \sum_{j=1}^{N} |\zeta_j| < R\} \subset G_R$$

Dans le cas $N = 2$, $H_R = G_R$; si $N \geq 3$, $H_R \subset G_R$, $H_R \neq G_R$

$$H_R \supset \{\zeta \in C^N \mid \|\zeta\| < \frac{R}{\sqrt{2}}\}$$

2.3.10. L'expression de la frontière de Bergman-Šilov montre que si f est une fonction entière dans C^N, homogène d'ordre m (i.e. $f(\alpha z) = \alpha^m f(z)$ pour tout $\alpha \in \mathbb{C}$) bornée en module par M sur la sphère réelle $\{x \in R^N \mid \|x\|=1\}$, alors on a :

$$|f(z)| \leq [\|z\|^2 + \sqrt{\|z\|^2 - |\Sigma z_j^2|^2}]^{\frac{m}{2}} M \quad (z \in C^N)$$

§ 2.4. PROLONGEMENT ANALYTIQUE DES FONCTIONS HARMONIQUES D'ORDRE QUELCONQUE.

Soit f une fonction analytique réelle dans un ouvert connexe $\Omega \in R^N$. D'après 1.1.2. il existe un ouvert connexe $\hat{\Omega} \subset C^N$, $\hat{\Omega} \cap R^N = \Omega$ et une fonction holomorphe $\hat{f}$ dans $\hat{\Omega}$ dont la restriction à Ω est égale à f. Le domaine $\hat{\Omega}$

dépend en général de f , mais dans certains cas il peut être invariant par rapport à une classe de fonctions analytiques réelles dans Ω . C'est le cas notamment si on considère la classe constituée par les fonctions harmoniques dans Ω . Dans ce chapitre on se propose d'étudier si l'ouvert connexe Ω de $\mathbb{R}^N$ étant donné, il existe un ouvert connexe $\hat{\Omega}$ de $\mathbb{C}^N$, $\hat{\Omega} \cap \mathbb{R}^N = \Omega$, tel que toute fonction harmonique dans Ω , soit la restriction à Ω d'une fonction holomorphe dans $\hat{\Omega}$. Précisément on veut savoir s'il existe un plus grand domaine de $\mathbb{C}^N$ possédant cette propriété. Si on se restreint uniquement aux domaines de $\mathbb{C}^N$, le problème ci-dessus n'a une réponse positive que dans des cas particuliers de Ω . (Par exemple, si Ω est une boule de $\mathbb{R}^N$) . Par contre comme cette étude est liée au prolongement analytique, pour obtenir un énoncé général, il faut considérer des domaines au dessus de $\mathbb{C}^N$ (surface de Riemann non ramifiée au dessus de $\mathbb{C}^N$) . Nous rappelons brièvement quelques notions concernant ces domaines.

2.4.1. Rappel

a) prolongement analytique

Un domaine au dessus de $\mathbb{C}^N$, ou surface de Riemann non ramifiée au dessus de $\mathbb{C}^N$, est un couple (X,Φ) composé d'un espace topologique séparé connexe X , et d'une application Φ de X dans un ouvert de $\mathbb{C}^N$ qui est un homéomorphisme local (i.e. chaque point $a \in X$ a un voisinage V_a tel que $\Phi|_{V_a}$ est un homéomorphisme de V_a sur ΦV_a) .

Si X' est un sous -espace topologique de X , le couple (X',Φ) est appelé un sous-domaine de (X,Φ) . Le domaine (X,Φ) au dessus de $\mathbb{C}^N$, est dit univalent (Schlicht), si Φ est un homéomorphisme globale. Un domaine Ω de $\mathbb{C}^N$ est identifié au domaine (Ω,id_Ω) au dessus de $\mathbb{C}^N$, où id_Ω est l'injection naturelle (identité) de $\Omega \hookrightarrow \mathbb{C}^N$. Plus généralement soit Ω un ouvert connexe non vide de $\mathbb{C}^N$, s'il existe un domaine (X,Φ) au dessus de $\mathbb{C}^N$ et un homéomorphisme j de Ω sur un ouvert de X tel que $\Phi \circ j = id_\Omega$, le domaine Ω sera identifié au sous-domaine $(j\Omega,\Phi)$ de (X,Φ) .

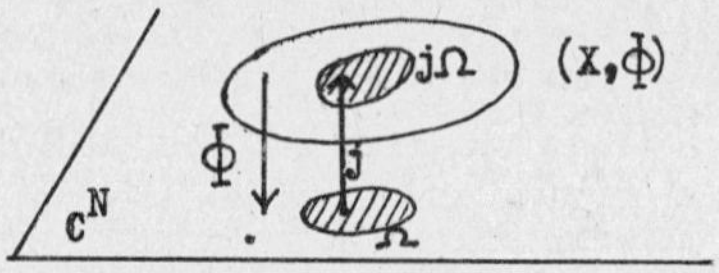

$\Omega = (j\Omega,\Phi) \subset (X,\Phi)$

b) Fonction holomorphe sur un domaine (X,Φ) au dessus de $\mathbb{C}^N$.

Une fonction continue $f : X \to \mathbb{C}$ est dite holomorphe au point $x \in X$ s'il existe un voisinage V_x de x tel que $\Phi|_{V_x}$ est un homéomorphisme de V_x sur son image $\Phi(V_x)$, et si, la fonction $f \circ (\Phi|_{V_x})^{-1}$ qui est définie sur $\Phi(V_x)$ est holomorphe au point $z = \Phi(x)$. La fonction f est dite holomorphe sur un sous domaine de (X,Φ) si ellle l'est en tout point de ce sous-domaine. Si f est holomorphe dans un domaine Ω de $\mathbb{C}^N$, elle sera holomorphe dans le domaine (Ω,id_Ω) au dessus de $\mathbb{C}^N$.

c) Application holomorphe. Soient $(X,\Phi),(X',\Phi')$ deux domaines au dessus de $\mathbb{C}^N$ une application continue $h : X \to X'$ est dite holomorphe, si pour tout ouvert $V' \subset X'$ et g' holomorphe dans (V',Φ') la fonction $g = g' \circ h$ est holomorphe dans $(h^{-1}(V'),\Phi)$. Si h est de plus un homéomorphisme de X sur X' , l'application réciproque h^{-1} est nécessairement holomorphe et h est un isomorphisme analytique. Les deux domaines (X,Φ) et (X',Φ') sont alors isomorphes $(\approx)$ et peuvent être identifiés. On remarquera que l'application $id_X : X \to X$ est holomorphe sur (X,Φ) .

d) Théorème fondamental du prolongement analytique.

Considérons un ouvert connexe Ω de $\mathbb{C}^N$ et une fonction holomorphe f dans Ω .

Le théorème fondamental du prolongement analytique exprime qu'il existe un domaine $\mathcal{H}_f = (X_f, \Phi)$ au dessus de $\mathbb{C}^N$ et une fonction holomorphe g sur (X_f, Φ) possédant les propriétés suivantes :

(i) Le domaine Ω s'identifie au sous-domaine $(j\Omega, \Phi)$ de (X_f, Φ) par un homéomorphisme $j : \Omega \to j\Omega \subset X_f$.

(ii) On a $g \circ j = f$ dans Ω .

(iii) pour tout $\{(X', \Phi'), j', g'\}$ possédant les propriétés (i) et (ii), il existe une application holomorphe et une seule $h : X' \to X_f$ telle que les diagrammes suivants soient commutatifs :

(1)

La condition (ii) exprime que f se prolonge en une fonction holomorphe g sur (X_f, Φ) ; g est nécessairement unique en vertu du principe du prolongement analytique dans Ω . La propriété (iii) implique l'unicité de (X_f, Φ) a un isomorphisme analytique près, c'est-à-dire si $\{(X_f, \Phi), j\}$ $\{(X_1, \Phi_1), j_1\}$ vérifient (i),(ii),(iii), ils sont nécessairement isomorphes. En effet, il existe $h : X_1 \to X_f$ et $k : X_f \to X_1$ holomorphes uniques, telles que

$$h \circ j_1 = j \quad \Phi \circ h = \Phi_1$$

$$k \circ j = j_1 \quad \Phi_1 \circ k = \Phi$$

L'application $\ell = h \circ k : X \to X$ vérifie

$$\ell \circ j = h \circ k \circ j = h \circ j_1 = j$$

$$\Phi \circ \ell = \Phi \circ h \circ k = \Phi_1 \circ k = \Phi$$

cela montre que les diagrammes suivants sont commutatifs :

$$\begin{array}{ccc} X_f & \xrightarrow{\text{hok}} & X_f \\ {\scriptstyle j}\nwarrow & & \nearrow{\scriptstyle j} \\ & \Omega & \end{array} \qquad\qquad \begin{array}{ccc} X_f & \overset{\text{hok}}{\longrightarrow} & X_f \\ {\scriptstyle \Phi}\searrow & & \swarrow{\scriptstyle \Phi} \\ & \mathbb{C}^N & \end{array}$$

mais l'application identité $id_X : X_f \to X_f$ est aussi holomorphe sur (X_f, Φ) et les deux derniers diagrammes restent commutatifs si on substitut $h \circ k$ par id_X. L'hypothèse de l'unicité de (iii) implique $h \circ k = id_{X_f}$. De même $k \circ h = id_{X_1}$ c'est à dire $k = h^{-1}$ et h est un isomorphisme analytique. D'où $(X, \Phi) \approx (X_1, \Phi_1)$.

e) La commutativité des diagrammes (1) montre en outre que si g_1 est un prolongement holomorphe de f sur (X_1, Φ_1) on a $g \circ h = g_1$. Ainsi parmi les domaines au dessus de $\mathbb{C}^N$ vérifiant (i),(ii) $\mathcal{H}_f$ est le plus grand (au sens de (iii)) ; $\mathcal{H}_f$ est appelé le domaine d'holomorphie de f construit au dessus de $\mathbb{C}^N$.

2.4.2. Existence des domaines d'holomorphies au dessus de $\mathbb{C}^N$.

Appelons élément analytique, un couple constitué par un point $z_o \in \mathbb{C}^N$ et une série $\Sigma\, a_n(z-z_o)^n$ convergente dans un polydisque centré au point z_o. Un élément analytique, est engendré par un autre, si la série définissant le premier élément, s'obtient, à partir de la série du second élément, par le procédé du prolongement analytique (dans $\mathbb{C}^N$), le long d'une suite finie de polydisques. Si f est holomorphe dans un ouvert connexe $\Omega \subset \mathbb{C}^N$, l'élément analytique de f au point $z_o \in \Omega$ est le couple (f_{z_o}, z_o) où f_{z_o} est la série de Taylor de f au point z_o.

Le domaine (X_f,Φ) au dessus de $\mathbb{C}^N$ est alors défini comme suit :

Les points de X_f sont des éléments analytiques. Un élément analytique appartient à X_f si et seulement si, il est engendré par un élément (f_{z_o},z_o) de $f(z_o \in \Omega)$. Deux points (f_z,z) et $(f_{z'},z')$ de X_f sont donc différents si et seulement si, leur projection z,z' sur $\mathbb{C}^N$ sont différentes ou bien, si $z = z'$, les séries $f_z, f_{z'}$ différent par un terme au moins. Evidemment pour tout $z_o \in \Omega$ on a $(f_{z_o},z_o) \in X_f$. La topologie de X_f est définie par la donnée en chaque point $m = (f_a,a) \in X_f$ d'un système fondamental de voisinage $\mathcal{V}_m$ de ce point.

Notons $((f_a)_b,b)$ l'élément analytique (f_b,b) engendré par (f_a,a). Une base V de m (i.e. un élément de $\mathcal{V}_m$) est constitué par l'ensemble des points $((f_a)_b,b)$ lorsque on fait varier b dans un polydisque $c(a,r)$ de $\mathbb{C}^N$, centré au point a, et dans lequel la série f_a converge uniformément.

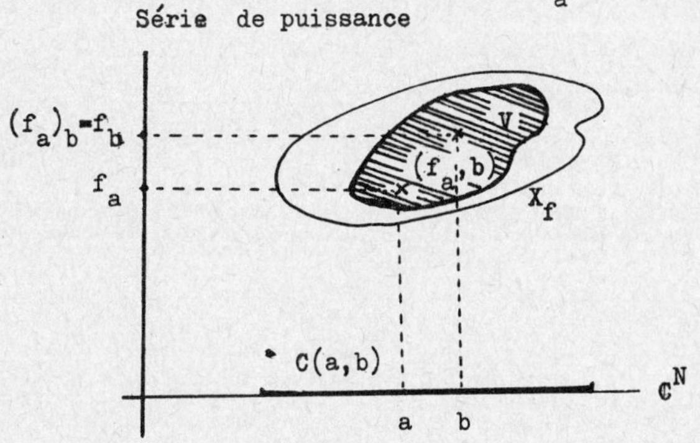

La topologie ainsi définie est séparée et X_f muni de cette topologie est connexe. L'homéomorphisme locale $\Phi : X_f \to \mathbb{C}^N$ est définie par $(f_a,a) \mapsto a$. La fonction holomorphe g sur (X_f,Φ) qui prolonge f est définie par

$$g[(f_a,a)] = f_a(a)$$

L'homéomorphisme $j : z \mapsto (f_z,z)$ de $\Omega \to j\Omega \subset X_f$ permet l'identification de Ω avec son image $j\Omega$. Si $z_o \in \Omega$ on a

$$(g \circ j)(z_o) = g[(f_{z_o},z_o)] = f_{z_o}(z_o) = f(z_o)$$

et g prolonge bien f .

2.4.3. <u>Points accessibles</u>. Soit f une fonction holomorphe dans un ouvert connexe non vide $\Omega \subset \mathbb{C}^N$. Disons qu'un point $z \in \mathbb{C}^N$ est f-accessible si quel que soit l'élément analytique (f_a, a) $(a \in \Omega)$, il existe un chemin dans $\mathbb{C}^N$ d'origine a et d'extrémité z le long duquel cet élément se prolonge analytiquement de a à z. L'ensemble des points f-accessibles est manifestement un domaine d'holomorphie de $\mathbb{C}^N$ qui contient Ω.

Le théorème de la <u>monodromie</u> affirme que si (f_a, a) peut être prolongé analytiquement le long de tout chemin (d'origine a et d'extrémité z) d'un domaine simplement connexe $\tilde{\Omega}_f$, il existe une fonction holomorphe $\tilde{f}$ dans $\tilde{\Omega}_f$ dont la restriction à Ω est égale à f.

Dans ce cas, les domaines au dessus de $\mathbb{C}^N$, (X_f, Φ) et $(\tilde{\Omega}_f, id)$ sont isomorphes. Les deux diagrammes (1) sont ici de la forme ci-dessous :

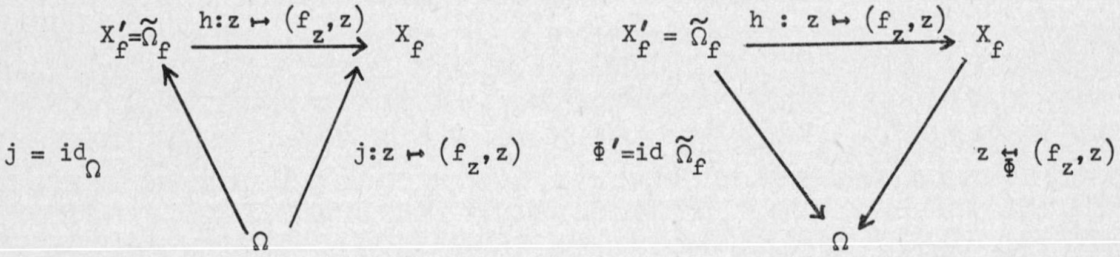

$$\mathcal{H}_f = (X_f, \Phi) \approx (\tilde{\Omega}_f, id_{\tilde{\Omega}_f}) \approx \tilde{\Omega}_f$$

$$\tilde{\Omega}_f \text{ simplement connexe}$$

on notera que $\tilde{\Omega}_f$ est univalent.

2.4.4. THÉORÈME.

<u>Soient</u> Ω <u>un ouvert connexe de</u> $\mathbb{R}^N$, $\partial\Omega \neq \emptyset$ <u>et</u> $\mathcal{H}(\Omega)$ <u>la cellule d'harmonicité de</u> Ω <u>dans</u> $\mathbb{C}^N$.

a - <u>Si</u> f <u>est une fonction harmonique dans</u> Ω, <u>son domaine d'holomorphie</u> $\mathcal{H}_f$ <u>au dessus de</u> $\mathbb{C}^N$, <u>contient un sous-domaine</u> (X_f, Φ) <u>tel que</u>

$\Phi(X_f) = \mathcal{H}(\Omega)$.

b - Si $N = 2p \geq 4$, ou bien si Ω est étoilé par rapport à un de ses points, toute fonction harmonique dans Ω a un prolongement holomorphe sur $\mathcal{H}(\Omega)$ (i.e. $(X_f,\Phi) = (\mathcal{H}(\Omega), id_{\mathcal{H}(\Omega)})$. En particulier toute fonction harmonique dans un ouvert convexe Ω de $\mathbb{R}^N$, est la restriction à Ω d'une fonction holomorphe sur $\mathcal{H}(\Omega)$.

c - Dans tous les cas, il existe un ouvert connexe $\hat{\Omega} \supset \Omega$ de $\mathbb{C}^N$ tel que toute fonction harmonique dans Ω a un prolongement holomorphe dans $\hat{\Omega}$. En particulier, toute fonction harmonique dans la boule $B(a,R) \subset \mathbb{R}^N$, a un prolongement holomorphe dans la boule $B(a,\frac{R}{\sqrt{2}})$ de $\mathbb{C}^N$.

Démonstration. Nous procédons comme dans [19] .

Soient $(K_m)_{m\geq 1}$ une suite d'exhaustive de compacts avec

$K_m \subset K_{m+1}$, $D_m = \overset{\circ}{K}_m$, $\bigcup\limits_{m=1}^{\infty} D_m = D$, ∂D_m continûment différentiable par morceaux

et
$$E(x) = \begin{cases} \dfrac{1}{(2-N)\omega_N(1)} \dfrac{1}{(x_1^2+\dots+x_N^2)^{\frac{N-2}{2}}} & \text{si } N \geq 3 \\ -\dfrac{1}{2\pi} \operatorname{Log}(x_1^2 + x_2^2) \quad \text{si } N = 2 \end{cases}$$

le noyau de la théorie du potentiel neutonien dans $\mathbb{R}^N (\omega_N(1)$ = l'aire de la sphère unité dans $\mathbb{R}^N = 2\pi^{\frac{N}{2}} / \Gamma(\frac{N}{2}))$. A partir de la représentation :

$$(14) \qquad I_m(f,x) = \int_{t\in\partial D_m} [f(t) \frac{\partial E(x-t)}{\partial n_{ext}} - E(x-t) \frac{\partial f(t)}{\partial n_{ext}}] d\omega_N(t) = \begin{cases} f(x) & \text{si } x \in D_m \\ 0 & \text{si } x \notin D_m \end{cases}$$

où f est harmonique dans Ω , on peut montrer que l'élément analytique (f_{a_o}, a_o) $(a_o \in \Omega)$ se prolonge analytiquement à tout point $z_o \in \mathcal{H}(D_m)$ le long de tout chemin polygonal γ_m dans $\mathcal{H}(D_m)$, d'origine a_o et d'extrémité z_o.

a) En effet, on recouvre l'image de γ_m par un nombre fini de domaines $\delta_{i,m}$ simplement connexe dans $\mathcal{H}(D_m)$ et sur chaque $\delta_{i,m}$ de proche en proche (à partir de a_o) on précise les branches de $E(z-t)$, $\frac{\partial E(z-t)}{\partial n_{ext}}$ $(z \in \mathbb{C}^N)$, holomorphes et bornées, quel que soit $t \in \partial D_m$.
Considérons alors la fonction $I_m(f,z)$ de (14) où on a substituer à la variable réelle x la varibale $z \in \delta_{i,m}$. La fonction $I_m(f,z)$ est de la forme :

$$\int K_1(z,t)d\mu_1(t) + \int K_2(z,t)d\mu_2(t)$$

$$\int |d\mu_1(t)|<\infty \quad \int |d\mu_2(t)|<\infty$$

et représente une fonction holomorphe dans $\delta_{i,m}$. Finalement $I_m(f,z)$ permet d'obtenir un élément analytique (f_{z_o}, z_o) prolongement analytique de (f_{a_o}, a_o) le long des $\delta_{i,m}$. Si maintenant $z \in \mathcal{H}(\Omega)$ et $\gamma, \gamma(0) = a_o, \gamma(1) = z$, est un chemin polygonal dont l'image $\gamma(I)$ comporte un nombre fini de côtés (donc compact) dans $\mathcal{H}(\Omega)$, on aura $\gamma(I) \subset \mathcal{H}(D_m)$ pour $m \geq m_o$ (2.1.8) et on peut trouver un élément analytique (f_z, z) prolongeant (f_{a_o}, a_o) le long de $\gamma(I)$.
Ainsi tous les points de $\mathcal{H}(\Omega)$ sont f-accessibles. Le domaine d'holomorphie $\mathcal{H}_f$ de f construit au dessus de $\mathbb{C}^N$ (§ 2.4.e)) contient donc un sous-domaine (X_f, Φ) avec $\Phi X_f = \mathcal{H}(\Omega)$.

b) Dans le cas $N = 2p \geq 4$, si $z \in \mathcal{H}(D_m)$, la fonction $\hat{f}_m(z)=I_m(f,z)$ est holomorphe, $\tilde{f}_{m|D_m} = f_{|D_m}$, $m \geq 1$, et pour $m \leq \ell$, $\tilde{f}_m = \tilde{f}_{\ell|\mathcal{H}(D_m)}$. La fonction $\hat{f}$ dont la restriction à chaque $\mathcal{H}(D_m)$ est égale à $\tilde{f}_m$ est une fonction holomorphe dans $\mathcal{H}(\Omega)$ avec $\tilde{f}|_{\Omega} = f$.

Si Ω est étoilé en un point (resp. convexe), $\mathcal{H}(\Omega)$ est étoilé (2.1.6) au même point (resp. convexe) donc simplement connexe. Il est aisé de voir que dans ce cas le théorème de la monodromie s'applique. D'où l'existence d'une fonction holomorphe dans $\mathcal{H}(\Omega)$ dont la restriction à Ω est égale à f.

c) En particulier si Ω est la boule $B(a,R)$, il existe une fonction holomorphe dans la boule de Lie $\mathcal{H}[B(a,R)] = \{\zeta \in \mathbb{C}^N | L(\zeta - a) < R\}$ (où L est la norme de Lie (2.3.1.) dans $\mathbb{C}^N$) dont la restriction à $B(a,R)$ est f. Par conséquent si Ω est un domaine quelconque de $\mathbb{R}^N$ $(\partial\Omega \neq \phi)$ le domaine :

$$\hat{\Omega} = \bigcup_{a\in\Omega} \mathcal{H}[B(a,r_a)]$$

où r_a = distance de a à $\partial\Omega$, répond à notre exigence.
Le dernier point résulte du fait que la cellule d'harmonicité de la boule $B(a,R)$ contient la boule $B(a,\frac{R}{\sqrt{2}}) \subset \mathbb{C}^N$.

2.4.5. Remarque importante. P. Lelong démontre dans [18], qu'il existe une fonction harmonique h dans Ω dont le domaine d'holomorphie $\mathcal{H}_h = (X_h, \Phi)$. Dans le cas particulier d'une boule voir [26].

2.4.6. Remarque. Soit h une fonction harmonique dans la boule $B(o,r)$ de $\mathbb{R}^N$ $(N\geq 2)$ et continue sur son adhérence. Le noyau du poisson

$$P_r(x,a) = \frac{1}{\omega_N(1)} \frac{r^2 - \|x\|^2}{r\|x-a\|^N} \qquad (\|a\| = r)$$

qui est harmonique dans $B(o,r)$ a un prolongement holomorphe dans la cellule d'harmonicité de $B(o,r)$ (et en particulier dans la boule $B(0,\frac{r}{\sqrt{2}})$ de $\mathbb{C}^N$) donnée par :

$$P_r(z,a) = \frac{1}{\omega_N(1)} \frac{r^2 - (z_1^2 + \dots + z_N^2)}{r[(z_1-a_1)^2 + \dots + (z_N-a_N)^2]^{N/2}}$$

où la branche du dénominateur a été choixi par la condition qu'elle coïncide avec $\|x-a\|^N$ si $z = x \in \mathbb{R}^N$. La fonction $\tilde{h}$

$$(15) \qquad \tilde{h}(z) = \int_{\|a\|=r} P_r(z,a)\, h(a)\, d\omega_N(a)$$

est alors holomorphe dans la cellule d'harmonicité de $B(o,r)$, (et en particulier dans la boule de $\mathbb{C}^N$ centrée en 0 et de rayon $\frac{r}{\sqrt{2}}$) et qui coïncide avec h dans $B(o,r)$.

La représentation (15) permet de majorer $|\tilde{h}(z)|$ connaissant une majoration de $|h(x)|$.

2.4.7. Exemples d'applications

PROPOSITION [21].

Soient $h(x)$ une fonction harmonique dans la boule $B(0,R) \subset \mathbb{R}^N$ et $\tilde{h}(z)$ son prolongement holomorphe dans la boule $B(0,\frac{R}{\sqrt{2}})$ de $\mathbb{C}^N$.

Posons $m(r) = \max\limits_{\|x\|=r} |h(x)| \quad (x \in \mathbb{R}^N)$

$$M(r') = \max_{\|z\|=r'} |\tilde{h}(z)| \quad z \in \mathbb{C}^N, \ r' < \frac{r}{\sqrt{2}} < \frac{R}{\sqrt{2}}$$

alors on a

$$(16) \qquad M(r') \leq 3.2^{\frac{N}{2}-1} \left(1-2\frac{r'^2}{r^2}\right)^{-\frac{N}{2}} m(r)$$

En effet, considérons la représentation (15) dans $B(o,r) \subset \mathbb{R}^N$ on a si $a_k = r\alpha_k$ $\|\alpha\|=r$,

$$|\Sigma(z_k-r\alpha_k)^2| \geq |\mathrm{Re}\ \Sigma(z_k-r\alpha_k)^2| \geq$$
$$\geq (r-r'\cos\varphi)^2 - r'^2\sin^2\varphi$$

avec $\|x\| = \|z\| \cos\varphi = r' \cos\varphi$, $\|z\| = r' < \frac{r}{\sqrt{2}}$

D'où $|\Sigma(z_k - r\alpha_k)^2| \geq \frac{r^2}{2} - r'^2$

et $|P_r(z,a)| \leq \frac{3r}{2\omega_N(1)(\frac{1}{2}r^2 - r'^2)^{N/2}}$ $(|z| \leq r' < \frac{r}{\sqrt{2}} < \frac{R}{\sqrt{2}})$

Donc, $|\tilde{f}(z)| \leq \frac{3r.2^{\frac{N}{2}-1}}{(r^2-2r'^2)^{\frac{N}{2}}} \frac{1}{\omega_N(1)} \int_{\|a\|=r} |h(a)| d\omega_N(a)$

$$\leq 3.2^{\frac{N}{2}-1} M(r). \frac{r^N}{(r^2-2r'^2)^{\frac{N}{2}}}$$

Finalement

$$M(r') \leq 3.2^{\frac{N}{2}-1} (1 - 2\frac{r'^2}{r^2})^{-\frac{N}{2}} m(r) \quad (r' < \frac{r}{\sqrt{2}} < \frac{R}{\sqrt{2}})$$

2.4. 8. COROLLAIRE [12]

<u>Soient</u> h <u>harmonique dans</u> $B(0,R) \subset \mathbb{R}^N$ <u>et</u>

(17) $$h(x) = \sum_{n \in \mathbb{N}^N} a_n x^n \qquad (x = x_1^{n_1} \ldots x_N^{n_N})$$

<u>la série de Taylor de</u> h <u>au point</u> 0 .

on a

(18) $$|a_n| \leq k(N) \left(\frac{2}{|n|}\right)^{\frac{|n|}{2}} (N+|n|)^{\frac{N+|n|}{2}} \frac{m(r)}{r^{|n|}}$$

<u>pour tout</u> $n \in \mathbb{N}^N$ <u>et</u> $r < R$; $m(r) = \underset{\|x\|=r}{\text{Max}} |h(x)|$, $k(N) = \frac{3}{2}(\frac{2}{N})^{\frac{N}{2}}$

En effet, le polydisque $|z_1| < \frac{r'}{\sqrt{N}}$,...., $|z_N| < \frac{r'}{\sqrt{N}}$ est contenu dans la boule $B(o,r') \subset B(o,\frac{r}{\sqrt{2}}) \subset \mathbb{C}^N$. D'après les inégalités de **Cauchy** appliquées à $\tilde{h}(z)$ et en tenant compte de (16) on obtient :

$$|a_n| \leq 3.2^{\frac{N}{2}-1} \left(1-\frac{2r'^2}{r^2}\right)^{-\frac{N}{2}} m(r).r'^{-|n|} \quad (r' \in]0, \tfrac{r\sqrt{2}}{2}[)$$

la fonction $g(r') = \left(1 - \frac{2r'^2}{r^2}\right)^{-\frac{N}{2}} r'^{-|n|}$ a un maximum dans $]0, \frac{r\sqrt{2}}{2}[$ au point

$$r' = \left[\frac{|n|}{2(N+|n|)} \right]^{\frac{1}{2}} r$$

D'où l'inégalité (18) . Le coefficient de $\frac{m(r)}{r^{|n|}}$ dans (18) est un $O(\sqrt{2}^{|n|}.|n|^{N/2})$ $(|n| \to \infty)$.

2.4.9. COROLLAIRE.

<u>Si</u> h <u>est harmonique dans tout</u> $\mathbb{R}^N$ <u>et de type exponentiel</u> (i.e. $|h(x)| \leq Ae^{B\|x\|}$) <u>ses coefficients de Taylors</u> $a_n = \frac{D^n h(o)}{n!}$ <u>vérifient :</u>

$$a_n = O((\sqrt{2}eB)^{|n|} \; |n|^{\frac{N}{2}-|n|}) \; (|n| \to \infty)$$

En effet, $\inf_{r>o} e^{Br}.r^{-|n|} = (eB)^{|n|} |n|^{-|n|}$

L'estimation 2.4. 9. peut être raffinée; on a en effet,

2.4. 1o. LEMME [2] .

<u>Si</u> h <u>est une fonction harmonique dans tout</u> $\mathbb{R}^N$ <u>vérifiant</u>

$$\lambda[|h|,o,r] = O(e^{Br}) \quad B > o \; , \; r \to \infty$$

<u>alors</u>

$$(19) \qquad a_n = \frac{D^n h(o)}{n!} = O\left(\frac{|n|^{N-\frac{3}{2}}}{n!} \; . \; B^{|n|}\right)$$

Il en résulte :

2.4.11. PROPOSITION.

Soit h une fonction harmonique dans tout $\mathbb{R}^N$ de type exponentiel : $|h(x)| \leq Ae^{B\|x\|}$. Alors pour tout $\varepsilon > 0$ il existe une constante $C(\varepsilon) > 0$ telle que le prolongement holomorphe $\tilde{h}$ de h dans $\mathbb{C}^N$ vérifie :

$$|\tilde{h}(z)| \leq C(\varepsilon) e^{(B+\varepsilon)(|z_1|+\dots+|z_N|)}$$

En effet, si $h(x) = \sum_{n\in\mathbb{N}^N} a_n x^n$, on a $\tilde{h}(z) = \sum_{n\in\mathbb{N}^N} a_n z^n$ $(z \in \mathbb{C}^N)$

où a_n vérifie (19). Donc, si $B' = B+\varepsilon$ et C est une constante convenable,

$$|\tilde{h}(z)| \leq C_1 \sum_{n\in\mathbb{N}^N} \frac{|n|^{N-\frac{3}{2}}}{n!} B^{|n|} |z|^n \leq$$

$$\leq \sum_{m>0}^{\infty} \Big[\sum_{|n|=m} |n|^{N-\frac{3}{2}} \left(\frac{B}{B'}\right)^{|n|} \frac{B'^{|n|}}{(|n|)!} \frac{|n|!\,|z|^n}{n!} \Big]$$

$$\leq \sum_{m=0}^{\infty} m^{N-\frac{3}{2}} \left(\frac{B}{B'}\right)^m \frac{B'^m}{m!} \Big(\sum_{|n|=m} \frac{|n|!}{n!} |z|^n \Big)$$

or, $m^{N-\frac{3}{2}} \left(\frac{B}{B'}\right)^m$ étant borné, il existe une constante $c_2(\varepsilon)$ telle que

$$|\tilde{h}(z)| \leq c_2(\varepsilon) \sum_{m=0}^{\infty} \frac{B'^m}{m!} (|z_1|+\dots+|z_N|)^m =$$

$$c_2(\varepsilon)\, e^{B'(|z_1|+\dots+|z_N|)}$$

2.4.12. Cas des fonctions harmoniques d'ordre infini

THÉORÈME.

Si u est une fonction harmonique d'ordre infini dans un ouvert étoilé

Ω (resp. convexe) de $\mathbb{R}^N$, il existe une fonction $\tilde{u}$ holomorphe dans la cellule d'harmonicité $\mathcal{H}(\Omega) \subset \mathbb{C}^N$ dont la restriction à Ω est égale à u.

Démonstration :

Soit $H(x,t)$ $(x,t) \in \Omega \times \mathbb{R}$ la fonction harmonique telle que $H(x,t)|_{t=o} = u(x)$ (1.2.4.). Il existe d'après 2.4.4. une fonction holomorphe $\tilde{H}(z,\zeta)$ dans $\mathcal{H}(\Omega \times \mathbb{R}) \subset \mathbb{C}^{N+1}$ telle que

$$\tilde{H}(z,\zeta)|_{\Omega \times \mathbb{R}} = H(x,t)$$

D'après 2.2.13, $\mathcal{H}(\Omega \times \mathbb{R}) \supset \mathcal{H}(\Omega) \times \{0\}$. Donc, $\tilde{u}(z) = \tilde{H}(z,o)|_{\Omega} = H(x,o) = u(x)$ et $\tilde{H}(z,o)$ est holomorphe de $z \in \mathcal{H}(\Omega)$.

En particulier :

2.4.13. COROLLAIRE.

Toute fonction harmonique d'ordre infini dans la boule $B(0,R) \subset \mathbb{R}^N$ est la restriction à $B(0,R)$ d'une fonction holomorphe dans la boule de Lie $BL(0,R) = \{z \in \mathbb{C}^N | L(z) < R\}$ (2.3.1).

CHAPITRE 3

APPLICATIONS

§ 3.1. FONCTIONS HARMONIQUES (D'ORDRE QUELCONQUE) ARITHMÉTIQUES.

Les développements récents des fonctions entières de type exponentiel et arithmétiques de plusieurs variables complexes, permettent d'obtenir des résultats analogues pour les fonctions harmoniques d'ordre quelconque et cela grâce au procédé de complexification de ces fonctions. Nous utilisons ici comme théorèmes de bases les énoncés démontrés dans [5] .

Une fonction entière $f(z)$, $z \in \mathbb{C}^N$ (resp. une fonction harmonique h d'ordre quelconque) est dite arithmétique si $f(\mathbb{N}^N) \subset \mathbb{Z}$ (resp. $h(\mathbb{N}^N) \subset \mathbb{Z}$) . On a :

3.1.1. (PROPOSITION 3.3.1. [5]).

Soit f une fonction entière dans $\mathbb{C}^N$ vérifiant :

(1) $$|f(z_1,\dots,z_N)| \leq M \exp(\alpha_1|z_1|+\dots+\alpha_N|z_N|)$$

si $0 \leq \alpha_j < \text{Log } 2$ $(1 \leq j \leq N)$ $M = \text{cte} > 0$, alors f se développe selon

(2) $$f(z) = \sum_{\nu\in\mathbb{N}^N} \Delta_\nu(f) \binom{z_1}{\nu_1} \cdots \binom{z_N}{\nu_N}$$

développement uniformément convergente sur tout compact de $\mathbb{C}^N$, où

$$\binom{z_j}{\nu_j} = \frac{z_j(z_j-1)\ldots(z_j-\nu_j+1)}{\nu_j!} \quad (1 \le j \le N)$$

$$\Delta_\nu(f) = \sum_{\substack{0\le\beta_j\le\nu_j \\ 1\le j\le N}} \binom{\nu}{\beta}(-1)^{|\nu|-|\beta|} f(\beta) \quad (\nu,\beta \in \mathbb{N}^N)$$

$$\binom{\nu}{\beta} = \binom{\nu_1}{\beta_1} \ldots \binom{\nu_N}{\beta_N} \quad , \; |\nu| = \nu_1 + \ldots + \nu_N .$$

<u>Pour tout</u> $\max_j \alpha_j < \alpha < \text{Log } 2$, <u>il existe une constante</u> $M_\alpha \ge 0$ <u>telle que</u>

$$|\Delta_\nu(f)| \le M_\alpha (e^\alpha - 1)^{|\nu|}, \; \nu \in \mathbb{N}^N$$

3.1.2. La proposition (3.1.1) montre que si f est de plus arithmétique, $\Delta_\nu(f)$ est un entier, dont la valeur absolue est < 1 dès que $M_\alpha(e^\alpha-1)^{|\nu|} < 1$, donc $\Delta_\nu(f) = 0$ à partir d'un certain ν_0 , et finalement f est un polynome ; si $f(\mathbb{N}^N) = 0$ alors $f = 0$.

3.1.3. (COROLLAIRE 3.2.9. [5])

<u>Soit</u> f <u>entière arithmétique vérifiant</u> (1)

a - Si $\alpha_j < |\text{Log } (\frac{3}{2} + i \frac{\sqrt{3}}{2})| = 0,7588\ldots$

f <u>est de la forme</u> :

$$(3) \quad f = \sum P_{k_1,\ldots,k_N}(z_1,\ldots,z_N)\, C_{1,k_1}^{z_1} \ldots C_{N,k_N}^{z_N} \quad \text{(somme finie)}$$

<u>avec</u> $C_{j,k_j} \in \{1,2\}$ <u>pour tout</u> j,k_j <u>et les</u> P_k <u>polynomes</u>.

b - Si $\alpha_j < \text{Log} 2$, $C_{j,k_j} = 1$ pour tout j, et f est un polynome.

On a alors l'analogue du théorème 3.1.1. avec les mêmes notations :

3.1.4. THÉORÈME.

Soit h une fonction harmonique dans tout $\mathbb{R}^N$ vérifiant :

(4) $|h(x)| \leq Ae^{B\|x\|} \quad (x \in \mathbb{R}^N)$

Si $B < \text{Log } 2$, alors f se développe selon

$$h(x) = \sum_{\nu \in \mathbb{N}^N} \Delta_\nu(h) \binom{x_1}{\nu_1} \cdots \binom{x_N}{\nu_N}$$

où la convergence est uniforme sur tout compact de $\mathbb{R}^N$.

En particulier :

3.1.5. COROLLAIRE.

Soit h harmonique dans tout $\mathbb{R}^N$ et vérifiant (4) avec $B < \text{Log } 2$.

Si h est arithmétique alors h est un polynome.

Si h est nulle sur le réseau $\mathbb{N}^N$ alors $h = o$ (i.e. le réseau $\mathbb{N}^N$ est un ensemble d'unicité pour la classe de fonctions harmoniques dans $\mathbb{R}^N$ vérifiant (4) avec $B < \text{Log } 2$) .

Le théorème 3.1.4. est la conséquence de 3.1.2. et de la proposition 2.4.11 .

3.1.6. Remarque.

On peut améliorer dans 3.1.5. la constante Log 2 . En effet :

3.1.7. THÉORÈME.

Soit h une fonction harmonique arithmétique dans tout R^N et vérifiant

$$|h(x)| \leq Ae^{B\|x\|}$$

Si $B < a_o = |\text{Log}(\frac{3}{2} + i\frac{\sqrt{3}}{2})| = 0{,}7588...$ alors h est un polynome .

Démonstration.

Remarquer que $a_o > \text{Log}\, 2 = 0{,}693...$. Il existe une fonction entière $\tilde{h}(z)$ vérifiant

$$|\tilde{h}(z)| \leq A' \, e^{B'(|z_1|+...+|z_N|)}$$

avec $B < B' < a_o$ et dont la restriction à R^N est égale à h (2.4.11.). L'énoncé 3.1.3. s'applique, $h(z)$ est de la forme :

$$\tilde{h}(z) = \sum_k P_{k_1,...,k_N}(z_1,...,z_N) \, c_{1,k_1}^{z_1} \cdots c_{N,k_N}^{z_N} \quad \text{(somme finie)}$$

avec $C_{j,k_j} \in \{1,2\}$ pour tout j,k . D'où

$$h(x) = \sum_k P_k(x) \, 2^{\alpha_{1,k_1}x_1+...+\alpha_{N,k_N}x_N} \quad (\alpha_{j,k_j} \in \{0,1\})$$

$$= \sum_k P_k(x) \, e^{\langle \beta_k, x\rangle} \quad (\beta_k = (\beta_{1,k_1},...,\beta_{N,k_N}) \, , \beta_{j,k_j} \in \{0,\text{Log}2\})$$

or la dernière somme est une exponentielle polynome ; celle-ci est harmonique si et seulement si, c'est un polynome harmonique. En effet

$$\Delta h(x) = \sum_k e^{\langle\beta_k, x\rangle} \left[\Delta P_k(x) + 2 \sum \frac{\partial}{\partial x_j} P_k(x)\, \beta_{j,k_j} + P_k(x)\|\beta_k\|^2\right]$$

le crochet est un polynome, et $\langle\beta_k, x\rangle - \langle\beta_{k'}, x\rangle$ non constante, en remarquant qu'une fonction analytique dans $\mathbb{C}^N$ et nulle sur $\mathbb{R}^N$ est identiquement nulle, le théorème de E. Borel [10] (qui est aussi valable dans $\mathbb{C}^N$ [24]) montre que $\Delta h \equiv 0$ dans $\mathbb{R}^N$ implique

$$\Delta P_k(x) + 2 \sum \frac{\partial}{\partial x_j} P_k \cdot \beta_{j,k} + P_k\|\beta_k\|^2 \equiv 0 .$$

Nécessairement $\|\beta_k\|^2 = 0$, $\beta_{j,k_j} = 0$ pour tout j, k_j et h est un polynome harmonique.

Soit $m \leq N$. Posons

$$\mathbb{N}_m = \{x = (x_1, \ldots, x_m , 0, \ldots, 0) \in \mathbb{R}^N | x_i \in \mathbb{N} \quad 1 \leq i \leq m\}$$

3.1.8. COROLLAIRE [1].

Soit h une fonction harmonique dans tout $\mathbb{R}^N$ vérifiant

$$|h(x)| \leq Ae^{B\|x\|}$$

Si $h(\mathbb{N}_m) \subset \mathbb{Z}$, $B < \text{Log } 2$, il existe un polynome $P(x_1, \ldots, x_m)$ dans $\mathbb{R}^m$ tel que

$$h(x_1, \ldots, x_m, 0, \ldots, 0) = P(x_1, \ldots, x_m) \quad (m \leqslant N)$$

Si $m < N$ la constante $\text{Log } 2$ est la meilleure possible.

En effet, la complexifiée $\tilde{h}$ de h vérifie (2.4.11)

$$|\widetilde{h}(z_1,\dots,z_N)| \leq A' e^{B'(|z_1|+\dots+|z_N|)}$$

$B < B' < \text{Log } 2$, $A' = \text{cte}$, et la fonction $(z_1,\dots,z_m) \mapsto \widetilde{h}|_{z_{m+1}=\dots=z_N} = 0$

est arithmétique on conclut par 3.1.2. Contrairement au cas $m = N$ (3.1.6.), l'exemple $h(x_1,\dots,x_N) = \mathcal{R}e\ 2^{x_1+ix_N}$ $m = 1$, montre que la constante Log 2 de l'énoncé est la meilleure possible.

3.1.9. THEOREME.

<u>Soit</u> $f \in C^\infty(\mathbb{R}^N)$ <u>vérifiant</u>

$$|\Delta^m f(x)| \leq A \frac{(2m)!}{m^{\alpha m}} e^{B\|x\|} \quad (x \in \mathbb{R}^N , A = \text{cte})$$

a) <u>Si</u> $\alpha > 2$, $B < \text{Log } 2$ <u>et</u> $f(\mathbb{N}^N) \subset \mathbb{Z}$, <u>alors</u> f <u>est un polynome</u>.

b) <u>Si</u> $\alpha = 2$, $B < \text{Log}[\frac{1}{2}(3+\sqrt{5})] - \frac{2}{e} = 0,224\dots$ et $f(\mathbb{Z}^N) \subset \mathbb{Z}$, <u>alors</u> f <u>est un polynome</u>.

<u>Démonstration</u>.

a. La fonction f est harmonique d'ordre infini dans $\mathbb{R}^N$. Considérons la fonction

$$H(x,t) = \sum_{m=0}^{\infty} (-1)^m \frac{\Delta^m f(x)}{(2m)!} t^{2m}$$

qui est harmonique dans $\mathbb{R}^{N+1}(x,t) = \mathbb{R}^N(x) \times \mathbb{R}^N(t)$ (1.2.4).

Si $\mathbb{N}_1^{N+1} = \{(x,t) \in \mathbb{R}^{N+1} | t=0 , x \in \mathbb{N}^N\}$

on a

$$(5) \quad H(\mathbb{N}_1^{N+1}) = f(\mathbb{N}^N) \subset \mathbb{Z}$$

D'autre part,

$$(6) \qquad |H(x,t)| \leq Ae^{B\|x\|}\sum_{m=o}^{\infty}\frac{t^{2m}}{m^{\alpha m}} = Ae^{B\|x\|}\Phi(t^2)$$

Or la fonction entière dans le plan complexe :

$$\Phi(z) = \sum_{m=o}^{\infty}\frac{z^m}{m^{\alpha m}}$$

est d'ordre $P = \frac{1}{\alpha}$ et de type $\tau = \frac{\alpha}{e}$ en vertu des égalités bien connues reliant l'ordre et le type d'une fonction entière à ses coefficients de Taylor. Donc, pour tout $\varepsilon > o$, il existe une constante $A_1(\varepsilon)$ tels que

$$|\Phi(z)| \leq A_1\, e^{(\frac{\alpha}{e}+\varepsilon)|z|^{\frac{1}{\alpha}}} \qquad (z \in \mathbb{C})$$

en particulier

$$|\Phi(t^2)| \leq A_1\, e^{(\frac{\alpha}{e}+\varepsilon)|t|^{\frac{2}{\alpha}}} \qquad (t \geq 0)$$

Si $\alpha > 2$, alors $\frac{2}{\alpha} < 1$ et il existe une constante $A_2(\varepsilon)$ telle que

$$|\Phi(t^2)| \leq A_2\, e^{\varepsilon|t|} \quad (t \in \mathbb{R})$$

et d'après (6) :

$$(7) \qquad |H(x,t)| \leq A_3(\varepsilon)e^{B\|x\|+\varepsilon|t|} \leq A_3 e^{(B+\varepsilon)\sqrt{x_1^2+\ldots+x_N^2+t^2}}$$

Les conditions du corollaire 3.1.8. sont vérifiées si ε est choisi de manière que $\beta + \varepsilon < \mathrm{Log}\ 2$. Donc $H(x,o) = f(x)$ est un polynome.

b. Dans le cas $\alpha = 2$, $|\Phi(t^2)| \leq A_1\, e^{(\frac{2}{e}+\varepsilon)|t|}$

et

$$(8) \qquad |H(x,t)| \leq A_4\, e^{(\beta+\frac{2}{e}+\varepsilon)\|(x,t)\|} \qquad .$$

Si ε est choisi de manière que $\beta + \varepsilon < \frac{1}{2}(3+\sqrt{5}) - \frac{2}{e}$, les conditions d'application du théorème 2 de [1] (i.e. si h harmonique dans $\mathbb{R}^N$ vérifie $|h(x)| \le Ae^{B\|x\|}$, $B < \frac{1}{2}(3+\sqrt{5})$ et $f(\mathbb{Z}_m) \subset \mathbb{Z}$, $\mathbb{Z}_m = \{x \in \mathbb{R}^N, x_i \in \mathbb{Z}$, $1 \le i \le m$, $x_i = o$ $m < i \le N\}$ alors $h(x_1, \ldots, x_m, 0, \ldots, 0)$ est un polynome) sont vérifiées et $H(x,o) = f(x)$ est un polynome.

Un théorème récent de D.H. Armitage [2] affirme que si une fonction harmonique h dans tout $\mathbb{R}^N$ vérifie (9) $\lambda[|h|,o,r] = O(e^{\alpha r})$ $o \le \alpha < 1$, $r \to \infty$, et $D^\nu h(o) \in \mathbb{N}$ pour tout $\nu \in \mathbb{N}^N$, alors h est un polynome. Ce résultat conduit à

3.1.10. PROPOSITION.

<u>Soit</u> $f \in C^\infty(\mathbb{R}^N)$ <u>vérifiant</u> :

$$|\Delta^m f(x)| \le A \frac{(2m)!}{m^{\alpha m}} e^{B\|x\|} \quad , \ x \in \mathbb{R}^N \ , \ A = \text{cte} \ , \ \alpha > o$$

a. <u>Si</u> $\alpha > 2$, $0 \le B < 1$, <u>et</u> $D^\nu f(0) \in \mathbb{N}^N$, <u>alors</u> f <u>est un polynome</u>.

b. <u>Si</u> $\alpha = 2$, $B + \frac{2}{e} < 1$, <u>et</u> $D^\nu f(0) \in \mathbb{N}$ <u>pour tout</u> $\nu \in \mathbb{N}^N$, <u>alors</u> f <u>est un polynome</u>.

En effet, la fonction $H(x,t)$ vérifie (9) dans les deux cas en vertu des inégalités (7) et (8) .

Comme $D^\nu H(0,0) = D^\nu f(0) \in \mathbb{N}$, $H(x,t)$ est un polynome en x,t et par conséquent $f(x) = H(x,0)$ l'est aussi.

BIBLIOGRAPHIE

[1] ARMITAGE D.H. On harmonic functions wich take integer values on integer lattices. Math. Proc. camp. Phil. soc. (1979).

[2] ARMITAGE D.H. On the derivatives at the origin of entire harmonic functions. Glasgow Math. J. 20 (1979).

[3] ARONZAJN N. Sur les décompositions des fonctions analytiques uniformes et sur leurs applications. Acta Math. 65 (1935).

[4] AVANISSIAN V. Distributions harmoniques d'ordre infini...
Lecture notes (Sém. Probabilités XI) 581 (1977).

[5] AVANISSIAN V. & GAY R. Sur une transformation des fonctionnelles analytiques... Bull. Soc. Math. France 103 (1975).

[6] AVANISSIAN V. Sur l'harmonicité des fonctions séparément harmoniques. Lectures notes (Sém. Probabilités) 39 (1967).

[7] AVANISSIAN V. & FERNIQUE X. Sur l'analycité des distributions harmoniques d'ordre infini. Ann. Inst. Fourier 18,2 (1968).

[8] BAOUENDI M.S. & GOULAOUI C. HANOUZET B.
Caractérisation de classes de fonctions C^∞ et analytiques. J. Math. pure et appl. 52 (1973).

[9] BAOUENDI M.S. & GOULAOUI C. Régularité analytique et itérés d'opérateurs elliptiques dégénérés (centre de Math. Ecole polytechniques) (séminaire d'analyse Janvier 1971 n° M 53.017)

[10] BOREL E. Sur les zéros des fonctions entières. Acta Math. 20 (1896-97).

[11] CARTAN E. Sur les domaines bornés homogènes. Abh. Math. Sém. Hamburg, 11 (1935) oeuvres, partie 1.

[12] HAYMANN W.K. Power series expensions for harmonic functions. Bull. London Math. Soc. 2 (1970).

[13] HUA L.K. Harmonic analysis of functions of several complex variables in classical domains Moscow (1959). Trans. of math. monog. vol. 6 A.M.S., Prov. Rhode Islind, 1963.

[14] JARNICKI M. Prace Mathematyczne, zeszyt 17 (1975).

[15] KISELMANN C.O. Prolongement des solutions d'une équation aux dérivées partielles à coeffs. constants. Bull. Soc. Math. France 97 (4) (1969).

[16] KORANYI A. & WOLF J. Realization of hermitian symetric Spaces as generalized Half. planes. **Ann. of Maths. Vol.81,No. 2, 1965.**

[17] LELONG P. Fonctions plurisousharmoniques et fonctions analytiques de variables réelles. Ann. inst. Fourier 11 (1961).

[18] LELONG P. Sur les singularités complexes d'une fonction harmonique. C.R. Acd. Sc. Paris t. 232 (1951).

[19] LELONG P. Prolongement analytique et singularité complexes des fonctions harmoniques. Bull. Soc. Math. Belgique (1954).

[20] LELONG P. Sur les fonctions indéfiniment dérivables... Duke Math. J. vol. 14 n° 1 (1947).

[21] LELONG P. Fonctions entières et fonctions plurisousharmoniques d'ordre fini dans C^n . J. d'analyse Math. Jérusalem t. 9 1964.

[22] LOWDENSLAGER D.B. Potential theory in bounded symetric homogenous complex domains. Ann. of math. vol. 67 (1958)

[23] MOREMOTO M. Analytic functionals on the Lie sphere. Tokyo Journal of Mathém. Vol. 3 n° 1 (1980).

[24] NARASIMHAN R. Un analogue holomorphe du théorème de Lindemann. Ann. Inst. Fourier 21,3 (1971).

[25] OVCARENKO On multiply superharmonic functions. Uspehi Math. Nauk. 16 n° 3 (99) (1961).

[26] SICIAK J. Holomorphic continuation of harmonic functions. Ann. Polonici Mathematici XXIX (1974).

Institut de Recherche Mathématique Avancée

7, rue René Descartes, 67084 Strasbourg Cédex

DÉVELOPPEMENTS ASYMPTOTIQUES DES FONCTIONS OBTENUES PAR INTÉGRATION SUR LES FIBRES*

par D. B A R L E T

Le but de cet exposé est de présenter quelques résultats sur le problème suivant :

Soit X un espace analytique complexe réduit et irréductible de dimension $n+1$. Soit $f : X \longrightarrow \mathbb{C}$ une fonction holomorphe sur X , non constante . Pour chaque $s \in f(X) = D$, $f^{-1}(s)$ est une hypersurface de X , et nous noterons par X_s le cycle de X sous-jacent (voir [B.0]) .

Pour φ forme différentielle C^{∞} à support f-propre de type (n,n) sur X , la fonction $F_\varphi : D \longrightarrow \mathbb{C}$ définie par

$$F_\varphi(s) = \int_{X_s} \varphi$$

est continue sur D (voir [B.1]) . Comme le montre l'exemple qui va suivre , cette fonction continue n'est pas , en général , différentiable :

<u>Exemple</u> :

On prend $n = o$, $X = \{(t,z) \in \mathbb{C}^2 \ / \ z^2 = t\}$, $f : X \longrightarrow \mathbb{C}$ donnée par $f(t,z) = t$ et $\varphi(t,z) = z\bar{z}$. On trouve alors $F_\varphi(t) = 2|t|$.

<u>Problème</u> :

Décrire les singularités possibles des fonctions continues de la forme F_φ et les relier à la géométrie de f .

I <u>Le théorème d'existence des développements asymptotiques</u>

Commençons par donner l'énoncé du théorème fondamental :

<u>Théorème 1</u>

Soit $D = \{s \in \mathbb{C} \ / \ |s| < 1\}$, et soit X un espace analytique complexe réduit et irréductible de dimension $n+1$. Soit $f : X \longrightarrow D$ une application holomorphe surjective , et notons par X_s le cycle $f^{-1}(s)$. Soit K un compact de X . Il existe des rationnels $r_1, \ldots, r_k$ dans $[o,2[$ tels que pour toute forme différentielle φ C^{∞} de type (n,n) sur X et à support dans K , la fonction F_φ admette , quand $s \longrightarrow o$, un développement asymptotique de la forme :

$$F_\varphi(s) \sim \sum_{r=r_1,\ldots,r_k} \sum_{j=o}^{n} \sum_{(m,m') \in \mathbb{N}^2} T^{r,j}_{m,m'}(\varphi) . s^m \bar{s}^{m'} |s|^r (\mathrm{Log}|s|)^j$$

*Conférence faite au Colloque de Wimereux en l'honneur du Professeur P.LELONG.

où $T^{r,j}_{m,m'}$ est un courant de type (1,1) sur X .

Ce théorème est assez proche de son analogue réel démontré par Maire dans sa thèse (voir [M] et [J]) . Sa démonstration (comme dans le cas réel) utilise de manière essentielle la résolution des singularités pour se ramener au cas où X est un polydisque de $\mathbb{C}^{n+1}$ et où $f(z) = z_o^{a_o} \ldots z_n^{a_n}$ avec $a_o, \ldots, a_n$ entiers positifs non tous nuls . Dans ce cas nous prouvons le théorème par des calculs élémentaires d'intégrales (*).

Ces calculs sont assez explicites pour permettre de décrire complètement les fonctions de la forme F_φ au voisinage de o dans cette situation particulière , grace à la

<u>Proposition</u> 1

Soit X un polydisque ouvert de centre o dans $\mathbb{C}^{n+1}$, et soit K un voisinage compact de o dans X . Soient $a_o \neq o$, $a_1, \ldots, a_n$ des entiers positifs , et soit $f : X \to \mathbb{C}$ donnée par

$$f(z) = z_o^{a_o} \ldots z_n^{a_n} .$$

Si l'entier $c > o$ est répété exactement j-fois dans la suite $(1, a_o, a_1, \ldots, a_n)$, pour tout $h \in [o, j-1]$ $(j \geqslant 1)$ il existe une forme C^∞ de type (n,n) et à support dans K $\varphi_{h,c}$ vérifiant :

$$F_{\varphi_{h,c}}(s) = |s|^{2/c} (\mathrm{Log}|s|)^h \quad \text{pour } |s| \text{ assez petit.}$$

complétée par le lemme suivant :

<u>Lemme</u> 1

Dans la situation du théorème , si $f^{-1}(o) \cap \mathring{K} \neq \emptyset$, il existe une forme ω C^∞ de type (n,n) sur X et à support dans K telle que $F_\omega(s) \equiv 1$ pour s assez voisin de o .

En particulier , si θ est une fonction C^∞ sur D , pour $\varphi = f^*(\theta).\omega$ on aura $F_\varphi \equiv \theta$ au voisinage de o .

Après ce théorème de développement asymptotique, deux questions viennent à l'esprit :

1) Donner une interprétation des rationnels intervenant dans ce développement .

2) Décrire les courants $T^{r,j}_{m,m'}$.

(*) voir "Développement asymptotique des fonctions obtenues par intégration sur les fibres I", preprint de l'Institut E. Cartan , Nancy, janvier 81 [B.3]

Pour ce qui est de la première question, il apparait dans la démonstration du théorème 1 que ces rationnels (exépté o) sont de la forme 2/c où c est la multiplicité d'une composante irréductible du diviseur à croisements normaux obtenu après désingularisation de X et de $f^{-1}(o)$. La proposition suivante, qui ne fait que reprendre dans ce cadre un calcul désormais classique, montre que les valeurs possibles de r , en supposant X lisse , sont de la forme $-2\lambda_o-2q$ où $q \in \mathbb{N}^*$, et λ_o est racine du polynôme de Berstein-Sato de f (voir [S]).

Proposition 2

Soit X un ouvert connexe de $\mathbb{C}^{n+1}$ contenant o , et soit $f : X \longrightarrow \mathbb{C}$ une application holomorphe non constante vérifiant $f(o) = o$. Supposons que l'on ait sur X un opérateur différentiel holomorphe dépendant polynomialement de $P(z,\frac{\partial}{\partial z},\lambda)$ à coefficients holomorphes , et un polynôme $b \in \mathbb{C}[\lambda]$, $b \neq o$, vérifiant :

$$P(z,\frac{\partial}{\partial z},\lambda).f^{\lambda+1} = b(\lambda).f^{\lambda} \quad \text{sur } X$$

(ceci est en fait une identité en $\lambda \in \mathbb{C}$ sur le revêtement universel de $X - \{f=o\}$) . Alors si $T^{r_o,j_o}_{m_o,m'_o} \neq o$, il existe λ_o racine de b telle que $\lambda_o + r_o/2 \in -\mathbb{N}^*$ (*) .

De plus la somme des ordres des racines de b vérifiant la relation (*) est au moins égale à $(j_o+1)/2$.

Je ne résiste pas au plaisir d'esquisser cette démonstration :

Comme les valeurs précises de m_o et m'_o n'interviennent pas dans la conclusion , nous pouvons supposer (m_o,m'_o) minimal pour l'ordre lexicographique tel que $T^{r_o,j_o}_{m_o,m'_o} \neq o$.

Soit φ une forme différentielle C^{∞} de type (n,n) et à support dans K telle que $T^{r_o,j_o}_{m_o,m'_o}(\varphi) = 1$.

Pour $Re(\lambda) > o$ soit $C(\lambda)$ le courant de type (n,n) sur X et dépendant holomorphiquement de λ , défini par

$$< C(\lambda), \theta > = \int_X |f|^{2\lambda} \varphi \wedge \theta$$

pour θ forme différentielle C^{∞} de type (1,1) sur X . Pour θ fixée la fonction holomorphe $<C(\lambda),\theta>$ admet un

prolongement méromorphe à $\mathbb{C}$ tout entier :
en effet l'équation

$$P(z,\frac{\partial}{\partial z},\lambda).f^{\lambda+1} = b(\lambda).f^{\lambda}$$

donne , en remplaçant λ par $\bar{\lambda}$ et en conjuguant :

$$\bar{P}(\bar{z},\frac{\partial}{\partial \bar{z}},\lambda).\bar{f}^{\lambda+1} = \bar{b}(\lambda).\bar{f}^{\lambda}$$

où $\bar{P}(\bar{z},\frac{\partial}{\partial \bar{z}},\lambda) = \sum \overline{C_{a,p}(z)}.\lambda^p \frac{\partial^a}{\partial \bar{z}^a}$ si on a posé

$$P(z,\frac{\partial}{\partial z},\lambda) = \sum C_{a,p}(z).\lambda^p \frac{\partial^a}{\partial z^a} \quad \text{et} \quad \bar{b}(\lambda) = \overline{b(\bar{\lambda})} .$$

On aura donc , puisque P et $\bar{P}$ sont respectivement holomorphe et antiholomorphe :

$$P.\bar{P}\, f^{\lambda+1}\, \bar{f}^{\lambda+1} = b(\lambda)\bar{b}(\bar{\lambda})\, |f|^{2\lambda}$$

ce qui donne l'égalité :

$$b(\lambda)\bar{b}(\bar{\lambda}) \int_X |f|^{2\lambda} \varphi\wedge\theta = \int_X \left[P.\bar{P}\ |f|^{2(\lambda+1)}\right] \varphi\wedge\theta$$

$$= \int_X |f|^{2(\lambda+1)} \left[\bar{Q}.Q(\varphi\wedge\theta)\right]$$

si Q et $\bar{Q}$ sont les adjoints de P et $\bar{P}$.

Ceci permet de prolonger analytiquement $< C(\lambda),\theta >$ à l'ouvert $\mathrm{Re}(\lambda) > -2$ quitte à avoir des pôles aux zéros de $b(\lambda)\ \bar{b}(\bar{\lambda})$. En itérant ce procédé (classique) on obtient un prolongement méromorphe de $C(\lambda)$ à $\mathbb{C}$ tout entier dont les pôles éventuels sont contenus dans les translatés par des entiers négatifs des zéros de $b(\lambda)\ \bar{b}(\bar{\lambda})$ contenus dans la bande $\mathrm{Re}(\lambda) < 0$ (*).

Prenons maintenant $\theta = f^{m'_o}\, \bar{f}^{m_o}\, df\wedge d\bar{f}$; on aura alors :

$$< C(\lambda),\theta > = \int_X |f|^{2\lambda} f^{m'_o}\, \bar{f}^{m_o}\ \varphi\wedge df\wedge d\bar{f}$$

$$= \int_{\mathbb{C}} |s|^{2\lambda} s^{m'_o}\, \bar{s}^{m_o}\, F_{\varphi}(s)\, ds\wedge d\bar{s} \qquad \text{par Fubini .}$$

En utilisant le développement asymptotique de F_{φ} en $s=0$

$$F_{\varphi}(s) = \sum_{m+m'+r\leq N} \sum_{j=0}^{n} T^{r,j}_{m,m'}(\varphi).s^m \bar{s}^{m'} |s|^r (\mathrm{Log}|s|)^j + O(|s|^N)$$

avec $N > m_o+m'_o+r_o$, et en multipliant les deux membres de cette égalité par une fonction C^{∞} à support compact ρ

(*) en fait pour le choix de θ que nous allons faire on ne peut avoir de singularités que pour $\mathrm{Re}(\lambda) \leq -1$.

radiale et valant identiquement 1 au voisinage du support de F_φ (qui est compact), on obtient :

$$< C(\lambda), \theta > = \sum_{m+m'+r \leq N} \sum_{j=o}^{n} T^{r,j}_{m,m'}(\varphi) \int_C |s|^{2\lambda+r+1} s^{m+m'_o} \bar{s}^{m'+m_o}$$

$$(\mathrm{Log}|s|)^j \rho(s)\, ds \wedge d\bar{s} + G_N(\lambda) ,$$

où G_N est analytique pour $2\mathrm{Re}(\lambda) + N+1 > -1$.

Comme ρ est radiale, le calcul en coordonnées polaires donne l'existence d'un pôle en $\lambda_o = -r_o/2 - m_o - m'_o - 1$ d'ordre égal à j_o+1 (les rationnels intervenant dans le développement asymptotique étant dans $[o,2[$, il ne peut y avoir superposition de pôles correspondant à des valeurs différentes de r) . Ceci achève notre esquisse de démonstration .

Il serait évidemment très intéressant de savoir si chaque zéro du polynôme de Berstein-Sato de f contribue effectivement au développement asymptotique du théorème 1 (cette question contient le fait que les zéros du polynôme de Bernstein-Sato de f sont des rationnels négatifs !)

Venons en à la question 2) en commençant par quelques remarques :

Si a et b sont des entiers , on a

$$f^a \bar{f}^b T^{r,j}_{m,m'} = T^{r,j}_{m-a,m'-b}$$

égalite entre courants sur X qui résulte immédiatement de l'identité $F_{f^a \bar{f}^b \varphi} = s^a \bar{s}^b F_\varphi$.

On a $d\, T^{r,j}_{m,m'} = o$ pour tous les r,j,m,m' puisque $\varphi = d\psi$ implique $F_\varphi = o$ d'après la formule de Stokes . De plus , il est clair que $T^{o,o}_{o,o}$ est le courant $[X_o]$ d'intégration sur le cycle X_o . On en déduit que $T^{o,o}_{m,m'}$ s'obtient en divisant le courant X_o par $f^m \bar{f}^{m'}$ et en conservant la relation de fermeture (*).

Ces courants $T^{o,o}_{m,m'}$ sont donc de support $|X_o|$ exactement et s'interprètent comme des dérivées normales le long de $|X_o|$. Plus géneralement , si k est la multiplicité d'une composante irréductible de X dans le cycle X_o , les courants $T^{r,j}_{m,m'}$ pour $j=o$ et $r.k \in 2\mathbb{N}$ peuvent s'interpréter comme

(*) L'existence d'une telle division avec conservation de la condition de fermeture ne semble pas évidente à priori .

des "dérivées normales multiformes" le long des composantes irréductibles de X_o ayant k comme multiplicité dans le cycle X_o (ou bien un multiple de k !) ; pour un peu plus de precision sur ces termes voir [B.2] et [B.3] .

Le théorème qui suit précise le support des autres courants, dans le cas où $|X_o|$ n'a pas de composante irréductible contenue dans le lieu singulier de X .

Théorème 2

On se place dans la situation du théorème 1 , et on pose $X_o = \sum k_i.C_i$ où $C_1,\ldots,C_I$ sont des sous-ensembles analytiques irréductibles de X . On suppose qu'aucun C_i n'est contenu dans le lieu singulier de X , noté $S(X)$. Soit Y la réunion de $|X_o| \cap S(X)$ et du lieu singulier de $|X_o|$. Alors on a

$$\mathrm{Supp}\; T^{r,j}_{m,m'} \subset Y$$

dès que $r.k_i \notin \mathbb{N}$ pour chaque i , ou dès que $j \neq o$.

Remarque :

Dans le cas où $(X_s)_{s\in D}$ est une famille (analytique) à un paramètre de cycles d'une variété analytique lisse Z donnée , les résultats précédents s'appliquent , mais on doit prendre pour X le graphe de la famille (c'est à dire la réunion dans $D \times Z$ des $\{s \times |X_s|\}$) . On peut alors caractériser en terme de déformation infinitésimale d'ordre 1 de la famille en $s = o$ le fait que X_o n'ait pas de composante irréductible contenue dans $S(X)$, et préciser l'ensemble $S(X) \cap |X_o|$ en terme de symbole de la déformation infinitésimale d'ordre 1 (voir [B.2] §3 et [B.3]) .

Considérons , toujours dans la situation du théorème 1 , un compact K de X fixé . Posons :

$$\widetilde{\mathcal{M}}_K = \{F_\varphi \text{ , pour } \varphi \in C_K^\infty(X) \text{ de type } (n,n)\} .$$

Alors $\widetilde{\mathcal{M}}_K$ est un module sur $C^\infty(D)$ par multiplication des fonctions , puisque si $\theta \in C^\infty(D)$ on a :

$$\theta . F_\varphi = F_{f*(\theta).\varphi} .$$

Notons par $\widetilde{\mathcal{M}}_K^\infty$ le sous-ensemble des F_φ admettant un développement asymptotique nul en o , et par I^∞ l'idéal de $C^\infty(D)$ formé des fonctions plates en o (c'est à dire ayant une série de Taylor nulle en s = o) . Alors on a

$$I^\infty \widetilde{\mathcal{M}}_K \subset \widetilde{\mathcal{M}}_K^\infty$$

ce qui montre que le module des développements asymptotiques $\mathcal{M}_K$ qui est , par définition , le quotient $\widetilde{\mathcal{M}}_K / \widetilde{\mathcal{M}}_K^\infty$ est un $C^\infty(D)/I^\infty \equiv \mathbb{C}[[s,\bar{s}]]$- module .

Avec ces notations , le théorème 1 admet le corollaire immédiat suivant :

<u>Corollaire</u> :

Le $\mathbb{C}[[s,\bar{s}]]$-module $\mathcal{M}_K$ des développements asymptotiques est de type fini .

En effet , le théorème 1 nous assure que $\mathcal{M}_K$ est un sous-module du module libre de type fini :

$$\bigoplus_{r,j} \mathbb{C}[[s,\bar{s}]] . |s|^r (\text{Log } |s|)^j .$$

Comme l'anneau $\mathbb{C}[[s,\bar{s}]]$ est noethérien , le corollaire s'en déduit .

Pour "remonter" des développements asymptotiques aux fonctions obtenues par intégration sur les fibres , nous utiliserons le résultat suivant qui résultera du lemme 1 et de l'étude de la dérivabilité terme à terme des dévelop--pements asymptotiques qui sera faite plus loin :

<u>Lemme</u> 2

Dans la situation du théorème 1 , si $\mathring{K} \cap f^{-1}(o) \neq \emptyset$, il existe $\omega \in C_K^\infty(X)$ de type (n,n) telle que :

$$\tilde{\mathcal{M}}^{\infty}_K \subset I^{\infty} F_{\omega} \quad .$$

En effet , en choisissant ω comme dans le lemme 1 , si $F_{\varphi} \in \tilde{\mathcal{M}}^{\infty}_K$, et si l'on sait que cette hypothèse implique que F_{φ} est C^{∞} au voisinage de o on aura :

$$F_{\varphi} = F_{f*(F_{\varphi}).\omega} \quad \text{au voisinage de } s = o \quad .$$

Mais il n'est pas évident que $F_{\varphi} \in \tilde{\mathcal{M}}^{\infty}_K$ implique que F_{φ} soit C^{∞} au voisinage de $s = o$.

Remarquons que le lemme 2 joint au corollaire précédent donne la finitude des germes en o d'éléments de $\tilde{\mathcal{M}}_K$ comme module sur l'anneau des germes de fonctions C^{∞} en $s = o$.

Comme application de la finitude du module des dévelop--pements asymptotiques , on obtient le résultat suivant :

<u>Proposition</u> 3

Dans la situation du théorème 1 , pour K compact fixé de X , il existe un entier k tel que pour tout $N \in \mathbb{N}$ et toute forme différentielle $\varphi \in C^{\infty}_K(X)$ de type (n,n) et vérifiant :

$$T^{r,j}_{m,m'}(\varphi) = o \quad \text{pour} \quad m+m'+r \leq N+k$$

il existe des formes différentielles $\varphi_{a,b}$ et $\emptyset$ dans $C^{\infty}_K(X)$ de type (n,n) vérifiant :

$$\varphi = \sum_{a+b = N} f^a \bar{f}^b \varphi_{a,b} + \emptyset$$

avec $F_{\emptyset} \equiv o$ au voisinage de $s = o$.

La démonstration consiste à appliquer le lemme d'Artin-Rees pour comparer les deux filtrations naturelles de $\mathcal{M}_K$: la filtration $\mathfrak{m}$-adique où $\mathfrak{m} = (s,\bar{s})$ est l'idéal maximal de $\mathbb{C}[[s,\bar{s}]]$, et la filtration par les sous-modules

$$\mathcal{M}^{\nu}_K = \left\{ F_{\varphi} \in \mathcal{M}_K / \ T^{r,j}_{m,m'}(\varphi) = o \quad \text{pour} \quad m+m'+r \leq \nu \right\} \quad .$$

On "remonte" ensuite à $\tilde{\mathcal{M}}_K$ en utilisant le lemme 2 .

Commençons l'étude de la dérivabilité terme à terme des développements asymptotiques par une remarque ; il suffit de traiter le cas où X est non singulier , en utilisant l'existence d'une désingularisation . De plus , un argument simple de partition de l'unité montre que le problème est local sur la désingularisée . Nous supposerons donc dans la suite que X est un polydisque ouvert de centre o dans $\mathbb{C}^{n+1}$. Comme les valeurs critiques d'une fonction holomorphe $f : X \longrightarrow \mathbb{C}$ sont isolées , on peut supposer (quitte à localiser encore) que $df \neq o$ si $o<|f|<\varepsilon$; ceci montre que , dans la situation du théorème 1 , pour K compact donné , il existe $\varepsilon > o$ tel que les fonctions de la forme F_φ pour $\operatorname{supp}\varphi \subset K$ soient C^∞ sur le disque pointé $o<|s|<\varepsilon$.

On a alors le résultat facile suivant :

Proposition 4

Dans la situation du théorème 1 si $\dfrac{\partial^{a+b}}{\partial s^a \partial \bar{s}^b} F_\varphi$ admet quand $s \longrightarrow o$ un développement asymptotique de la forme

$$\sum a^{r,j}_{m,m'}\, s^m \bar{s}^{m'} |s|^r (\operatorname{Log}|s|)^j$$ avec les restrictions suivantes : r et j prennent un nombre fini de valeurs dans $[o,2[$ et $\mathbb{N}$ respectivement , et $(m,m')\in \mathbb{Z}^2$ avec $m \geq -p$ et $m' \geq -p$ où p est un entier donné , alors ce développement asymptotique s'obtient en dérivant terme à terme le développement asymptotique de F_φ donné par le théorème 1 .

La preuve est immédiate en intégrant terme à terme le développement asymptotique donné dans l'hypothèse et en utilisant l'unicite du développement asymptotique de F_φ (*).

Nous en arrivons maintenant au résultat clé de la derivation des développements asymptotiques donnés par le théorème 1 :

(*) L'intégration terme à terme ne pose guère de problème contrairement à la dérivation terme à terme !

<u>Théorème</u> 3

Dans la situation du théorème 1 , il existe un entier $l \geqslant 1$ tel que l'on ait :

$$s^l \frac{\partial}{\partial s} \mathcal{M}_K \subset \mathcal{M}_K$$

(et aussi $\bar{s}^l \frac{\partial}{\partial \bar{s}} \mathcal{M}_K \subset \mathcal{M}_K$ puisque la situation est auto-conjuguée).

Donnons la démonstration dans le cas (auquel on se ramène comme ci-dessus) où X est un ouvert de $\mathbb{C}^{n+1}$, où K est un polydisque compact de X , et où $df = o$ implique $f = o$. Soit $\mathcal{K}$ le faisceau cohérent sur X quotient de Ω_X^{n+1} par le sous-faisceau $\Omega_X^n \wedge df$ des formes holomorphes de degré $n+1$ qui s'écrive localement $\alpha \wedge df$. Comme on a $\Omega_X^{n+1} = \Omega_X^n \wedge df$ dès que $df \neq o$, on aura :

$$\operatorname{supp} \mathcal{K} \subset \{df = o\} \subset \{f = o\} .$$

D'après le Nullstellensatz il existe $l \in \mathbb{N}$ tel que f^l annule $\mathcal{K}$ au voisinage de K . Ceci signifie en particulier qu'il existe une forme holomorphe de degré n au voisinage de K vérifiant

$$f^l \, dz_o \wedge \ldots \wedge dz_n = \omega \wedge df \qquad (*)$$

Pour $\varphi \in C_K^\infty(X)$ de type (n,n) , on peut écrire

$$d'\varphi = \psi \wedge dz_o \wedge \ldots \wedge dz_n$$

avec $\psi \in C_K^\infty(X)$ de type (o,n) ; on aura alors :

$$f^l \, d'\varphi = \chi \wedge df$$

avec $\chi \in C_K^\infty(X)$ de type (n,n) .

Mais $d'F_\varphi$ au sens des distributions , qui coïncide avec $\frac{\partial}{\partial s} F_\varphi . ds$ pour $o < |s| < \varepsilon$ (puisque F_φ est C^∞ sur ce disque pointé), se calcule de la manière suivante : puisque $F_\varphi = f_*(\varphi)$ (dans le sens que la fonction continue F_φ représente la distribution image directe de φ par f), on a :

$$s^l \, d'F_\varphi = f_*(f^l \, d'\varphi)$$

(*) on remarquera que l'existence d'une telle écriture globale au voisinage de K utilise le théorème B de Cartan !

et donc pour tout $\theta \in C^{\infty}(D)$ de type (o,1) :

$$\langle s^{l}\, d'F_{\varphi}, \theta \rangle = \int_X f^{l} d'\varphi \wedge f^{*}(\theta)$$

$$= \int_X \chi \wedge df \wedge f^{*}(\theta)$$

$$= \int_{\mathbb{C}} F_{\chi} \wedge ds \wedge \theta \quad \text{par Fubini} .$$

On a donc l'égalité au sens des distributions :

$$s^{l} \frac{\partial}{\partial s} F_{\varphi} = F_{\chi} \qquad \text{d'où le résultat} .$$

<u>Corollaire</u> 1

Le développement asymptotique donné par le théorème 1 est indéfiniment dérivable terme à terme en s et $\bar{s}$.

<u>Corollaire</u> 2

Si F_{φ} a un développement asymptotique nul en o , alors F_{φ} est C^{∞} au voisinage de $s = o$ (et plate en $s = o$!) .

BIBLIOGRAPHIE

[B.0] Espace analytique réduit... Lecture Notes 482 p. 1-163

[B.1] Convexité de l'espace des cycles Bull. Soc. Math. France 1978.

[B.2] Déformation d'ordre 1... Revue de l'Institut E. Cartan n° 1, 1981.

[B.3] Développement asymptotiques des fonctions... I Préprint Nancy 1981.

B. JORK : Rings of differential operators - North Holland 1979

[J] JEANQUARTIER : Note au CRAS (1973)

[M] MAIRE : thèse Genève 1974.

The Operator $(dd^c)^n$ on Complex Spaces

Eric Bedford*

Dedicated to P. Lelong

1. Introduction

Let X be a complex space, which we will assume to be reduced, paracompact, and of pure dimension n; and let P(X) be the plurisubharmonic (p.s.h.) functions on X. A reduced complex space is a Hausdorff space that is locally given as a complex variety in an open subset of complex Euclidean space, and a function on X is p.s.h. if it is locally the restriction of a p.s.h. function in the ambient space. By a theorem of Fornaess and Narasimhan [9] a function ψ is a p.s.h. if it is uppersemicontinuous (u.s.c.) and if $\psi(f)$ is subharmonic on the unit disk $\Delta \subset \mathbb{C}$ for all holomorphic mappings $f: \Delta \longrightarrow X$.

Here we wish to study $P(X) \cap L^\infty(X,\mathrm{loc})$ in the tradition of Lelong [13,14]: we consider nonnegative currents on the singular space X. Our method is to work with the operator

$$(dd^c)^n: P(X) \cap L^\infty(X,\mathrm{loc}) \longrightarrow M_{n,n}(X)$$

where $M_{n,n}(X)$ denotes the Radon measures on X. For $\psi \in P(X) \cap L^\infty(X,\mathrm{loc})$ $(dd^c\psi)^n$ is defined in the usual way on Reg(X) (see [4]), and it is given the extension "by zero" to X, i.e. by definition

$$(1) \qquad \int_E (dd^c\psi)^n = \int_{E \cap \mathrm{Reg}(X)} (dd^c\psi)^n .$$

For $E \subset \Omega^{\mathrm{open}} \subset X$ we set

$$(2) \qquad C(E,\Omega) = \sup \left\{ \int_E (dd^c v)^n : v \in P(\Omega),\ 0 < v < 1 \right\} .$$

*Sloan Fellow.

(Note that the definition (1) spares us the problem of defining d and d^c on Sing (X).)

In this paper we show that the methods and ideas of [1,2,5] may be extended to singular spaces. Our main attention is given to the Comparison Theorem of [4]; we show that it holds in complex spaces and give some applications.

In the first application we discuss the equivalence of locally and globally pluripolar sets. This is an old conjecture of Lelong and a deep fact about p.s.h. functions. This equivalence was first shown for $\mathbb{C}^n$ by Josefson [12], and the extension to Stein manifolds and spaces (which is in fact trivial, assuming the result in $\mathbb{C}^n$) has been noted by several authors (see [1,2,7,17]).

We recall that the original proof of Josefson [12] and its refinement by El Mir [7] both involve a careful study of p.s.h. functions of the form c log |f|. On more general spaces, where this result still holds, the Hartogs functions (i.e. the family generated by upper and lower envelopes of c log |f|) do not span P(X). The equivalence result obtained below (Theorem 5.3) works in a more general context: by an example of Grauert [24], Theorem 5.3 applies to certain complex maifolds which possess no nonconstant holomorphic functions.

Section 6 gives a simple application of Josefson's Theorem to show that the polynomial hulls of certain submanifolds of $\mathbb{C}^n$ are small.

Our second application of the Comparison Theorem is based on the fact (Theorem 4.4) that $\psi \in P(X) \cap L^{\infty}(X,loc)$ is a free upper envelope of p.s.h. functions off of supp $(dd^c\psi)^n$. With this we give a version of the classic Evans' Law, which says that the discontinuities of a subharmonic function lie in supp $\Delta\phi$. (The analogous statement is false in the p.s.h. case since there is no local regularity for $(dd^c)^n$: e.g. if $\psi = \psi(z_1,\ldots,z_{n-1})$ is independent of z_n, then $(dd^c\psi)^n = 0$.) The p.s.h. version of Evans' Law says roughly that if $\psi \in P(X) \cap L^{\infty}(X,loc)$ is continuous on supp $(dd^c\psi)^n$ *and on* ∂X, then $\psi \in C(\text{Reg}(X))$. We do not know what happens on Sing (X).

2. Smoothing: Example of Fornaess

A frequent situation in trying to establish the results of [1,2,4,5] has been that the theorems are essentially trivial when the functions involved are p.s.h. and class C^2. The difficulty is to consider a sequence of smooth, p.s.h. functions decreasing to a given one, and then to pass to the limit. In a Stein manifold, such global smoothings are possible. We present here an example of J.E. Fornaess to show that global smoothings are not always possible on domains in $\mathbb{C}^2$.

Let us consider the domain

$$\Omega = \{(z,w) \in \mathbb{C}^2 : |z| < 2,\ |z| \neq 1, |w| < 1\} \cup \bigcup_{n=1}^{\infty} \mathcal{O}_n$$

where

$$\mathcal{O}_n = \{(z,w) : |z| = 1,\ |w - \tfrac{1}{n}| < e^{-e^n}\} .$$

It follows that Ω is a connected open set containing the disks $D_n = \{|z| < 2,\ w = \frac{1}{n}\}$ which "converge" to the split disk

$$D_o' = \{|z| < 2,\ |z| \neq 1,\ w = 0\} .$$

If we set

$$\psi(w) = \sum_{n=2}^{\infty} \frac{2^{-n}}{\log n} \log |w - \tfrac{1}{n}| ,$$

then $\psi(w)$ is subharmonic on $\mathbb{C}$ and $\psi(0) = -\frac{1}{2}$. Further, the $\mathcal{O}_n$ are chosen sufficiently small that the function defined by

$$\Phi(z,w) = \begin{cases} \max\,(\psi(w),-1) & \text{for } |z| < 1 \\ -1 & (z,w) \in \Omega,\ |z| \geq 1 \end{cases}$$

is p.s.h. on Ω.

We note that the behavior of $\Phi|_{D_n}$ is "discontinuous", i.e. $\Phi|_{D_n} = -1$ for $n = 1,2,3,\ldots$ and

$$\Phi|_{D_o'} = \begin{cases} -\tfrac{1}{2} & \text{for } |z| < 1 \\ -1 & \text{for } 1 < |z| < 2 \end{cases} .$$

From this apparent violation of the maximum principle we conclude that <u>there exists no sequence</u> $\{\Phi_j\} \subset P(\Omega) \cap C(\Omega)$ <u>decreasing pointwise to</u> Φ. The reason for this is that if $\Phi \in P(\Omega) \cap C(\Omega)$, then the maximum principle holds on $\Phi|_{D_n}$, and thus taking the limit as $n \longrightarrow \infty$, we have

(3) $$\Phi(0,0) \leq \sup_{|z|=3/2} \Phi(z,0).$$

Thus (3) also holds for any pointwise decreasing sequence in $C(\Omega) \cap P(\Omega)$.

Let us note that if there is a bounded, continuous, strongly p.s.h. function on a complex space Ω, then by a theorem of Richberg [14], every $\Phi \in C(\Omega) \cap P(\Omega)$ is the uniform limit of a sequence in $C^{\infty}(\Omega) \cap P(\Omega)$.

We note also that there are always semiglobal smoothings (i.e. $\{\psi_j\} \subset C^{\infty}(\Omega') \cap P(\Omega')$ for $\Omega' \subset\subset \Omega$) on any open subset Ω of a Stein manifold. The question seems to be open, however, whether there exist semiglobal smoothings on more general complex spaces. Since a semiglobal smoothing was used to obtain the Comparison Theorem of [4], we will need here to use slightly different smoothings and slightly different arguments.

A form of smoothing that will be adequate for our purposes is as follows. Let $\psi \in P(X)$ be given, and let $\{U_j\}$ be an open covering of X such that for each j there is an open set $\hat{U}_j \subset \mathbb{C}^{n_j}$, U_j is a subvariety of $\hat{U}_j$, and there exists $\hat{\psi}_j \in P(\hat{U}_j)$ with $\hat{\psi}_j|_{U_j} = \psi|_{U_j}$. And let $\{\chi_j\}$ be a partition of unity subordinate to $\{U_j\}$. Now by convolution with a usual radial smoothing kernel T_ε on $\mathbb{C}^{n_j}$, we have a smoothing $\psi_j^\varepsilon = T_\varepsilon * \psi_j$, which is p.s.h. in a neighborhood of supp χ_j for $\varepsilon > 0$ small.

Finally, our smoothing is obtained as the sum

$$\psi^\varepsilon = \sum \chi_j \psi_j^\varepsilon . \tag{4}$$

It is easily seen that ψ^ε decreases monotonically to ψ. In general, however, $\psi^\varepsilon \notin P(X)$, but ψ^ε is sufficiently close to being p.s.h. that we will be able to discuss $(dd^c\psi^\varepsilon)^n$.

3. Integration at the Singular Set

By definition, Sing (X) has zero capacity. Now we show that it has outer capacity zero.

Lemma 3.1. If X is a complex space and if $\Omega \subset X$ is an open subset, then $C^*((\text{Sing } X) \cap \Omega, \Omega) = 0$.

Proof: Since $C^*(E,\Omega)$ is countably subadditive in E and monotone decreasing in Ω, it is sufficient to show that

$$C^*((\text{Sing } X) \cap \Omega_j, \Omega_j) = 0$$

for an open cover $\{\Omega_j\}$ of Ω. Let us choose Ω_j small enough that it may be realized as a variety in $\Delta^n \times \mathbb{C}^{n_j}$, where $\Delta^n = \{(z_1,\ldots,z_n) \in \mathbb{C}^n: |z_1|,\ldots,|z_n| < 1\}$ is the unit polydisk. Further, we may assume that Ω_j is a λ-sheeted branched cover of Δ^n. Let $\pi: \Omega_j \longrightarrow \Delta^n$ denote projection.

It is sufficient to find open sets $\mathcal{O}_k$, Sing (X) $\subset \mathcal{O}_k \subset \Delta^n$, such that

$$\lim_{k \to \infty} C(\pi^{-1}(\mathcal{O}_k), \Omega_j) = 0. \tag{5}$$

Now for $\psi \in P(\Omega_j)$, we consider

$$\tilde{\psi}(z) = \sum_{p \in \pi^{-1}(z)} \psi(p)$$

for $z \in \mathbb{C}^n \backslash \pi(\text{Sing } X)$. Since ψ is bounded above, so is $\tilde{\psi}$, and thus $\tilde{\psi} \in P(\Delta^n)$. Further, if $0 < \psi < 1$, then $0 < \tilde{\psi} < \lambda$.

For $z_0 \in \Delta^n \backslash \pi(\text{Sing } X)$, there exists $\varepsilon > 0$ and holomorphic functions

$$\phi_\mu: \{|z-z_0| < \varepsilon\} \longrightarrow \mathbb{C}^{n_j},$$

$1 \leq \mu \leq \lambda$, which are local inverses of π. Since dd^c is invariant under holomorphic mappings

$$\int_{\pi^{-1}\{|z-z_0|<\varepsilon\}} (dd^c\psi)^n = \sum_{\mu=1}^{\lambda} \int_{\{|z-z_0|<\varepsilon\}} (dd^c\psi(\phi_\mu(z)))^n$$

$$\leq \int_{\{|z-z_0|<\varepsilon\}} (dd^c(\sum_{\mu=1}^{\lambda} \psi(\phi_\mu)))^n = \int_{\{|z-z_0|<\varepsilon\}} (dd^c\tilde{\psi})^n \quad .$$

Thus it follows that

$$\int_{\pi^{-1}(O_k)} (dd^c\psi)^n \leq \int_{O_k} (dd^c\tilde{\psi})^n$$

which yields $C(\pi^{-1}(O_k),\Omega_j) \leq \lambda^n\, C(O_k,\Delta^n)$ for any open $O_k \subset \Delta^n$. Now $\pi(\mathrm{Sing}\,(X))$ is a subvariety of Δ^n, and so by [1,2] we may find open sets $\mathrm{Sing}\, X \subset O_k \subset \Delta^n$ with $C(O_k,\Delta^n) \longrightarrow 0$. Thus (5) holds, which completes the proof.

From Lemma 3.1 we conclude several things. First, the definition (1) in fact defines $(dd^c\psi)^n$ as a Radon measure on X. Second, all the local analysis of [5] extends over Sing X. For instance, $\psi \in P(X)$ is quasi continuous, and all the "obvious" algebraic identities hold.

Let us define $A^k(X)$ to be the linear span of all of the k-forms on Reg X which are representable by integration and which are obtained as wedge products of

a) smooth forms θ on X

b) du_j, d^cu_j, or dd^cu_j ,

where $u_j \in P(X) \cap L^\infty(X,\mathrm{loc})$.

We say that a sequence $\{f_j\}$ of Borel functions on X converges <u>quasi uniformly</u> to f if the f_j are uniformly bounded, $f_j \longrightarrow f$ a.e. and for $\varepsilon > 0$, there is an open $\mathcal{O} \subset X$ such that $C(\mathcal{O},X) < \varepsilon$ and $f_j \longrightarrow f$ uniformly on $X \setminus \mathcal{O}$.

Another consequence of [5] is that if $\{f_j\}$ converges quasi uniformly to f, and if $\{\eta_j\} \subset A^{2n}(X)$ is a sequence such that the forms θ_j in a) converge uniformly and the p.s.h. functions u_j in b) converge monotonically a.e., then

$$f_j\eta_j \longrightarrow f\eta$$

where the convergence is in the weak sense of measures on X.

Let us recall the method of smoothing of Section 2, equation (4). We note that

$$\begin{aligned} dd^c\psi^\varepsilon &= \sum((dd^c\chi_j)\psi_j^\varepsilon + d\chi_j \wedge d^c\psi_j^\varepsilon \\ &\quad + d\psi_j^\varepsilon \wedge d^c\chi_j + \chi_j dd^c\psi_j^\varepsilon) \end{aligned}$$

and thus $(dd^c\psi^\varepsilon)^k \in A^{2k}(X)$. By our remark above, we see that the smoothings $(dd^c\psi^\varepsilon)^k$ converge weakly to $(dd^c\psi)^k$ as $\varepsilon \longrightarrow 0$.

<u>Lemma</u> 3.2 (Stokes' Theorem). <u>If</u> $\eta \in A^{2n-1}(X)$ <u>has compact support, then</u> $\int_X d\eta = 0$.

<u>Proof</u>: We may assume that η is a wedge product of terms of the form a) and b). As in Section 2, we may take a smoothing u_j^ε of u_j and denote the corresponding smooth form by η^ε. By the remarks on smoothing,

$$\lim_{\varepsilon \to 0} \int_X d\eta^{\varepsilon} = \int_X d\eta \quad .$$

Now the Lemma follows since

$$\int_X d\eta^{\varepsilon} = 0 \ ,$$

i.e. Stokes' Theorem holds for smooth forms on analytic spaces (see Bungart [6] and Herrera [11]).

4. Comparison Theorem

Lemma 4.1 *Let $\chi \in C^1(\mathbb{R})$ be a monotone decreasing function such that $\chi(t) = 1$ for $t \leq a$ and $\chi(t) = 0$ for $t \geq b$. Let $h \in C^1(X)$ and $\theta \in A^{2n-1}(X)$ be given, and assume that either $h^{-1}(-\infty,b] \subset\subset X$ or θ has compact support. Then*

$$
(6) \qquad \begin{aligned} -\int \chi'(t)dt \int_{\{h<t\}} d\theta &= \\ \int \chi(h)\, d\theta &= \int d\chi(h)\wedge\theta \\ &= \int \chi'(h)\, dh\wedge\theta \; . \end{aligned}
$$

Proof: If we show that the first equality holds, then the second follows by Lemma 3.2 and the third is immediate.

The first equality, however, is clear if h is a simple function, i.e.

$$h = \sum_{j=1}^{N} c_j \, 1_{S_j}$$

where $\{S_1,\ldots,S_N\}$ is a Borel partition of X. For in this case we may write the measure $d\theta$ as

$$d\theta = \sum \mu_j$$

where μ_j is the restriction of $d\theta$ to S_j. Thus it suffices to show that

$$-\int \chi'(t)dt \int_{\{h<t\}} \mu_j = \int \chi(h)\mu_j \; .$$

But this is seen because both sides may be computed directly to be $\chi(c_j)\mu_j(S_j)$.

<u>Lemma</u> 4.2 <u>Let</u> $u,v \in P(X) \cap L^{\infty}(X,\mathrm{loc})$ <u>and</u> $\theta \in A^{2n-1}(X)$ <u>be given. If</u> $a < b$ <u>and</u> $\{u-v < b\} \subset\subset X$, <u>then</u>

$$(7) \qquad \int_a^b dt \int_{\{u-v<t\}} d\theta = \int_{\{a<u-v<b\}} d(u-v)\wedge\theta \quad .$$

<u>Proof</u>: First we show that (6) holds with $h = u-v$. To see this, we replace θ by $\theta' = \phi\theta$ where $\rho \in C_0^{\infty}(X)$. If we set $h = u^{\varepsilon} - v^{\delta}$, where u^{ε} , v^{δ} are smoothings as in (4) of Section 2, then by Lemma 4.1,

$$(8) \qquad -\int \chi'(t)dt \int_{\{h<t\}} d\theta' = -\int \chi'(h)\, dh\wedge\theta' \quad .$$

Letting first $\varepsilon \searrow 0$ and then $\delta \searrow 0$, we see that the LHS of (8) holds with $h = u-v$.

To evaluate the convergence of the right hand side we first use the fact that $u^{\varepsilon} \searrow u$ and $v^{\varepsilon} \searrow v$ quasi uniformly to conclude that

$$\chi'(h)dh\wedge\theta' \text{ converges.}$$

Now we may choose $\rho \in C_0^{\infty}(X)$ such that $\{\rho = 1\} \supset \{u-v < b\} \supset \mathrm{supp}\, \chi'(u-v)$. Thus (8) holds with θ' replaced again by θ. Finally, we may choose a sequence χ_j with $0 \geq \chi_j' \geq -2$ such that

$$\lim_{j\to\infty} \chi_j'(t) = \begin{cases} -1 & \text{if} \quad t \in (a,b) \\ 0 & \text{if} \quad t \notin (a,b) \end{cases}$$

Both sides of (8) will then converge, as $j \longrightarrow \infty$, to the equality (7).

Theorem 4.3 (Comparison). Let $u,v \in P(X) \cap L^{\infty}(X,\mathrm{loc})$ be given such that $\{u \leq v\} \subset\subset X$. Then

$$\int_{\{u<v\}} (dd^c u)^n \geq \int_{\{u<v\}} (dd^c v)^n. \tag{9}$$

Proof: Let us set

$$\theta = d^c(u-v)\wedge[(dd^c u)^{n-1} + (dd^c u)^{n-2}\wedge dd^c v + \dots + (dd^c v)^{n-1}]. \tag{10}$$

Now by Lemma 4.2, for $a < b < 0$

$$\int_a^b dt \int_{\{u-v<t\}} d\theta = \int_{\{a<u-v<b\}} d(u-v)\wedge\theta \quad . \tag{11}$$

It is easily seen that

$$d(u-v)\wedge\theta \geq 0$$

and thus the right hand integral above is ≥ 0. Thus it follows that for a dense subset $\{t_1, t_2, \dots\} \subset (-\infty, 0)$, we have

$$\int_{\{u-v<t_j\}} d\theta \geq 0 \ .$$

If $t_j \nearrow 0$, then $\cup\{u-v < t_j\} = \{u-v < 0\}$. Thus $\int_{\{u-v<0\}} d\theta \geq 0$. Since $d\theta = (dd^c u)^n - (dd^c v)^n$, we have the desired result.

Theorem 4.4 (Envelopes). *Suppose that* $\Omega \subset\subset X$ *and that there exists* $\phi \in P(\Omega) \cap L^{\infty}(\Omega,\mathrm{loc})$ *such that*

$$\int_{\mathcal{O}} (dd^c \phi)^n > 0$$

for all open $\mathcal{O} \subset \Omega$. *Then for each* $\psi \in P(\Omega) \cap L^{\infty}(\Omega,\mathrm{loc})$,

(12) $$\psi(z) = \sup \{v(z) : v \in P(\Omega),\ v \leq \psi \text{ on } \operatorname{supp}(dd^c\psi)^n \text{ and } \limsup_{\zeta \to \partial\Omega} (v(\zeta) - \psi(\zeta)) \leq 0\} .$$

Proof: It suffices to show that each element v of the family above satisfies $v \leq \psi$ on Ω.

In particular, it suffices to show that for $\varepsilon > 0$,

$$\omega = \{\psi + \varepsilon < v\} = \emptyset.$$

We may assume that $\phi \leq 0$. If $\omega \neq \emptyset$, then we may choose $\delta > 0$ small enough that

$$\omega' = \{\psi + \varepsilon < v + \delta\phi\} \neq \emptyset .$$

On the other hand, $\omega' \subset \omega \subset\subset \Omega$ and so by the comparison theorem

$$\int_{\omega'} (dd^c(\psi+\varepsilon))^n \geq \int_{\omega'} (dd^c(v+\delta\phi)^n \geq \delta^n \int_{\omega'} (dd^c\phi)^n.$$

Since $\omega' \cap (\operatorname{supp} dd^c\psi)^n) = \emptyset$, the left hand integral vanishes. But ω' is essentially open (by quasicontinuity) and so the right hand integral is strictly positive. By this contradiction, we see that $\omega = \emptyset$.

<u>Remark</u>: We cannot simply drop the assumption of ϕ in Theorem 4.4. Otherwise, we can set $X = \mathbb{C} \times M$, where M is any compact Riemann surface. If $\Omega = \{|z| < 1\} \times M$, then a subharmonic function $\psi \in P(\Omega)$ depends on z alone and always satisfies $(dd^c\psi)^2 = 0$. But ψ is only given as the envelope in (12) if $\Delta\psi = 0$, i.e. ψ is harmonic.

5. Globally Pluripolar Sets

As our first application of the Comparison Theorem, we will give a generalization of Josefson's Theorem. As usual, we work with the extremal functions

$$u(E,\Omega,z) = \sup \{v(z) : v \in P(\Omega),\ v < 0,\ v \leq 01 \text{ on } E\}$$
$$v < 0,\ v \leq -1 \text{ on } E\}$$
$$u(E,\Omega,z)^{*} = \limsup_{\zeta \to z} u(E,\Omega,\zeta).$$

Since there will be a large overlap with the arguments of Theorem 1.5 of [1], we only sketch the details where they are different.

In this Section, we let $\psi \in P(X)$ be an exhaustion function, i.e. $\{\psi < c\} \subset\subset X$ for all $c \in \mathbb{R}$. We will also use the notation

$$\Omega(c) = \{\psi < c\} .$$

For any $\lambda_1 \in \mathbb{R}$, we may choose

$$\lambda_2 > \sup \ \{\psi(z) : z \in \Omega(\lambda_1)\}$$

and thus $\Omega(\lambda_1) \subset\subset \Omega(\lambda_2)$.

<u>Lemma</u> 5.1. <u>If</u> $K \subset \Omega(\lambda_1)$ <u>is compact, then</u>

$$\int_K (dd^c u(K,\Omega(\lambda_1))^{*})^n = C(K,\Omega(\lambda_1)). \tag{13}$$

<u>Proof:</u> We let $v \in P(\Omega(\lambda_1))$, $-1 < v < 0$, be given. For $0 < \varepsilon < 1$ the functions

$$v' = \begin{cases} \max \{(1-\varepsilon)v, c(\psi-\lambda_1)\} & \text{on } \Omega(\lambda_1) \\ c(\psi-\lambda_1) & \text{on } \Omega(\lambda_2)\setminus\Omega(\lambda_1) \end{cases}$$

$$u' = \begin{cases} \max\ u(K,\Omega(\lambda_1))^*, c(\psi-\lambda_1) & \text{on } \Omega(\lambda_1) \\ c(\psi-\lambda_1) & \text{on } \Omega(\lambda_2)\setminus\Omega(\lambda_1) \end{cases}$$

are uppersemicontinuous on X and thus $u', v' \in P(X)$. Since $\{u' < v'\} \subset \Omega(\lambda_1) \subset\subset \Omega(\lambda_2)$, we may apply the Comparison Theorem to deduce that

$$\int_{\{u' < v'\}} (dd^c u')^n \geq \int_{\{u' < v'\}} (dd^c v')^n \ .$$

Furthermore, by [1,2], $F = \{u' > -1\} \cap K$ is a locally pluripolar set, and thus

$$\int_F (dd^c w)^n = 0$$

for $w \in P \cap L^\infty_{loc}$. Taking $c > 0$ sufficiently large, we have $u' = u(K,\Omega(\lambda_1))^*$ and $v' = (1-\varepsilon)v$ in a neighborhood of K. Thus we have

$$\int_{\{u' < v'\}} (dd^c u')^n = \int_{\{u' < v'\}} (dd^c u(K,\Omega(\lambda_1))^*)^n$$
$$= \int_{K\setminus F} (dd^c u(K,\Omega(\lambda_1))^*)^n =$$
$$\int_K (dd^c u(K,\Omega(\lambda_1))^*)^n \geq \int_{\{u' < v'\}} (dd^c v')^n.$$

Now for any $\delta > 0$, we may choose an open set $\mathcal{O} \subset \Omega(\lambda_1)$ with $C(\mathcal{O},\Omega(\lambda_1)) < \delta$ and such that u' and v' are continuous on $\Omega(\lambda_1)\setminus\mathcal{O}$. Thus $\mathcal{O} \cup \{u' < v'\} \supset K$ and so

$$\int_{\{u'<v'\}} (dd^c v')^n \geq \int_K (dd^c v')^n - (1-\varepsilon)^n C(\mathcal{O},\Omega(\lambda_1))$$

$$\geq \int_K (dd^c v)^n - \delta(1-\varepsilon)^n.$$

Since $\delta > 0$ may be taken arbitrarily small, $\geq$ holds in (13). Now the inequality $\leq$ in (13) is obvious, so (13) holds. This completes the proof.

Now if $\mathcal{O} \subset\subset \Omega(\lambda_1)$ is an open set, and if $K_1 \subset K_2 \subset \ldots$ are compacts with $\cup K_j = \mathcal{O}$ we have

$$\lim_{j\to\infty} u(K_j,\Omega)^* = u(\mathcal{O},\Omega) = u(\mathcal{O},\Omega)^*.$$

Thus it follows that

$$C(\mathcal{O},\Omega) = \int_{\partial\mathcal{O}} (dd^c u(\mathcal{O},\Omega))^n.$$

Repeating the proof of Proposition 6.5 of [5], we have

$$C^*(E,\Omega(\lambda_1)) = \int (dd^c u(E,\Omega(\lambda_1))^*)^n$$

for arbitrary $E \subset\subset \Omega(\lambda_1)$.

<u>Lemma</u> 5.2. <u>If</u> $E \subset\subset \Omega(\lambda_1)$, <u>and if</u> $u(E,\Omega(\lambda_1))^* < 0$, <u>then either</u> $C^*(E,\Omega(\lambda_1)) > 0$, <u>or</u>

$$\int_{\Omega(\lambda_1)} (dd^c \phi)^n = 0 \text{ for all } \phi \in P(\Omega(\lambda_2)) \cap L^\infty(\Omega(\lambda_2)).$$

<u>Proof</u>: Without loss of generality we may assume that $\psi - \lambda_1 \leq -1$ on E. It follows, then, that

$$0 > u(E,\Omega(\lambda_1))^* \geq \psi - \lambda_1$$

holds on $\Omega(\lambda_1)$. Since ψ is uppersemicontinuous, $\psi - \lambda_1 \geq 0$ on $\partial\Omega(\lambda_1)$, and thus

$$\psi' = \begin{cases} \max \{u(E,\Omega(\lambda_1))^*, \psi-\lambda_1\} & \text{on } \Omega(\lambda_1) \\ \psi - \lambda_1 & \text{on } \Omega(\lambda_2)\setminus\Omega(\lambda_1) \end{cases}$$

belongs to $P(\Omega(\lambda_2))$.

Let us suppose that there exists $\phi \in P(\Omega(\lambda_2)) \cap L^\infty(\Omega(\lambda_2))$ such that

$$\int_{\Omega(\lambda_1)} (dd^c\phi)^n > 0 .$$

Thus for some $\varepsilon > 0$,

$$\int_{\Omega(\lambda_1-\varepsilon)} (dd^c\phi)^n > 0. \tag{14}$$

Now we may choose $a > 0$ such that

$$0 > \phi'(z) = a\phi(z) - \varepsilon/2 > -\varepsilon$$

holds for $z \in \Omega(\lambda_2)$.

By simple inequalities

$$\Omega(\lambda_1-\varepsilon) \subset \{\psi' < \phi'\} \subset \Omega(\lambda_1), \tag{15}$$

and by the choice of λ_2,

$$\Omega(\lambda_1) \subset\subset \Omega(\lambda_2).$$

Now we may apply the Comparison Theorem to obtain

$$\int_{\{\psi' < \phi'\}} (dd^c\psi')^n \geq \int_{\{\psi' < \phi'\}} (dd^c\phi')^n .$$

Since $\psi' = u(E,\Omega(\lambda_1))^*$ on $\Omega(\lambda_1)$, (14) and (15) yield

$$\int_{\{\psi' < \phi'\}} (dd^c u(E,\Omega(\lambda_1))^*)^n \geq \int_{\Omega(\lambda_1-\varepsilon)} (dd^c\phi') > 0 .$$

The Lemma now follows from Lemma 5.1.

Theorem 5.3. Let X be a complex space. If there is a p.s.h. exhaustion function on X, and if

(16) there exists $\phi \in P(X) \cap L^\infty(X,\text{loc})$ with $0 < \int_X (dd^c\phi)^n \leq +\infty$,

then for every locally pluripolar subset E of X, there exists $p \in P(X)$ such that $E \subset \{p = -\infty\}$.

Proof: The proof follows the lines of Theorem 1.5 of [1]. That is, we may assume that $E = \cup E_j$, and that E_j is strictly pluripolar in a small domain ω_j, i.e. $u(E_j,\omega_j)^* = 0$. Since we may choose ω_j to be Stein, it follows $C^*(E_j,\omega_j) = 0$.

Now by (16), we may choose λ_0 large enough that

$$\int_{\Omega(\lambda_0)} (dd^c\phi)^n > 0.$$

Since $C^*(E_j,\Omega(\lambda_j)) = 0$ for any $\Omega(\lambda_j) \supset \omega_j$, it follows by Lemma 5.2 that

$$u(E_j,\Omega(\lambda_j))^* = 0$$

if $\lambda_j \geq \lambda_0$.

Thus we may take

$$p = \sum p_j$$

where $p_j = \begin{cases} \max\{h_j, \psi - \lambda_j\} & \text{on } \Omega(\lambda_j) \\ \psi - \lambda_j & \text{on } X \setminus \Omega(\lambda_j) \end{cases}$

$h_j \in P(\Omega(\lambda_j))$, $h_j < u(E_j, \Omega(\lambda_j))$, and h_j is sufficiently small in $L^1(\Omega(\lambda_j))$.

Remark 1. We note that if the equivalence of locally- and globally-pluripolar sets holds on a space X, then it holds on any open subspace $Y \subset X$. Thus the assumption of a p.s.h. exhaustion in Theorem 5.3 is not natural. It is only there because we do not yet understand $C^*(E,\Omega)$ when Ω is not pseudoconvex.

Remark 2. The assumption (16) is in principle the same as assuming that the elements of $P(X) \cap L^\infty(X,\text{loc})$ are not functionally dependent. If (16) does not hold, then

(17) $$d\phi_1 \wedge \ldots \wedge d\phi_n = 0$$

holds for all $\phi_1, \ldots, \phi_n \in P(X) \cap L^\infty(X,\text{loc})$. Indeed, for $\phi_1, \ldots, \phi_n$ we may consider

$$\Phi = e^{\phi_1} + \ldots + e^{\phi_n} .$$

By Section 3,

$$\begin{aligned} (dd^c\Phi)^n &= (e^{\phi_1}(dd^c\phi_1 + d\phi_1 \wedge d^c\phi_1) + \ldots + e^{\phi_n}(dd^c\phi_n + d\phi_n \wedge d^c\phi_n))^n \\ &\geq (e^{\phi_1} d\phi_1 \wedge d^c\phi_1 + \ldots + e^{\phi_n} d\phi_n \wedge d^c\phi_n)^n \\ &\geq n!\, e^{\phi_1 + \ldots + \phi_n} d\phi_1 \wedge d^c\phi_1 \wedge \ldots \wedge d\phi_n \wedge d^c\phi_n . \end{aligned}$$

Thus if (16) does not hold, then (17) holds.

6. Polar Manifolds and Polynomial Hulls

An interesting question that arises from Josefson's Theorem is to know how far a given polar set is from being "complete". That is, if $E \subset \mathbb{C}^n$ is polar, we may consider the new polar set

$$E^* = \{z \in \mathbb{C}^n : p(z) = -\infty \text{ for all } p \in P(\mathbb{C}^n) \text{ such that } p|_E = -\infty\} .$$

Although E^* may be characterized in other ways (see [10]), the relation between E and E^* is not clear. A nontrivial example has been given by Sadullaev [18], who showed that a smoothly bounded complex disk

$$E = \{(z,w) \in \mathbb{C}^2 : w = f(z),\ |z| < 1\}$$

can satisfy $E^* = E$.

If K is compact, then the polynomial $\hat{K}$ coincides with the hull with respect to $P(\mathbb{C}^n)$. Thus $K \subset \hat{K} \subset K^*$. Although local and global polynomial convexity are not the same, the relationship with polar sets will give a method which sometimes showed that global hulls are small.

Let M be a C^1 submanifold of $\mathbb{C}^n$. For $p \in M$, we let H_pM denote the largest $\mathbb{C}$-linear subspace of $T_pM \subset \mathbb{C}^n$. M is said to be CR if H_pM has constant dimension, and M is <u>generating</u> at p if there is no $\mathbb{C}$-linear subspace L with

$$T_pM \subset L \subsetneq \mathbb{C}^n .$$

We remark that if a smooth manifold is generating at any point then it is not pluripolar. We are interested in the converse.

<u>Theorem</u> 6.1. <u>If</u> $M \subset \mathbb{C}^n$ <u>is a real analytic submanifold which is nowhere generating, then</u> M <u>is pluripolar</u>.

<u>Proof</u>: By Josefson's Theorem, it suffices to show that M is a union of locally pluripolar subsets. First we may stratify $M = M_0 \cup \ldots \cup M_k$ into a union of semianalytic subsets with the property that $M_j \subset \tilde{M}_j$, where $\tilde{M}_j$ is a real analytic CR manifold which is nowhere generating.

Let us write $M = \tilde{M}_j$ and work locally near $0 \in M$. We may rotate coordinates so that $\mathbb{C}^m \times \{0\} \subset \mathbb{C}^n$ is the smallest $\mathbb{C}$-linear subspace of $\mathbb{C}^n$ containing T_0M. If $\pi(z_1,\ldots,z_n) = (z_1,\ldots,z_m)$, then $\pi|_M: M \longrightarrow \pi M$ is a CR diffeomorphism near zero. Thus

$$f = \pi^{-1}: \pi\,(M) \longrightarrow M$$

is also a CR mapping. By a theorem of Tomassini [21] f extends to a holomorphic mapping $F(z) = (z_1,\ldots,z_m,F_{m+1}(z),\ldots,F_n(z))$ in a neighborhood U of $\pi(M)$ in $\mathbb{C}^m$. It follows now that

$$M \subset \{(z_1,\ldots,z_n) \in \mathbb{C}^n:$$
$$\pi(z) \in U,\ z_j = F_j(\pi(z)) \text{ for } m+1 \le j \le n\}\ .$$

Thus M is pluripolar, which completes the proof.

<u>Corollary</u> 6.2. <u>If</u> $M \subset \mathbb{C}^n$ <u>is a real analytic subset that is nowhere generating, then</u> $\hat{M}$ <u>is a polar subset of</u> $\mathbb{C}^n$. <u>In particular</u>, $\hat{M}$ <u>has no interior</u>.

If M is allowed to be generating, then M may be locally polynomially convex but $\hat{M}$ may contain an open set, e.g. $\widehat{(\partial\Delta)^n} = \overline{\Delta^n}$.

7. Evans' Law

We have tried to present some parallels between the operator $(dd^c)^n$ and the Laplacian Δ. Evans' Law, which says that the discontinuities of a function ψ lie inside the support of $\Delta\psi$, does not generalize exactly. For instance, by an example at the end of Section 3 of [1], the function $u(K,\Omega)^*$ is not necessarily continuous on $\Omega\setminus K$.

Let us assume that X is a Stein space, and that $\Omega \subset\subset X$ is an open subset. It follows, then, that there is a finite number of holomorphic vector fields $Z_1,\ldots,Z_m$ on X such that for each $p \in \overline{\Omega} \cap \mathrm{Reg}(X)$ $\{Z_1(p),\ldots,Z_m(p)\}$ spans T_pX. (See, for instance, Fischer [8].) We may assume that the Z_j all vanish on Sing X. Thus any integral curve of Re Z_j which starts in Reg (X) will remain in Reg (X). This gives us a mapping $p \longrightarrow T^s_{Z_j}(p)$ obtained by starting at p and following the integral curve of Re Z_j for time s, $|s| \leq s(p)$. If $\Omega \subset\subset X$, then for $|s| \leq s(\Omega)$

$$T^s_{Z_j} : \Omega \longrightarrow T^s_{Z_j}(\Omega) \subset X$$

is a biholomorphic mapping.

Given a function on $\Omega \subset\subset X$, we may define the modulus

$$\omega(f,Z,\delta) = \sup \{|f(p)-f(T^s_Z\, p)| : p \in \Omega,\ 0 \leq |s| \leq \delta \leq s(\Omega)\ ,$$

$$Z = \sum a_j Z_j,\ \sum |a_j|^2 \leq 1, \text{ and } T^s_Z p \in \Omega\}.$$

We will say that $\omega(f,Z,f)$ is a modulus of continuity if $\lim_{\delta\to 0} \omega(f,Z,\delta) = 0$.

An alternative would be to take a Stein neighborhood Ω', $\Omega \subset\subset \Omega' \subset\subset X$ and imbed Ω' properly into $\mathbb{C}^N$. We could give Ω the imbedded metric, and thus obtain another modulus of continuity $\tilde{\omega}(f,\delta)$. It follows, then, that

$$\omega(f,Z,\delta) \leq \tilde{\omega}(f,C\delta)$$

for some constant $C > 0$. Conversely, if $\lim_{\delta \to 0} \omega(f,Z,\delta) = 0$ then f is continuous on Reg X, but we cannot conclude anything about the behavior of f on Sing X.

Theorem 7.1. *Let X be Stein, $\Omega \subset\subset X$, and $f \in C(\Omega)$ be given. If $\psi \in P(\Omega) \cap L^\infty(\Omega,\text{loc})$, $\psi \leq f$, and $\psi(p) = f(p)$ for $p \in \text{supp}\,(dd^c\psi)^n$ and for p in a neighborhood of $\partial\Omega$, then $\omega(\psi,Z) \leq \omega(f,Z)$. In particular, $\psi \in C(\text{Reg}\,(\Omega))$. If, in addition, $f \in C^{1,\alpha}(\Omega)$, for some $0 < \alpha \leq 1$, then $\psi \in C^{1,\alpha}(\text{Reg}\,(\Omega))$.*

Proof: By Theorem 4.4, ψ is given as an upper envelope, which may as well be written

$$\psi(z) = \sup\,\{v(z)\colon v \in P(\Omega),\ v \leq f\}$$

by the hypotheses on f. For $Z = \sum a_j Z_j$, with $\sum |a_j|^2 \leq 1$ we consider

$$\psi_s(p) = \psi(T_Z^s p)\ .$$

Repeating now the proof of Theorem 6.2 of [4], making the appropriate modifications, as in Theorem 3 of [3], we see that

$$|\psi(p)-\psi_s(p)| \leq \omega(f,Z,s).$$

This gives the desired continuity result.

For the other conclusions, we recall that if $f \in C^{1,\alpha}$, then

$$|f_s - 2f + f_{-s}| \leq C\,|s|^{1+\alpha}$$

(see Stein [20]). Now we repeat the proof of Theorem 3 of [3], applied to ψ_s, instead of translates, to see that $\psi \in C^{1,\alpha}(\operatorname{Reg}\Omega)$. This completes the proof.

We point out that for $n = 1$, a sharper result has been obtained by Lewy [15]. For other results on the continuity of envelopes of p.s.h. functions, see [19,22,23].

We will say that $\psi \in P(\Omega)$ is continuous at $\partial\Omega$ if there is a continuous function $f \in C(\partial\Omega)$ such that

$$f(z) = \limsup_{\substack{\zeta \to z \\ \zeta \in \Omega}} \psi(\zeta)$$

holds for $z \in \partial\Omega$.

Theorem 7.2 (Evans' Law). *Let* $\Omega = \{\phi < 0\}$, *where* $\phi \in C^2(\overline{\Omega})$ *and* ϕ *is strongly p.s.h. Let* $\psi \in P(\Omega) \cap L^\infty(\Omega, \mathrm{loc})$ *be continuous on the set*

$$\partial\Omega \cup \operatorname{supp}(dd^c\psi)^n.$$

Then $\psi \in C(\overline{\Omega} \cap \operatorname{Reg} X)$.

Proof: By the continuity of ψ at $\partial\Omega \cup \operatorname{supp}(dd^c\psi)^n$ and the uppersemicontinuity of ψ on Ω, there is a continuous function $f \in C(\overline{\Omega})$ with $f \geq \psi$ and such that $f = \psi$ on $\partial\Omega \cup \operatorname{supp}(dd^c\psi)^n$.

Thus by Theorem 4.4 ψ is given as the envelope of the p.s.h. functions dominated by f. The proof now follows by repeating Lemma 1 of Walsh [22], with translates replaced by $\psi(T_z^s p)$.

The method of looking at upper envelopes yields a procedure for smoothing solutions of the Dirichlet problem.

Theorem 7.3. Let X be a Stein manifold, and let $\Omega = \{\phi < 0\} \subset\subset X$ where $\phi \in C^2(\overline{\Omega})$ is strongly p.s.h. If $g \in C(\partial\Omega)$ and $v \in P(\Omega)$ is the solution to

$$(dd^c v)^n = 0 \quad \text{on } \Omega$$
$$v = g \quad \text{on } \partial\Omega$$

then there exists a sequence $\{v_j\}$ of functions which are p.s.h., $C^{1,1}$, and satisfy $(dd^c v_j)^n = 0$ in a neighborhood of $\overline{\Omega}$, and which converge uniformly to v on $\overline{\Omega}$.

Proof: We may approximate g uniformly by a smooth function and add $C\phi$, so without loss of generality we may assume that $g \in C^2(X)$ and is strongly p.s.h. in a neighborhood of $\overline{\Omega}$. Next we consider domains $\Omega_j = \{\phi < \frac{1}{j}\}$ with boundary values $g_j = g\big|_{\partial\Omega_j}$, and the corresponding Dirichlet solutions $\tilde{v}_j$ converge uniformly to v on $\overline{\Omega}$.

Now we let

$$G_j = g + C_j\,(\phi - \tfrac{1}{j})^2$$

be defined on the region $A_j = \{z: \frac{1}{2j} < \phi(z) < \frac{3}{2j}\}$. We may choose C_j large enough that $G_j \geq \|g\|_{L^\infty}$ on $\{\phi = \frac{1}{2j}\}$. We define

$$v_j(z) = \sup\{w(z) : w \in P\{\phi < \frac{3}{2j}\},\ w \le G_j \text{ on } A_j\}\ .$$

Since the function which is equal to g on $\{\phi < \frac{1}{j}\}$ and G_j on $\{\frac{1}{j} < \phi < \frac{3}{2j}\}$ is p.s.h., we see that $v_j = G_j$ in a neighborhood of $\{\phi = \frac{3}{2j}\}$. Clearly $v_j \le \tilde{v}_j$ on $\{\phi < \frac{1}{j}\}$, and so $G_j > v_j$ near $\{\phi = \frac{1}{2j}\}$. Thus we may apply Theorem 7.1 and conclude that $v_j \in C^{1,1}$. Also, $(dd^c v_j)^n = 0$ in a neighborhood of $\overline{\Omega}$. Finally, it is clear that $\{v_j\}$ converges uniformly to v on $\overline{\Omega}$, which gives the desired approximation.

Similar arguments yield the following.

Corollary 7.4. Let $\Omega_0 \subset\subset \Omega_1 \subset\subset X$, $\Omega = \Omega_1 \setminus \overline{\Omega}_0$, let X be a Stein manifold and let Ω_0, Ω_1 be strongly pseudoconvex with smooth boundary. The solution $u \in P(\Omega) \cap C(\overline{\Omega})$ of

$$(dd^c u)^n = 0 \quad \text{on } \Omega$$

$$u = \sigma \quad \text{on } \partial\Omega_\sigma,\ \sigma = 0,1$$

may be approximated uniformly on $\overline{\Omega}$ by $C^{1,1}$ solutions of $(dd^c)^n = 0$ in a neighborhood of $\overline{\Omega}$.

References

1. E. Bedford, Extremal plurisubharmonic functions and pluripolar sets in $\mathbb{C}^2$. Math. Ann. 249 (1980), 205-233.

2. ________, Envelopes of continuous, plurisubharmonic functions. Math. Ann. 251 (1980), 175-183.

3. ________, Stability of envelopes of holomorphy and the degenerate Monge-Ampère equation.

4. E. Bedford and B.A. Taylor, The Dirichlet problem for a complex Monge-Ampère equation. Inventiones Math. 37 (1976), 1-44.

5. ________________________, Some potential theoretic properties of plurisubharmonic functions.

6. L. Bungart, Integration on real analytic varieties II. Stokes formula. J. of Math. and Mech. 15 (1966), 1047-1054.

7. H. El Mir. Fonctions plurisousharmoniques et ensembles polaires, in Seminaire Lelong-Skoda (Analyse) Annees 1978/79, Springer Lecture Notes 822 (1980), 61-76.

8. G. Fischer, Complex Analytic Geometry. Springer Lecture Notes, 538 (1976).

9. J.E. Fornaess and R. Narasimhan, The Levi problem on complex spaces with singularities. Math. Ann. 248 (1980), 47-72.

10. T. Gamelin and N. Sibony, Subharmonicity for Uniform Algebras, J. of Functional Analysis, 35 (1980), 64-108.

11. M. Herrera, Integration on a semianalytic set. Bull. Soc. Math. France 94 (1966), 141-180.

12. B. Josefson, On the equivalence between locally polar and globally polar sets for plurisubharmonic functions on $\mathbb{C}^n$, Ark. Math. 16 (1978), 109-115.

13. P. Lelong, Fonctions Plurisousharmoniques et formes Différentielles Positives, Gordon and Breach, 1968.

14. ________, Fonctionelles Analytiques et Fonctions Entières (n Variables) Les Presses de l' Université de Montreal, 1968.

15. H. Lewy, On a refinement of Evans' law in potential thoery. Rendiconti Accad. Naz. dei Lincei, 48 (1970), 1-9.

16. R. Richberg, Stetige streng pseudokonvexe Funktionen. Math. Annalen 175 (1968), 257-286.

17. A. Sadullaev, The operator $(dd^c)^n$ and condenser capacities, Doklady Nauk USSR, 251 (1980), 44-47.

18. A. Sadullaev, P-regularity of sets in $\mathbb{C}^n$. in Analtyic Functions, Kozubnik, 1979, Springer Lecture Notes 798 (1980), 402-408.

19. J. Siciak, Extremal plurisubharmonic functions in $\mathbb{C}^n$. Proceedings of the first Finnish-Polish Summer School in Complex Analysis, (1977), 115-152.

20. E. Stein, Singular Integrals and Differentiability Properties of Functions. Princeton U. Press, 1970.

21. G. Tomassini, Tracce delle funzioni olomorfe sulle sottovarieta analitiche reali d'una varietà complessa, Ann. Scuola Norm. Sup. Pisa, 20 (1966), 31-43.

22. J.B. Walsh, Continuity of envelopes of plurisubharmonic functions, J. Math. Mech. 18 (1968), 143-148.

23. V. Zaharjuta, Extremal plurisubharmonic functions ..., Ann. Pol. Math. 33 (1976), 137-148.

24. H. Grauert, Bemerkenswerte pseudokonvexe Mannigfaltigkeiten, Math. Z. 81 (1963), 377-391.

Princeton University
Princeton, New Jersey 08544

STABILITÉ DU NOMBRE DE LELONG PAR RESTRICTION A UNE SOUS-VARIÉTÉ

Christer O. Kiselman

Étude dédiée à Pierre Lelong à l'occasion de sa promotion au grade de docteur honoris causa le 5 juin 1981.

Resumo. La nombro de Lelong estas mezurilo por maskoncentraĵoj ĉe kurentoj aperantaj en la kompleksa analitiko, do ankaŭ por plursubharmonaj funkcioj, kaj pliĝeneraligas tiel la klasikan koncepton de obleco. El ĝiaj ecoj menciindas la stabileco je malplivastigo al rektaj linioj: se f estas plursubharmona, $f|L$ kaj f havas la saman nombron de Lelong en la origino por ĉiu kompleksa rekto L entenanta la nulpunkton escepte se L apartenas al malgranda (loke polusa) aro en la projektiva spaco. La celo de tiu ĉi noto estas pliĝeneraligo de tiu stabileco: L estas anstataŭigita per parametrigita analitika subaro X. La teoremo 4.1 montras ke por holomorfaj mapoj h, la nombro de Lelong de $f \circ h$ kaj de f estas la sama krom por h en polusa aro. La analitika aro X do estas la bildo de h; tia parametrigo permesas senperan difinon kaj de la nombro de Lelong de $f|X$ kaj de la esceptara karaktero.

1. *Introduction*

Le nombre de Lelong sert à mesurer la concentration de masse des courants qui apparaissent en analyse complexe : c'est une généralisation de la notion de multiplicité. Le nombre de Lelong $\nu_f(x)$ d'une fonction plurisousharmonique f est le nombre de Lelong du courant $\frac{i}{\pi}\partial\bar\partial f$ ou la densité de la mesure $\frac{1}{2\pi}\Delta f$. Il a des propriétés remarquables, dont une propriété de stabilité que voici : si f est plurisousharmonique dans un ouvert Ω de $\mathbb{C}^n$, la restriction $f|L$ a le même nombre de Lelong que f au point $x \in \Omega$ pour toute droite complexe passant par x sauf quand L appartient à une partie localement polaire dans l'espace projectif, voir [7, proposition 15]. Grâce à ce résultat, Lelong a pu donner [7, proposition 14, b] une définition très convenable du nombre de Lelong $\nu_f(x)$ d'une fonction plurisousharmonique f définie sur un espace E de dimension infinie : on pose simplement

$$\nu_f(x) = \inf_L \nu_{f|L}(x)$$

où L est une droite passant par x, et on sait que la borne inférieure est atteinte sous certaines hypothèses sur E.

De façon analogue, si f est plurisousharmonique au voisinage de $0 \in \mathbb{C}^n$ et L est un sous-espace vectoriel de dimension m avec $1 \leq m \leq n$, alors $\nu_{f|L}(0) = \nu_f(0)$ sauf quand L appartient à un sous-ensemble exceptionnel dans la grassmannienne $G_m(\mathbb{C}^n)$ de tous les sous-espaces de dimension m. Pour ce résultat, voir [3, théorème 4.1]. Or, pourquoi exiger que L soit un sous-espace linéaire de $\mathbb{C}^n$? Evidemment on doit chercher un énoncé qui est invariant par applications biholomorphes : on doit montrer que

$$\nu_{f|X}(0) = \nu_f(0)$$

pour tout sous-espace analytique X de $\mathbb{C}^n$ contenant l'origine, sauf quand X appartient à un petit ensemble exceptionnel. Voilà le but de la présente note dont le résultat principal est le théorème 4.1. Pour donner un sens élémentaire à l'équation $\nu_{f|X}(0) = \nu_f(0)$ on supposera que X est paramétrisé par m paramètres complexes et alors il est immédiat de définir $\nu_{f|X}$ aussi bien que le caractère de l'ensemble exceptionnel. Comme l'espace de toutes les sous-variétés analytiques est de dimension infinie, il faut préalablement étudier les fonctions plurisousharmoniques sur de tels espaces ; c'est ce que nous ferons au paragraphe 3. Dans le paragraphe 2 nous résumerons les définitions principales dont nous aurons besoin et nous verrons que le nombre de Lelong peut être défini à l'aide du maximum de la fonction sur des boules.

2. Le nombre de Lelong

Soit μ une mesure de Borel positive dans $\mathbb{R}^m$ et notons $x + rB$ la boule fermée dans $\mathbb{R}^m$ de centre x et de rayon r. La masse portée par cette boule sera notée $\mu(x + rB)$ et on pourrait appeler densité moyenne k-dimensionnelle le quotient

$$\theta_k(x,r) = \frac{\mu(x + rB)}{\lambda_k(rB \cap \mathbb{R}^k)} \quad , \quad x \in \mathbb{R}^m, \ r > 0 ,$$

où λ_k est la mesure de Lebesgue dans $\mathbb{R}^k$ (on suppose que $0 \leq k \leq m$ et que $m \geq 1$). Si la limite

$$\theta_k(x) = \lim_{r \to 0} \theta_k(x,r) \ , \quad x \in \mathbb{R}^m ,$$

existe, elle a droit d'être nommée la densité k-dimensionnelle de μ en x. Par exemple, si μ provient d'une fonction convexe g,

$$\mu = \Delta g = \frac{\partial^2 g}{\partial x_1^2} + \ldots + \frac{\partial^2 g}{\partial x_m^2} ,$$

on obtient (en ne trichant que très peu)

$$\mu(x + rB) = \int_{x+rB} \Delta g = \int_{x+rS} \frac{\partial g}{\partial r} = \frac{\partial \hat{g}}{\partial r} r^{m-1} \int_{S^{m-1}} 1$$

où $\hat{g}(x,r)$ est la valeur moyenne de g sur la sphère $x + rS$ de rayon r et centre x, et

$$\lambda_{m-1}(rB \cap \mathbb{R}^{m-1}) = r^{m-1} \int_{B^{m-1}} 1 .$$

Donc

$$\theta_{m-1}(x,r) = c_m \frac{\partial \hat{g}}{\partial r} (x,r) \ , \quad x \in \mathbb{R}^m, \ r > 0 ,$$

et la densité $(m-1)$-dimensionnelle existe comme $\hat{g}$ est fonction convexe croissante de $r > 0$ pour x fixé. (La dérivée $\partial\hat{g}/\partial r$ doit être interprétée comme dérivée à droite.) Cette densité correspond au nombre de

Lelong, mais elle n'admet aucune stabilité par restriction.

Si μ provient d'une fonction plurisousharmonique non identiquement $-\infty$ dans un voisinage de $x \in \mathbb{C}^n$ où $n = m/2$, plus précisement si

$$\mu = \frac{1}{2\pi} \Delta g ,$$

on aura

$$\mu(x + rB) = \frac{1}{2\pi} \frac{\partial \hat{g}}{\partial r} r^{2n-1} \int_{S^{2n-1}} 1$$

et

$$\lambda_{2n-2}(rB \cap \mathbb{C}^{n-1}) = r^{2n-2} \int_{B^{2n-2}} 1 .$$

Or, cette fois c'est $u(x,y) = \hat{g}(x,e^y)$ qui est fonction convexe croissante de y et il vient, avec $r = e^y$, $dy/dr = 1/r$:

$$\theta_{2n-2}(x,e^y) = \frac{\partial u}{\partial y}(x,y) \frac{\int_{S^{2n-1}} 1}{2\pi \int_{B^{2n-2}} 1} = \frac{\partial u}{\partial y}(x,y) ,$$

et la densité ponctuelle est

$$\nu_g(x) = \theta_{2n-2}(x) = \lim_{y\to-\infty} \theta_{2n-2}(x,e^y) = \lim_{y\to-\infty} \frac{\partial u}{\partial y}(x,y) = \lim_{y\to-\infty} \frac{u(x,y)}{y} .$$

<u>Le nombre de Lelong de</u> g est par définition cette densité $(2n-2)$-dimensionnelle de $\mu = \frac{1}{2\pi} g$. (Les astuces pour cacher le facteur 2π varient un peu dans la littérature, cf. [2, 6, 9].) Si g est $-\infty$ identiquement la même formule servira : on posera $\nu_{-\infty}(x) = +\infty$.

Tout cela est bien classique [4, 5, 6] et nous n'avons évoqué ces définitions que pour la raison suivante : dans plusieurs calculs il est plus pratique d'utiliser, au lieu de la valeur moyenne $u(x,y)$ sur la sphère $x + e^y S$, le maximum de la fonction g. Il apparaît alors que le maximum peut servir aussi bien que la valeur moyenne pour définir le nombre de Lelong, à savoir :

Proposition 2.1. (cf. AVANISSIAN [10]). - Soit g plurisousharmonique dans un ouvert Ω de $\mathbb{C}^n$ et posons pour $(x,y) \in \Omega \times \mathbb{R}$ tels que $x + e^y B \subset \Omega$:

$$u(x,y) = \text{valeur moyenne de } g \text{ sur } x + e^y S = \int_{x+e^y S} g \Big/ \int_{x+e^y S} 1 ,$$

et

$$v(x,y) = \sup_{x+e^y S} g .$$

Alors

$$\nu_g(x) = \lim_{y\to -\infty} u(x,y)/y = \lim_{y\to -\infty} v(x,y)/y , \quad x \in \Omega .$$

Démonstration. - On a toujours $u \leq v$. D'autre part, pour montrer une inégalité dans l'autre sens, on pourra supposer pour simplifier que $g \leq 0$ dans Ω. Etant données deux boules, l'une contenue dans l'autre :

$$x' + r'B \subset x'' + r''B \subset \Omega ,$$

on a une inégalité dans le sens inverse

$$\int_{x''+r''B} g \leq \int_{x'+r'B} g , \quad r''^{2n} \underset{x''+r''B}{M} g \leq r'^{2n} \underset{x'+r'B}{M} g$$

où M désigne "valeur moyenne". Cela donne

$$r''^{2n} g(x'') \leq r''^{2n} \underset{x''+r''B}{M} g \leq r'^{2n} \underset{x'+r'B}{M} g \leq r'^{2n} \underset{x'+r'S}{M} g$$

et, en faisant varier x'' dans la boule $x + rB = x + e^y B$,

$$v(x,y) \leq u(x',y')\left(\frac{r'}{r''}\right)^{2n} = u(x',y')e^{2n(y'-y'')}$$

où $y' = \log r'$, $y'' = \log r''$. Ici les seules conditions sont que $|x'-x| + r + r' \leq r''$ et que la boule $x + r''B$ soit contenue dans Ω . Si on choisit $x' = x$ et $r'' = r + r'$ on obtient

$$v(x,y) \leq u(x,y')\left(\frac{r'}{r + r'}\right)^{2n} = u(x,y') \frac{1}{(1 + e^{y-y'})^{2n}} .$$

Donc, si $y, y' < 0$,

$$\frac{u(x,y)}{y} \geq \frac{v(x,y)}{y} \geq \frac{u(x,y')}{y'} \cdot \frac{y'}{y} \cdot (1 + e^{y-y'})^{-2n} .$$

Il convient maintenant de choisir par exemple $y' = y + \sqrt{-y}$ de façon que $y - y' = -\sqrt{-y} \to -\infty$ et $y'/y \to 1$ quand $y \to -\infty$. Cela montre bien la proposition.

J'ignore si v donne aussi le nombre de Lelong dans le cas d'un espace de Banach.

3. Le principe de minimum et la transformation de Legendre

A la base de notre démonstration se trouve le résultat suivant.

Théorème 3.1 (Le principe de minimum). - Soient E un espace vectoriel complexe et F un espace vectoriel réel et notons $F_{\mathbb{C}}$ son complexifié. Soient Ω un ouvert pseudoconvexe dans $E \times F_{\mathbb{C}}$ et u plurisousharmonique dans Ω et supposons que, si $(x,y) \in \Omega$ et si y et y' ont même partie réelle, alors $(x,y') \in \Omega$ et $u(x,y) = u(x,y')$. Supposons aussi que $\Omega_x = \{y \in F_{\mathbb{C}} \,;\, (x,y) \in \Omega\}$ est connexe pour tout $x \in E$. Alors

$$v(x) = \inf_{y \in \Omega_x} u(x,y) , \quad x \in \omega ,$$

est plurisousharmonique dans ω , l'ensemble des $x \in E$ tels que Ω_x soit non vide.

Pour la démonstration de ce théorème, voir [1].

Si $f : \mathbb{R} \to [-\infty,+\infty]$ est une fonction numérique définie sur l'axe réel on définit sa transformée de Legendre $\tilde{f} : \mathbb{R} \to [-\infty,+\infty]$ par

$$\tilde{f}(\eta) = \sup_{y \in \mathbb{R}} (y\eta - f(y)) , \quad \eta \in \mathbb{R} .$$

On a toujours $\tilde{\tilde{f}} \leq f$ et l'égalité a lieu si et seulement si f est convexe, semicontinue inférieurement et admet la valeur $-\infty$ seulement si elle est la constante $-\infty$. Si f a ces trois propriétés on a donc

$$f(y) = \tilde{\tilde{f}}(y) = \sup_{\eta \in \mathbb{R}} (y\eta - \tilde{f}(\eta)), \quad y \in \mathbb{R},$$

formule qui représente f comme l'enveloppe supérieure d'une famille de fonctions affines, à savoir ses tangents. Si on sait seulement que f est convexe à valeurs dans $]-\infty,+\infty]$ on a $\tilde{\tilde{f}} = f$ dans l'intérieur et dans l'extérieur de l'intervalle I où f est finie, mais pas forcément dans ∂I. Nous renvoyons à Rockafellar [8] pour la théorie des fonctions convexes.

Soit maintenant f convexe et croissante sur $\mathbb{R}$, à valeurs dans $]-\infty,+\infty]$. Il existe donc un intervalle $]-\infty,a]$ ou $]-\infty,a[$ où f est finie, et on pourrait appeler nombre de Lelong de f la quantité

$$\nu_f = \lim_{t \to -\infty} f'(t) = \lim_{t \to -\infty} f(t)/t .$$

Il est facile d'exprimer cette pente à l'infini à l'aide de la transformée de Legendre :

Lemme 3.2. - Soit $f : \mathbb{R} \to]-\infty,+\infty]$ convexe, croissante et non identiquement $+\infty$. Alors ν_f est l'extrémité gauche de l'intervalle de finitude de $\tilde{f}$; en particulier $\tilde{f}(\eta) = +\infty$ si $\eta < \nu_f$. Si, en plus, f croît plus vite que toute fonction linéaire, alors $\tilde{f}(\eta) < +\infty$ pour tout $\eta > \nu_f$.

Soient maintenant E, F, Ω et u comme dans le théorème 3.1, mais supposons pour simplifier que $F = \mathbb{R}$, d'où $F_{\mathbb{C}} = \mathbb{C}$. On peut définir la transformée de Legendre partielle $\tilde{u}$ de u par

$$\tilde{u}(x,\eta) = \sup_{y \in \mathbb{R}} (y\eta - u(x,y)), \quad (x,\eta) \in E \times \mathbb{R}.$$

Alors, comme $u(x,\cdot)$ est convexe, on a

$$u(x,y) = \tilde{\tilde{u}}(x,y) = \sup_{\eta \in \mathbb{R}} (\mathrm{Re}\, y\eta - \tilde{u}(x,\eta)), \quad (x,y) \in \Omega$$

et le principe de minimum nous garantit que dans cette représentation de u comme enveloppe de fonctions affines en y chaque terme est aussi une fonction plurisousharmonique des variables $(x,y) \in \Omega$. En particulier, si $u(x,y)$ est croissante en y, on a

$$(3.1) \qquad \lim_{y \to -\infty} \frac{u(x,y)}{y} = \inf(\eta;\ -\tilde{u}(x,\eta) > -\infty)$$

et si $u(x,y) = +\infty$ pour y assez grand on a aussi

$$(3.2) \qquad \lim_{y \to -\infty} \frac{u(x,y)}{y} = \sup(\eta;\ -\tilde{u}(x,\eta) = -\infty)\ .$$

Le fait que $x \mapsto -\tilde{u}(x,\eta)$ est plurisousharmonique nous fournira des renseignements sur le nombre de Lelong dans le paragraphe suivant.

4. *Le résultat principal*

Théorème 4.1. - Soient ω un ouvert de $\mathbb{C}^m$ contenant l'origine, $m \geq 1$, et E l'espace de Fréchet de toutes les applications holomorphes h de ω dans $\mathbb{C}^n$, $n \geq 1$, telles que $h(0) = 0$. Si f est plurisousharmonique au voisinage de l'origine de $\mathbb{C}^n$, alors $\nu_{f \circ h}(0) = \nu_f(0)$ pour toute $h \in E$ sauf si h appartient à un sous-ensemble polaire de E.

Démonstration. - On va mesurer la croissance de $h \in E$ par

$$p(h,y) = \sup_{|z| \leq e^{\mathrm{Re}\, y}} \log|h(z)|, \quad (h,y) \in E \times \mathbb{C},$$

et celle de f par

$$u(t) = \sup_{|z| \leq e^t} f(z), \quad t \in \mathbb{R},$$

de façon que

$$f(z) \leq u(\log|z|)\ , \quad z \in \mathbb{C}^n ,$$

et

$$(4.1) \qquad f(h(z)) \leq u(p(h,\ \log|z|))\ , \quad (h,z) \in E \times \mathbb{C}^m .$$

(On convient de poser $u(t) = +\infty$ si f n'est pas définie dans toute la boule $|z| \leq e^t$ et de façon analogue pour $p(h,y)$.) Sans restreindre la généralité nous pourrons supposer que ω contient la boule unité fermée de $\mathbb{C}^m$, c'est-à-dire que $p(h,0) < +\infty$ pour toute $h \in E$, et que f est définie et négative dans la boule unite fermée de $\mathbb{C}^n$, c'est-à-dire que $u(0) \leq 0 < +\infty$. Soient

$$\Omega_0 = \{(h,y) \in E \times \mathbb{C}\ ;\ \mathrm{Re}\, y < 0\}$$

et

$$\Omega = \{(h,y) \in \Omega_0\ ;\ p(h,y) < 0\} .$$

L'hypothèse déjà faite que $p(h,0) < +\infty$ nous montre que $p < +\infty$ et par conséquent plurisousharmonique dans le domaine convexe Ω_0. Donc Ω est pseudoconvexe. Pour étudier le comportement de $f \circ h$ nous définissons

$$v(h,y) = \sup_{|z| \leq e^{\mathrm{Re}\, y}} f(h(z)) = \sup_{|z| \leq 1} f(h(ze^y)) \quad \text{si} \quad (h,y) \in \Omega ,$$

$$v(h,y) = +\infty \quad \text{si} \quad (h,y) \in E \times \mathbb{C} \setminus \Omega .$$

Vu (4.1) il s'en suit que $v \leq u \circ p$ dans Ω. Le fait que toute $h \in E$ s'annule à l'origine entraîne que

$$(4.2) \qquad p(h,y) \leq y + p(h,0)\ , \quad (h,y) \in \Omega_0\ , \ y \in \mathbb{R} .$$

Si $y < 0$ et $t = p(h,y) \leq y + p(h,0) < 0$, en d'autre termes si $(h,y) \in \Omega$ et $y \in \mathbb{R}$, on peut donc minorer v/y comme suit:

$$(4.3) \qquad \frac{v(h,y)}{y} \geq \frac{u(p(h,y))}{y} = \frac{u(t)}{t} \cdot \frac{t}{y} \geq \frac{u(t)}{t}\left(1 + \frac{p(h,0)}{y}\right) .$$

Grâce à la proposition 2.1 les nombres de Lelong de f et $f \circ h$ sont donnés, respectivement, par

$$\nu_f(0) = \lim_{t \to -\infty} u(t)/t$$

et

$$\nu_{f \circ h}(0) = \lim_{y \to -\infty} v(h,y)/y .$$

En faisant tendre y vers $-\infty$ dans (4.3) on obtient le résultat attendu :

$$\nu_{f \circ h}(0) \geq \nu_f(0) , \quad h \in E .$$

Comme v est plurisousharmonique dans le domaine pseudoconvexe Ω et comme

$$\{y \in \mathbb{C} ; (h,y) \in \Omega\}$$

est un demi-plan, donc connexe, on peut appliquer le principe de minimum (voir le théorème 3.1) et affirmer que $h \mapsto -\tilde{v}(h,\eta)$ est plurisousharmonique dans E pour tout $\eta \in \mathbb{R}$, où

$$\tilde{v}(h,\eta) = \sup_{y \in \mathbb{R}} (y\eta - v(h,y)) , \quad (h,\eta) \in E \times \mathbb{R} ,$$

est la transformée de Legendre partielle de v. Donc

$$E(\eta) = \{h \in E ; \ -\tilde{v}(h,\eta) = -\infty\}$$

est soit polaire soit égal à E tout entier. D'autre part on sait que

$$\nu_{f \circ h}(0) > \eta \Rightarrow -\tilde{v}(h,\eta) = -\infty \Rightarrow \nu_{f \circ h}(0) \geq \eta$$

(voir le lemme 3.2 et (3.1), (3.2); on note que $v(h,y) = +\infty$ si $y \geq 0$). D'où, en posant $\alpha = \nu_f(0)$,

$$\{h \in E ; \nu_{f \circ h}(0) \neq \alpha\} = \{h \in E ; \nu_{f \circ h}(0) > \alpha\} = \bigcup_1^\infty E(\alpha + 1/j) .$$

Par un résultat bien connu en dimension finie (cf. [7]), il existe une application linéaire $h_0 \in E$ telle que $\nu_{f \circ h_0}(0) = \nu_f(0)$. Donc $-\tilde{v}(h_0, \alpha + 1/j) > -\infty$ et, par conséquent, il existe des nombres $\varepsilon_j > 0$ tels que

$$\Sigma \varepsilon_j(-\tilde{v}(h_0, \alpha + 1/j)) > -\infty .$$

Nous allons poser

$$(4.4) \qquad w(h) = -\Sigma \varepsilon_j \tilde{v}(h,\alpha + 1/j) .$$

Nous savons donc que $w(h_0) > -\infty$ et nous allons assujettir les nombres ε_j à encore une condition, à formuler, pour que w soit bien définie et plurisousharmonique.

En transformant l'inégalité $v \leq u \circ p$ on obtient, vu (4.2), si $(h,y) \in \Omega$ et $y \in \mathbb{R}$,

$$-\tilde{v}(h,\eta) \leq u(p(h,y)) - y\eta \leq u(y + p(h,0)) - y\eta .$$

Ici on peut choisir, étant donnés $h \in E$ et $\eta \in \mathbb{R}$, n'importe quel $y < 0$ satisfaisant à $p(h,y) < 0$. Prenons

$$y = -p(h,0) - \delta < 0 \quad \text{si} \quad p(h,0) \geq 0$$

et

$$y = -\delta \quad \text{si} \quad p(h,0) < 0 .$$

D'où

$$(4.5) \qquad -\tilde{v}(h,\eta) \leq u(0) + \eta\, p(h,0)^+ , \quad (h,\eta) \in E \times \mathbb{R} ,$$

et

$$w(h) = -\Sigma\varepsilon_j \tilde{v}(h,\alpha + 1/j) \leq u(0)\, \Sigma \varepsilon_j + \Sigma \varepsilon_j(\alpha + 1/j)p(h,0)^+ .$$

On voit finalement qu'il suffit de supposer $\varepsilon_j > 0$ et $\Sigma\, \varepsilon_j < +\infty$ pour pouvoir conclure que w est bien définie par (4.4) et admet la majoration

$$w(h) \leq A_0 + A_1\, p(h,0)^+$$

avec des constantes A_0 et A_1. On note que $h \mapsto \exp p(h,0)$ est une seminorme continue sur E et que w est donc localement bornée supérieurement ; la représentation

$$w(h) = \lim_{q\to+\infty} \sum_{j=1}^{q} \varepsilon_j(-\tilde{v}(h,\alpha+1/j) - u(0) - (\alpha+1/j)p(h,0)^+)$$

$$+ \sum_{j=1}^{\infty} \varepsilon_j(u(0) + (\alpha+1/j)p(h,0)^+)$$

nous donne w comme limite d'une suite décroissante de fonctions semicontinues supérieurement, donc elle même semicontinue supérieurement. Cela dit, il est clair que w est plurisousharmonique et nous avons déjà vu que $w(h_0) > -\infty$. Donc $\bigcup_1^\infty E(\alpha+1/j)$ est polaire dans E : le théorème 4.1 est démontré.

Je ne sais pas si une réunion dénombrable d'ensembles polaires dans un espace de Fréchet est polaire. Dans le cas traité ici la conclusion dépend des majorations très particulières (4.5) pour les fonctions définissant les ensembles $E(\alpha+1/j)$.

Étant donnée une partie P de E on peut se demander s'il existe f plurisousharmonique au voisinage de l'origine telle que $\nu_{f\circ h}(0) > \nu_f(0)$ si $h \in P$. Si $m \geq n \geq 1$ il est clair que l'ensemble des $h \in E$ telle que $\nu_{f\circ h}(0) > \nu_f(0)$ est contenu dans l'ensemble algébrique où le rang de $h'(0)$ est inférieur à n. Si $1 \leq m < n$ on note le résultat particulier suivant. Soit $E_0 = L(\mathbb{C}^m, \mathbb{C}^n)$ le sous-espace de E formé des applications linéaires, et soit P une partie de E_0 telle que l'ensemble des droites contenues dans l'image d'une $h \in P$ soit localement polaire dans l'espace projectif de dimension $n-1$. Alors il existe f plurisousharmonique dans $\mathbb{C}^n$ telle que $\nu_{f\circ h}(0) = +\infty > \nu_f(0) = 1$ pour toute $h \in P$, et par suite g telle que $\nu_{g\circ h}(0) = b > \nu_g(0) = a$, a et $b > a \geq 0$ étant donnés. Cela est une conséquence du théorème 4.4 de [3].

Bibliographie

1. Kiselman, C.O., The partial Legendre transformation for plurisubharmonic functions. Invent. Math. 49 (1978), 137-148.

2. Kiselman, C.O., Densité des fonctions plurisousharmoniques. Bull. Soc. Math. France 107 (1979), 295-304.

3. Kiselman, C.O., The growth of restrictions of plurisubharmonic functions. Mathematical Analysis and Applications, ed. L. Nachbin; Advances in Mathematics Supplementary Studies vol. 7B, 435-454. Academic Press 1981.

4. Lelong, P., Propriétés métriques des variétés analytiques complexes définies par une équation. Ann. Sci. École Norm. Sup. (3) 67 (1950), 393-419.

5. Lelong, P., Intégration sur un ensemble analytique complexe. Bull. Soc. Math. France 85 (1957), 239-262.

6. Lelong, P., Fonctions plurisousharmoniques et formes différentielles positives. Gordon & Breach; Dunod, 1968.

7. Lelong, P., Plurisubharmonic functions in topological vector spaces: polar sets and problems of measure. Proceedings on Infinite Dimensional Holomorphy. Lecture Notes in Mathematics 364, 58-68. Springer-Verlag 1974.

8. Rockafellar, R.T., Convex analysis. Princeton University Press, 1970.

9. Skoda, H., Sous-ensembles analytiques d'ordre fini ou infini dans $\mathbb{C}^n$. Bull. Soc. Math. France 100 (1972), 353-408.

10. Avanissian, V., Fonctions plurisousharmoniques et fonctions doublement sousharmoniques. Ann. E.N.S., t. 78 (1961), 101-161.

Christer O. Kiselman
Université d'Uppsala
Département de Mathématiques
Thunbergsvägen 3
S-752 38 Uppsala, Suède

CAPACITY, TCHEBYCHEFF CONSTANT, AND TRANSFINITE HYPERDIAMETER ON COMPLEX PROJECTIVE SPACE

By

Robert E. Molzon and Bernard Shiffman*

Introduction

In studying the growth of the hyperplane sections of an analytic variety in $\mathbb{C}^n$, one uses the currents

$$\frac{\sqrt{-1}}{\pi}\partial\bar{\partial}\log\frac{|Z\cdot W|}{|W|}$$

of integration over the hyperplanes dual to the points W of $\mathbb{P}^{n-1}$. This expression suggests the kernel on $\mathbb{P}^{n-1}$:

$$\log\frac{|Z||W|}{|Z\cdot W|}.$$

In a joint work with N. Sibony [6], we used this kernel to define the capacity $C(E)$ (for compact subsets $E\subset\mathbb{P}^{n-1}$), which allowed us to extend a result of L. Gruman [4] on average growth estimates for hyperplane sections of affine analytic varieties. H. Alexander [2] then obtained the inequalities

$$c\,\rho(E)\le\exp(-1/C(E))\le\rho(E)$$

* Research partially supported by a National Science Foundation grant.

for compact $E \subset \mathbb{P}^{n-1}$, where $\rho(E)$ is the generalized Tchebycheff constant for E (see Definition 1) and c is a positive constant depending only on n . The capacity and Tchebycheff constant on $\mathbb{P}^{n-1}$ reduce to the classical elliptic capacity and elliptic Tchebycheff constant [7, pp.90-92] respectively, when $n = 2$.

In this work we define the <u>transfinite hyperdiameter</u> $\tau(E)$ and the <u>energy capacity</u> $C(E)$ (for compact $E \subset \mathbb{P}^{n-1}$) , which reduce to the elliptic transfinite diameter and elliptic capacity when $n = 2$. Although the classical identity

$$\exp(-1/C(E)) = \rho(E) = \tau(E)$$

does not generalize to the case $n > 2$, we are able to prove in this paper the identity $\tau(E) = \exp(-1/\tilde{C}(E))$ (Theorem 3) and the inequality $\rho(E)^{n-1} \leq \tau(E)$ (Theorem 2) for general n . Thus $C(E) \leq (n-1)\tilde{C}(E)$. The question whether there is an upper bound for $\tilde{C}(E)$ in terms of $C(E)$ remains unsolved.

1. Notation and definitions.

We let $Z = (z_1, \ldots, z_n)$ denote a point in either $\mathbb{C}^n$ or $\mathbb{P}^{n-1}$. For $Z = (z_1, \ldots, z_n)$, $W = (w_1, \ldots, w_n)$, we let

$$|Z| = (|z_1|^2 + \ldots + |z_n|^2)^{\frac{1}{2}},$$

$$Z \cdot W = z_1 w_1 + \ldots + z_n w_n .$$

For a k-vector $X \in \Lambda^k \mathbb{C}^n$, we let $|X|$ denote the inner-product norm of X induced from the above Euclidean norm. For a compact subset E of $\mathbb{P}^{n-1}$, we let $\mathcal{P}(E)$ denote the set of positive Borel measures μ on E with $\mu(E) = 1$, and we call these measures probability measures.

We first recall the definition from [6] of capacity in $\mathbb{P}^{n-1}$:

DEFINITION 1. For $\mu \in \mathcal{P}(\mathbb{P}^{n-1})$ the potential function u_μ on $\mathbb{P}^{n-1}$ is given by

$$u_\mu(Z) = \int_{\mathbb{P}^{n-1}} \log \frac{|Z|\,|W|}{|Z \cdot W|} \, d\mu(W) , \tag{1.1}$$

for $Z \in \mathbb{P}^{n-1}$. The capacity $C(E)$ of a compact subset E of $\mathbb{P}^{n-1}$ is defined by

$$C(E)^{-1} = \inf_{\mu \in \mathcal{P}(E)} \sup_{Z \in \mathbb{P}^{n-1}} u_\mu(Z) . \tag{1.2}$$

The capacity $C(E)$ of Definition 1 is related by Alexander's inequality to the Tchebycheff constant $\rho(E)$. (See Definition 2 and Theorem 1 below.) We do not work directly with $C(E)$ in this paper; instead we use another capacity $\hat{C}(E)$ (see Definition 4). Some conditions for the positivity of $C(E)$ are given in [6, Theorem 2.1] and [2, Theorem 3.3].

For k a positive integer and E a compact set in $\mathbb{P}^{n-1}$, put

$$(1.3) \qquad r_k(E) = \inf_{\{A_1,\dots,A_k\}\subset\mathbb{P}^{n-1}} \left(\sup_{Z\in E} \prod_{j=1}^{k} \frac{|Z\cdot A_j|}{|Z||A_j|} \right)^{1/k} .$$

Exactly as in the classical definition of the Tchebycheff constant (see [7, pp.72-73]), one obtains the following result:

LEMMA 1. If $E \subset \mathbb{P}^{n-1}$ is compact, then $\lim_{k\to\infty} r_k(E)$ exists and equals $\inf_k r_k(E)$.

Thus we can state the following definition:

DEFINITION 2. The Tchebycheff constant $\rho(E)$ of a compact set $E \subset \mathbb{P}^{n-1}$ is given by

$$\rho(E) = \lim_{k\to\infty} r_k(E) .$$

(We could let E be an arbitrary subset of $\mathbb{P}^{n-1}$ in Definition 2, but then $\rho(E) = \rho(\bar{E})$.)

This definition of the Tchebycheff constant was first given by N. Sibony (unpublished). H. Alexander [2] recently obtained the following result establishing the equivalence of $\rho(E)$ and $C(E)$:

THEOREM 1 (Alexander). Let E be a compact subset of $\mathbb{P}^{n-1}$. Then

$$(\exp a_n)\, \rho(E) \le \exp(-1/C(E)) \le \rho(E) ,$$

where a_n is the average value of $\log|z_1|$ on the unit sphere in $\mathbb{C}^n$.
(The Tchebycheff constant $\rho(E)$ is denoted by $\mathcal{S}$-cap(E) in [2].)

For $A_1,\dots,A_n \in \mathbb{P}^{n-1}$, we write

$$(1.4) \qquad \Phi(A_1,\dots,A_n) = \frac{|A_1 \wedge \dots \wedge A_n|}{|A_1| \dots |A_n|} \in [0,1] .$$

For $E \subset \mathbb{P}^{n-1}$, we let $d_k(E)$ denote the supremum of the product

(1.5) $$\prod_{1 \leq i_1 < \ldots < i_n \leq k} \Phi(A_{i_1},\ldots,A_{i_n})^{1/\binom{k}{n}}$$

over all $\{A_1,\ldots,A_k\} \subset E$, for $k \geq n$.

DEFINITION 3. For a compact set $E \subset \mathbb{P}^{n-1}$, we define the <u>transfinite hyperdiameter</u> $\tau(E)$ of E by

$$\tau(E) = \lim_{k\to\infty} d_k(E) .$$

Definition 3 (which coincides with the transfinite diameter in the Riemann sphere if $n = 2$) is a special case of a general definition given by F. Leja [5].

LEMMA 2 (Leja). For $E \subset \mathbb{P}^{n-1}$

$$d_k(E) \leq d_{k-1}(E) , \text{ for } k > n .$$

<u>Proof</u> (based on the proof in [5]): Since $d_k(E) = d_k(\bar{E})$, assume E is compact. Choose $A_1,\ldots,A_k \in E$ so that $d_k(E)$ equals the product (1.5). Then

(1.6) $$d_k(E)^{\binom{k}{n}} = \prod_{j\notin\{i_1,\ldots,i_n\}} \Phi(A_{i_1},\ldots,A_{i_n}) \prod_{j\in\{i_1,\ldots,i_n\}} \Phi(A_{i_1},\ldots,A_{i_n})$$

for $1 \leq j \leq k$. The first product in (1.6) is $\leq d_{k-1}(E)^{\binom{k-1}{n}}$. By multiplying the resulting inequalities for $j = 1,\ldots,k$, we obtain

$$d_k(E)^{\binom{k}{n}k} \leq d_{k-1}(E)^{\binom{k-1}{n}k} d_k(E)^{\binom{k}{n}n} ,$$

which yields Lemma 2.

In the classical potential theory of one complex variable, the reciprocal of the capacity can be given as the infimum of an "energy" integral. We now give a definition of an energy integral for measures on $\mathbb{P}^{n-1}$ and define a new capacity in terms of this integral. We shall then show that this capacity is given by the transfinite hyperdiameter (Theorem 3).

Given a positive Borel measure μ on $\mathbb{P}^{n-1}$, we associate with μ the <u>energy integral</u> $\mathcal{E}(\mu)$ given by

$$(1.7) \qquad \mathcal{E}(\mu) = \int_{(\mathbb{P}^{n-1})^n} \log \frac{1}{\Phi} \, d\mu^n$$

where μ^n denotes the product measure $\mu \times \ldots \times \mu$.

DEFINITION 4. Suppose E is a compact subset of $\mathbb{P}^{n-1}$. We define the energy potential $\mathcal{E}(E)$ of E by

$$\mathcal{E}(E) = \inf_{\mu \in \mathcal{P}(E)} \mathcal{E}(\mu)$$

and the energy capacity $C(E)$ by

$$\tilde{C}(E) = 1/\mathcal{E}(E) .$$

Note that if E is contained in a countable union of projective hyperplanes, then it is easy to see that $\tilde{C}(E) = 0$. We shall show in the next section that $\tilde{C}(E) = 0$ if and only if $\tau(E) = 0$.

2. Main Results.

In this section, we explore the relationships between the Tchebycheff constant, transfinite hyperdiameter, and the energy capacity. We begin with a linear algebra lemma.

LEMMA 3. For $X_1, \dots, X_n \in \mathbb{C}^n$, we have

$$|X_1 \wedge \dots \wedge X_n|^{n(n-2)} |X_1| \dots |X_n| \leq \prod_{j=1}^{n} |X_1 \wedge \dots \wedge \hat{X}_j \wedge \dots \wedge X_n|^{n-1} .$$

(Here ^ means deletion.)

Proof: Consider the $n \times n$ matrix

$$A = \begin{bmatrix} X_1 \\ \cdot \\ \cdot \\ \cdot \\ X_n \end{bmatrix} .$$

We may assume by homogeneity of the inequality that $\det A = 1$ and hence $|X_1 \wedge \dots \wedge X_n| = |\det A| = 1$. Consider the isometry (Hodge *-operator)

$$* : \Lambda^{n-1}\mathbb{C}^n \to \mathbb{C}^n$$

given by

$$*(e_1 \wedge \dots \wedge \hat{e}_j \wedge \dots \wedge e_n) = (-1)^j\, e_j$$

where $\{e_1, \dots, e_n\}$ is the standard basis for $\mathbb{C}^n$. The transpose inverse of A is then

$$^tA^{-1} = \begin{bmatrix} Y_1 \\ \cdot \\ \cdot \\ \cdot \\ Y_n \end{bmatrix}$$

where

$$(2.1)\qquad Y_j = (-1)^j * (X_1 \wedge \dots \wedge \hat{X}_j \wedge \dots \wedge X_n) .$$

Since ${}^t({}^tA^{-1})^{-1} = A$, we conclude that

$$(2.2)\qquad X_j = (-1)^j * (Y_1 \wedge \dots \wedge \hat{Y}_j \wedge \dots \wedge Y_n) .$$

Thus, by (2.2) $|X_j| \leq \prod_{i \neq j} |Y_i|$ and

$$|X_1| \dots |X_n| \leq |Y_1|^{n-1} \dots |Y_n|^{n-1} . \qquad \square$$

If E is a compact subset of $\mathbb{P}^1$, then $\rho(E) = \tau(E)$ (see [7, Theorem III.51]). For E in a higher dimensional projective space, we have the following inequality for the Tchebycheff constant and transfinite hyperdiameter:

THOEREM 2. For E a compact subset of $\mathbb{P}^{n-1}$, we have

$$\rho(E)^{n-1} \leq \tau(E) .$$

<u>Proof</u>: We write $d_k = d_k(E)$, $r_k = r_k(E)$. Fix $k \leq n$ and choose $A_1, \dots, A_k \in E$ such that

$$(2.3)\qquad \binom{k}{n} \log d_k = \frac{1}{n!} \sum{}' \log \Phi (A_{\alpha_1}, \dots, A_{\alpha_n})$$

where $\sum'$ denotes summation over distinct α_i . Fix α_1 ; then

$$(2.4)\qquad \sum_{\alpha_2 \dots \alpha_n}{}' \log \Phi (A_{\alpha_1}, A_{\alpha_2}, \dots, A_{\alpha_n}) = \sup_{Z \in E} \sum_{\alpha_2 \dots \alpha_n}{}' \log \Phi (Z, A_{\alpha_2}, \dots, A_{\alpha_n})$$

where $\sum'_{\alpha_2 \dots \alpha_n}$ denotes summation over $\alpha_2, \dots, \alpha_n$ distinct from each other and from α_1 . Let

$$B_{\alpha_2 \ldots \alpha_n} = * (A_{\alpha_2} \wedge \ldots \wedge A_{\alpha_n})$$

(where $*$ is as in the proof of Lemma 3). Then

$$\text{(2.5)} \qquad \log \Phi (Z, A_{\alpha_2}, \ldots, A_{\alpha_n}) = \log \frac{|Z \cdot B_{\alpha_2 \ldots \alpha_n}|}{|Z||A_{\alpha_2}| \ldots |A_{\alpha_n}|}$$

$$= \log \frac{|Z \cdot B_{\alpha_2 \ldots \alpha_n}|}{|Z||B_{\alpha_2 \ldots \alpha_n}|} + \log \frac{|A_{\alpha_2} \wedge \ldots \wedge A_{\alpha_n}|}{|A_{\alpha_2}| \ldots |A_{\alpha_n}|} .$$

By definition,

$$\text{(2.6)} \qquad \sup_{Z \in E} \sum_{\alpha_2 \ldots \alpha_n}{}' \log \frac{|Z \cdot B_{\alpha_2 \ldots \alpha_n}|}{|Z||B_{\alpha_2 \ldots \alpha_n}|} \geq (n-1)! \binom{k-1}{n-1} \log r_{\binom{k-1}{n-1}} .$$

Hence by (2.4), (2.5), and (2.6),

$$\text{(2.7)} \qquad \sum_{\alpha_2 \ldots \alpha_n}{}' \log \Phi (A_{\alpha_1}, \ldots, A_{\alpha_n}) \geq (n-1)! \binom{k-1}{n-1} \log r_{\binom{k-1}{n-1}}$$

$$+ \sum_{\alpha_2 \ldots \alpha_n}{}' \log \frac{|A_{\alpha_2} \wedge \ldots \wedge A_{\alpha_n}|}{|A_{\alpha_2}| \ldots |A_{\alpha_n}|} .$$

Summing (2.7) over $1 \leq \alpha_1 \leq k$ and recalling (2.3), we obtain

$$\text{(2.8)} \qquad \binom{k}{n} \log d_k = \frac{1}{n!} \sum_{\alpha_1} \sum_{\alpha_2 \ldots \alpha_n}{}' \log \Phi (A_{\alpha_1}, \ldots, A_{\alpha_n}) \geq \binom{k}{n} \log r_{\binom{k-1}{n-1}}$$

$$+ \frac{k-n+1}{n} \sum_{\alpha_2 < \ldots < \alpha_n} \log \frac{|A_{\alpha_2} \wedge \ldots \wedge A_{\alpha_n}|}{|A_{\alpha_2}| \ldots |A_{\alpha_n}|} .$$

By Lemma 3

$$n(n-2) \log \Phi (A_{\alpha_1}, \ldots, A_{\alpha_n}) \leq (n-1) \sum_{j=1}^{n} \log \frac{|A_{\alpha_1} \wedge \ldots \wedge \hat{A}_{\alpha_j} \wedge \ldots \wedge A_{\alpha_n}|}{|A_{\alpha_1}| \ldots |\hat{A}_{\alpha_j}| \ldots |A_{\alpha_n}|}$$

and therefore

$$(2.9) \qquad \binom{k}{n} \log d_k = \sum_{\alpha_1 < \ldots < \alpha_n} \log \Phi (A_{\alpha_1}, \ldots, A_{\alpha_n})$$

$$\leq \frac{(n-1)(k-n+1)}{n(n-2)} \sum_{\alpha_2 < \ldots < \alpha_n} \log \frac{|A_{\alpha_2} \wedge \ldots \wedge A_{\alpha_n}|}{|A_{\alpha_2}| \ldots |A_{\alpha_n}|} .$$

Combining inequalities (2.8) and (2.9), we obtain

$$(2.10) \qquad \log d_k \geq \log r_{\binom{k-1}{n-1}} + \frac{n-2}{n-1} \log d_k .$$

If $d_k > 0$, we conclude from (2.10) that

$$(2.11) \qquad (n-1) \log r_{\binom{k-1}{n-1}} \leq \log d_k .$$

If $d_k = 0$, then (2.11) is valid with E replaced by a larger sets E' with $d_k(E') > 0$. Letting $E' \searrow E$, we then conclude that $r_{\binom{k-1}{n-1}} = 0$, verifying (2.11) for the case $d_k = 0$. The conclusion of the theorem then follows by letting $k \to \infty$ in (2.11). □

We note in Section 3 that the exponent $n - 1$ in Theorem 2 is sharp and that the reverse inequality does not hold. Theorem 2 tells us that if $\tau(E) = 0$, then $\rho(E) = 0$; the converse is unkown. Theorem 1 of H. Alexander implies that $\rho(E) = 0$ if and only if $C(E) = 0$. The following result similarly implies that $\tau(E)$ vanishes if and only if $\tilde{C}(E)$ vanishes.

THEOREM 3. For E a compact subset of $\mathbb{P}^{n-1}$, we have

$$\exp(-1/\tilde{C}(E)) = \tau(E) .$$

<u>Proof</u>: We must show that $\mathcal{E}(E) = -\log \tau(E)$. The inequality

$$\mathcal{E}(E) \geq -\log \tau(E) \tag{2.12}$$

is elementary and is verified as follows: Let $\mu \in \mathcal{P}(E)$ and let $k \geq n$ be arbitrary. Then

$$\binom{k}{n} \log d_k(E) = \sup_{\{X_1,\dots,X_k\} \subset E} \left(\sum_{\alpha_1 < \dots < \alpha_n} \log \Phi(X_{\alpha_1}, \dots, X_{\alpha_n}) \right)$$

$$\geq \int \sum \log \Phi(X_{\alpha_1},\dots,X_{\alpha_n})\, d\mu^k(X_1,\dots,X_k)$$

$$= \binom{k}{n} \int \log \Phi \, d\mu^n = -\binom{k}{n} \mathcal{E}(\mu) ,$$

and thus

$$\mathcal{E}(\mu) \geq -\log d_k(E) .$$

Letting $k \to +\infty$ and then taking the infimum over $\mu \in \mathcal{P}(E)$, we obtain (2.12).

Thus we must show that $\mathcal{E}(E) \leq -\log \tau(E)$. For $A \in \mathbb{P}^{n-1}$ and $\delta > 0$, we let

$$B(A,\delta) = \{Z \in \mathbb{P}^{n-1} : |Z \wedge A|/(|Z||A|) \leq \delta\} \tag{2.13}$$

denote the δ-ball about A . Let λ denote the invariant probability measure on $\mathbb{P}^{n-1}$. (Thus λ is given by $(\frac{\sqrt{-1}}{\pi} \partial \bar{\partial} \log |Z|)^{n-1}$.) For $k > 1$, let $\delta_k \in \mathbb{R}^+$ be determined by

$$\lambda(B(A,\delta_k)) = 1/k \ . \tag{2.14}$$

Then

$$k \cdot \delta_k^{2n-2} \to 1 \quad \text{as} \quad k \to +\infty \ , \tag{2.15}$$

since $\lambda(B(A,\delta)) \approx \delta^{2n-2}$ for δ small.

Let D be an arbitrary open set containing E. Note that if $\{K_\nu\}$ is a sequence of compact sets decreasing to E, that is $K_{\nu+1} \subset K_\nu$ and $E = \bigcap K_\nu$, then $\mathcal{E}(K_\nu) \nearrow \mathcal{E}(E)$. (This fact is verified by choosing $\mu_\nu \in \mathcal{P}(K_\nu)$ such that $\mathcal{E}(K_\nu) = \mathcal{E}(\mu_\nu)$; passing to a subsequence so that $\mu_\nu \to \mu_0 \in \mathcal{P}(E)$, it then follows that $\mathcal{E}(E) \leq \mathcal{E}(\mu_0) = \lim \mathcal{E}(\mu_\nu)$.) Thus it suffices to show that

$$\mathcal{E}(\overline{D}) \leq -\log \tau(E) \ .$$

Let $k \geq n$ be fixed. Assume k is sufficiently large so that $\delta_k < \operatorname{dist}(E,\partial D)$. As in the proof of Theorem 2, choose $A_1, \ldots, A_k \in E$ such that (2.3) is satisfied. Let $U_j = B(A_j,\delta_k)$ for $j = 1, \ldots, k$; hence $\lambda(U_j) = 1/k$ and $\bigcup U_j \subset D$. Let

$$\mu = \Big(\sum_{j=1}^{k} \chi_{U_j}\Big)\lambda \in \mathcal{P}(\overline{D}) \ , \tag{2.16}$$

where χ_S denotes the characteristic function of S. Then

$$\mathcal{E}(\mu) = \sum_{\alpha_1\ldots\alpha_n} I_{\alpha_1\ldots\alpha_n} \tag{2.17}$$

where

$$I_{\alpha_1\ldots\alpha_n} = -\int_{U_{\alpha_1} \times \ldots \times U_{\alpha_n}} \log \Phi \, d\lambda^n \ . \tag{2.18}$$

Let $\Psi = -\log \Phi \geq 0$. We claim that

$$(2.19)\qquad \int_{U_j} \Psi(X_1, \ldots, X_n)\, d\lambda(X_1) \leq k^{-1}\, \Psi(A_j, X_2, \ldots, X_n) + o(k^{-1})$$

where $o(k^{-1})$ depends only on k and n . To prove (2.19), assume without loss of generality that $A_j = (1, 0, \ldots, 0)$, and write $X_1 = (1, z_2, \ldots, z_n)$. For $X_1 \in U_j$, we have

$$(2.20)\qquad \log |X_1| \leq -\frac{1}{2}\log(1-\delta_k^{\,2}) \leq 0(\delta_k^{\,2}) \leq o(1) .$$

Since

$$\Psi(X_1, \ldots, X_n) = \log|X_1| + \ldots + \log|X_n| - \log|X_1 \wedge \ldots \wedge X_n| ,$$

(2.19) then follows from (2.20) and the plurisubharmonicity in $(z_2, \ldots, z_n)$ of $\log|X_1 \wedge \ldots \wedge X_n|$.

By iterating (2.19), one easily obtains the estimate

$$(2.21)\qquad I_{\alpha_1 \ldots \alpha_n} = \int_{U_{\alpha_1} \times \ldots \times U_{\alpha_n}} \Psi\, d\lambda^n \leq k^{-n}[\Psi(A_{\alpha_1}, \ldots, A_{\alpha_n}) + o(1)] .$$

We need one more estimate:

$$(2.22)\qquad I_{\alpha_1 \ldots \alpha_n} = 0(k^{-n}\log k) \qquad \text{(depending only on } k \text{ and } n) .$$

(We shall use (2.22) when the α_j are not all distinct; in this case, the right hand side of (2.21) is $+\infty$.) To prove (2.22), we consider $A'_j \in U_{\alpha_j}$ (to be specified below) and let

$$U'_j = B(A'_j, 2\delta_k) \supset U_{\alpha_j}$$

for $j = 1, \ldots, n$. We may assume k is so large that the U'_j are all in D. Then by the proof of (2.21),

$$(2.23) \qquad I_{\alpha_1 \ldots \alpha_n} \leq \int_{U'_1 \times \ldots \times U'_n} \Psi \, d\lambda^n$$

$$\leq \left[\prod_{j=1}^{n} \lambda(U'_j)\right][\Psi(A'_1, \ldots, A'_n) + o(1)] .$$

Since $\lambda(U'_j) = O(k^{-1})$, to complete the proof of (2.22), we must choose the A'_j such that

$$(2.24) \qquad \Psi(A'_1, \ldots, A'_n) = O(\log k) .$$

We may assume without loss of generality that

$$A_{\alpha_1} = (1, 0, \ldots, 0)$$

and

$$A_{\alpha_j} = (a_j^1, \ldots, a_j^j, 0, \ldots, 0)$$

with $|A_{\alpha_j}| = 1$ for $2 \leq j \leq n$. Let $A'_1 = A_{\alpha_1}$ and

$$A'_j = \begin{cases} A_{\alpha_j} & \text{if } |A_j^j| \geq \delta_k/3 \\ (a_j^1, \ldots, a_j^{j-1}, \delta_k/3, 0, \ldots, 0) & \text{if } |a_j^j| < \delta_k/3 \end{cases}$$

for $2 \leq j \leq n$. Hence $1 \leq |A'_j| \leq (1 + \delta_k^2/9)^{\frac{1}{2}}$. Since

$$\frac{|A_{\alpha_j} \wedge A'_j|}{|A_{\alpha_j}|\,|A'_j|} \leq |A_{\alpha_j} \wedge A'_j| < \tfrac{2}{3}\delta_k ,$$

we have $A_j' \in U_{\alpha_j}$ as desired. Furthermore

$$|A_1' \wedge \ldots \wedge A_n'| \geq (\delta_k/3)^{n-1}$$

and thus

$$\Psi(A_1', \ldots, A_n') \leq (n-1) \log (1/\delta_k) + O(1) . \tag{2.25}$$

The estimate (2.24) then follows from (2.15) and (2.25).

We now complete the proof of the theorem. We write Σ' to denote summation over $\alpha_1, \ldots, \alpha_n$ all distinct and Σ'' to denote summation over the remaining $\alpha_1, \ldots, \alpha_n$. Then by (2.21) and (2.22),

$$\begin{aligned}\mathcal{E}(\mu) &= \Sigma' I_{\alpha_1 \ldots \alpha_n} + \Sigma'' I_{\alpha_1 \ldots \alpha_n} \\ &\leq k^{-n} \Sigma' \Psi(A_{\alpha_1}, \ldots, A_{\alpha_n}) + \frac{k!}{(k-n)!} o(k^{-n}) + p_n(k)\, O(k^{-n} \log k)\end{aligned}$$

where

$$p_n(k) = k^n - k(k-1) \ldots (k-n+1) = O(k^{n-1}) .$$

By (2.3)

$$\Sigma' \Psi(A_{\alpha_1}, \ldots, A_{\alpha_n}) = \frac{k!}{(k-n)!} \log (1/d_k)$$

and thus (noting that $k^{-n} \cdot k!/(k-n)! \leq 1$)

$$\mathcal{E}(\mu) \leq - \log d_k(E) + o(1) . \tag{2.26}$$

Letting $k \to \infty$ in (2.26), we obtain

$$\mathcal{E}(\bar{D}) \leq \mathcal{E}(\mu) \leq - \log \tau(E) . \qquad \square$$

3. Remarks

We now show that the inequality in Theorem 2 cannot be replaced by an equality. First, we observe that the exponent $n-1$ in Theorem 2 is sharp since one easily checks that

$$\rho(B(A,t)) \sim t \quad , \quad \tau(B(A,t)) \sim t^{n-1}$$

as $t \to 0$, where $B(A,t)$ is given by (2.13) (and $A \in \mathbb{P}^{n-1}$ is arbitrary). The following example shows that the reverse inequality is false.

EXAMPLE. Let $0 < \varepsilon < 1$ be arbitrary. We shall give a compact subset E of $\mathbb{P}^2$ with

$$\rho(E) \leq \varepsilon \,, \tag{3.1}$$

$$\tau(E) \geq c\sqrt{\varepsilon} \tag{3.2}$$

where c is a positive constant (independent of ε). Let

$$E_0 = \{(1,a,b) \in \mathbb{P}^2 : 1 \leq a \leq 2 \,,\; 0 \leq b \leq \varepsilon^2\} \,.$$

We let $\sigma : \mathbb{P}^2 \to \mathbb{P}^2$ be the transposition given by

$$\sigma(z_1, z_2, z_3) = (z_1, z_3, z_2) \,.$$

We then let

$$E = E_0 \cup \sigma(E_0) \,.$$

To verify (3.1) we let $B_1 = (0, 1, 0)$, $B_2 = (0, 0, 1)$. Then

$$\rho(E) \leq r_2(E) \leq \sup_{Z\in E} \left(\frac{|Z\cdot B_1||Z\cdot B_2|}{|Z|^2|B_1||B_2|} \right)^{\frac{1}{2}}$$

$$= \sup_{(1,z_2,z_3)\in E_o} \left(\frac{|z_2 z_3|}{1+|z_2|^2+|z_3|^2} \right)^{\frac{1}{2}} \leq \varepsilon \ .$$

We now verify (3.2): Let $m \geq 3$ be fixed. Choose $t_1, \ldots, t_m$ in the unit interval $[0,1]$ so that the product

$$p_m = \prod_{ij}{}' |t_i - t_j|$$

is maximal, where Π' denotes the product over distinct indices between 1 and m. Recall that

$$p_m^{1/m(m-1)} \to \delta > 0 \tag{3.3}$$

where δ is the transfinite diameter of $[0,1]$ (see [7]).

Let

$$A_i = (1,\ 1+t_i,\ t_i^2\varepsilon^2) \quad , \quad A_{m+i} = \sigma(A_i) \ ,$$

for $1 \leq i \leq m$. Then

$$\begin{aligned} |A_i \wedge A_j \wedge A_k| &= \varepsilon^2|t_j-t_i||t_k-t_i||t_k-t_j| \ , \\ |A_i \wedge A_j \wedge A_{m+k}| &= |(1 + t_k + \varepsilon^2 t_j t_i)(t_j-t_i) + \varepsilon^2(1-t_k^2\varepsilon^2)(t_j^2-t_i^2)| \\ &\geq |t_j-t_i| \end{aligned} \tag{3.4}$$

for $1 \leq i, j, k \leq m$. Thus, we set

$$\pi_1 = \Pi'_{ijk} |A_i \wedge A_j \wedge A_k| = \varepsilon^{2m(m-1)(m-2)} p_m^{3(m-2)}$$

(3.5)

$$\pi_2 = \Pi_k \Pi'_{ij} |A_i \wedge A_j \wedge A_{m+k}| \geq p_m^{\ m} .$$

Since $|A_\alpha| < \sqrt{6}$ for $1 \leq \alpha \leq 2m$, we have

$$d_{2m}(E) \geq 6^{-3/2}(\pi_1^{\ 2}\pi_2^{\ 6})^{1/2m(2m-1)(2m-2)} . \tag{3.6}$$

Combining (3.5) with (3.6) and letting $m \to \infty$, we obtain (3.2) with $c = (\delta/6)^{3/2}$.

An open question is whether we have an inequality of the form

$$\tau(E) \leq c_1 \rho(E)^{c_2} \tag{3.7}$$

for some positive constants c_1, c_2 depending only on n . If (3.7) were true, then by Theorems 1, 2, and 3, the vanishing of any one of the quantities $C(E)$, $\tilde{C}(E)$, $\rho(E)$, $\tau(E)$ would imply the vanishing of all of them. In order to explore this question, one could attempt to compare $C(E)$ and $\tilde{C}(E)$ using analytical methods. Along these lines, we define the <u>energy potential</u> v_μ of a positive Borel measure μ on $\mathbb{P}^{n-1}$ by

$$v_\mu(Z) = \int_{(\mathbb{P}^{n-1})^{n-1}} \Psi(X_1, \ldots, X_{n-1}, Z)\, d\mu^{n-1}(X_1, \ldots, X_{n-1}) , \tag{3.8}$$

where we write $\Psi = -\log \Phi$ as before. We then have

$$\mathcal{E}(\mu) = \int_{\mathbb{P}^{n-1}} v_\mu d\mu , \tag{3.9}$$

justifying the name "potential". Continuing the analogy with the classical case, we say that $\mu \in \mathcal{P}(E)$ is an <u>equilibrium measure</u> for a compact

$E \subset \mathbb{P}^{n-1}$ if $\mathcal{E}(\mu) = \mathcal{E}(E)$; i.e. (recalling Definition 4) if $\mathcal{E}(\mu)$ is minimal for $\mu \in \mathcal{P}(E)$. In the classical case, if μ is the equilibrium measure for $E \subset \mathbb{P}^1$, then $v_\mu \leq \mathcal{E}(E)$ on all of $\mathbb{P}^1$, with equality holding almost everywhere on E (see [3] or [7, Ch. III]). We do not know if this is true in $\mathbb{P}^{n-1}$. However, following the method of the classical proof, we obtain the following partial result:

PROPOSITION 1. Suppose μ is an equilibrium measure for a compact set $E \subset \mathbb{P}^{n-1}$. If $\mathcal{E}(\mu)$ is finite, then $v_\mu = \mathcal{E}(\mu)$ μ-almost everywhere on E.

<u>Proof:</u> We first show that $v_\mu \geq \mathcal{E}(\mu)$ μ-almost everywhere. Suppose on the contrary that $v_\mu < \mathcal{E}(\mu)$ on a set with positive μ measure. Then there exists $\varepsilon > 0$ and a compact set $E_1 \subset E$ with $\mu(E_1) > 0$ and

$$v_\mu(Z) \leq \mathcal{E}(\mu) - 2\varepsilon \, , \quad \text{for } z \in E_1 \, . \tag{3.10}$$

By (3.9), there exists a point A in the support of μ such that $v_\mu(A) \geq \mathcal{E}(\mu)$. By the lower semicontinuity of v_μ, we have $v_\mu > \mathcal{E}(\mu) - \varepsilon$ on a neighborhood N_A of A. (Thus $N_A \cap E_1 = \emptyset$.) Since $A \in \operatorname{Supp}\mu$, $\mu(N_A) > 0$. Let $K = \mu(N_A)/\mu(E_1)$ and let σ be the measure on E given by

$$\sigma = K\mu|E_1 - \mu|N_A \, . \tag{3.11}$$

Then $\sigma(E) = 0$. We consider the measures

$$\mu_t = \mu + t\sigma \in \mathcal{P}(E) \quad \text{for} \quad 0 < t < 1 \, .$$

Then

$$(3.12) \qquad \mathcal{E}(\mu_t) = \int_{E^n} \Psi d\mu^n = \mathcal{E}(\mu) + \sum_{k=1}^{n} t^k \binom{n}{k} \int_{E^n} \Psi d(\sigma^k \times \mu^{n-k}) .$$

Each integral in the above sum is convergent since $\mathcal{E}(\mu)$ is assumed finite and $|\sigma| \leq (K+1)\mu$ and thus $|\sigma^k \times \mu^{n-k}| \leq (K+1)^k \mu^n$. Thus recalling (3.8),

$$(3.13) \qquad \mathcal{E}(\mu_t) = \mathcal{E}(\mu) + tn\int_E v_\mu d\sigma + O(t^2) .$$

Now

$$\int_E v_\mu d\sigma = K\int_{E_1} v_\mu d\mu - \int_{N_A} v_\mu d\mu < K[\mathcal{E}(\mu) - 2\varepsilon]\mu(E_1) - [\mathcal{E}(\mu) - \varepsilon]\mu(N_A)$$

$$= -\varepsilon\mu(N_A) < 0 .$$

Hence by (3.13), $\mathcal{E}(\mu_t) < \mathcal{E}(\mu)$ for t sufficiently small, contradicting the assumption that μ is an equilibrium measure.

Finally, we show that $v_\mu \leq \mathcal{E}(\mu)$ μ-almost everywhere. If not, there exists $\varepsilon > 0$ and $E_2 \subset E$ with $\mu(E_2) > 0$ and $v_\mu \geq \mathcal{E}(\mu) + \varepsilon$ on E_2. Since we already know that $v_\mu \geq \mathcal{E}(\mu)$ μ-almost everywhere, we have

$$\mathcal{E}(\mu) = \int_{E_2} v_\mu d\mu + \int_{E-E_2} v_\mu d\mu \geq [\mathcal{E}(\mu)+\varepsilon]\mu(E_2) + \mathcal{E}(\mu)\mu(E-E_2)$$

$$= \mathcal{E}(\mu) + \varepsilon\mu(E_2) > \mathcal{E}(\mu) ,$$

a contradiction. □

References

1. H. Alexander, Projective capacity, Conference on Several Complex Variables, Ann. Math. Studies 100, Princeton Univ. Press, 1981, pp.3-27.

2. H. Alexander, A note on projective capacity, (manuscript).

3. L. Carleson, Selected Problems on Exceptional Sets, Van Nostrand, Princeton, N.J., 1967.

4. L. Gruman, La géométrie globale des ensembles analytiques dans $\mathbb{C}^n$, Séminaire Pierre Lelong-Henri Skoda (Analyse) Années 1978/79, Lecture Notes in Mathematics 822, Springer-Verlag, Berlin, 1980, pp.90-99.

5. F. Leja, Une généralisation de l'écart et du diamètre transfini d'un ensemble, Ann. Soc. Polonaise Math. 22 (1949), 35-42.

6. R.E. Molzon, B. Shiffman, and N. Sibony, Average growth estimates for hyperplane sections of entire analytic sets, to appear in Math. Annalen.

7. M. Tsuji, Potential Theory in Modern Function Theory, Maruzen Co., Tokyo, 1959.

University of Kentucky, Lexington, Kentucky

The Johns Hopkins University, Baltimore, Maryland

SEVERAL COMPLEX VARIABLES IN MATHEMATICAL PHYSICS

V.S.Vladimirov(Moscow, Steklov Institute of Mathematics).

Last 20-25 years the theory of analytic functions of several complex variables had many applications in Mathematical physics, especially in quantum field theory. On the other hand the quantum field theory found oneself as a source of many nontrivial problems in the theory of analytic functions and essentially influenced its development. This situation reminds somewhat one of 40-50 years ago when rapid development of the hydro-aerodynamics and the theory of elasticity stimulated a progress of the theory of analytic functions of one complex variable .

In quantum field theory (as in many other branches of Mathematical physics) the physical quantities arise as boundary values of some classes of analytic functions holomorphic in some "primitive" domains defined by axioms. The probleme is to construct the envelope of holomorphy for "primitive" domains and the corresponding integral representations which would evaluate values of holomorphic function by means of its values on the "essential" part of the boundary. By such a way it is possible in principle to obtaine so-called (manydimensional) dispersion relations between quantities observed in experiments. Realization of this programme in the frame of some system of axioms would give firstly a possibility to verify experimentaly the consistency of the system of axioms considered and secondly would lead to an analytical approach wich would capable to predict results of experiments.

In this talk I do not have a possibility to expose sufficiently completely this line of problems. I only expose here briefly several main results from the theory of holomorphic functions of several complex variables which serve as a mathematical tool in many problems of mathematical physics. I keep in mind the following four problems.

1. The "edge of the wedge" theorem by Bogoliubov;
2. The " C - convex hull" theorem (or the "double cone" theorem);
3. The "finite covariance" theorem;
4. Holomorphic functions with positive real part in tube domains over proper cones.

We denote points of $\mathbb{C}^n = \mathbb{R}^n + i\mathbb{R}^n$ by $z = x + iy = (z_1, z_2, \ldots, z_n)$; $H(\mathcal{D})$ is the space of functions holomorphic in a domain $\mathcal{D}$ with the topology of uniform convergence on each compact subset of $\mathcal{D}$; $\mathcal{H}(\mathcal{D})$ is the envelope of holomorphy of a domain $\mathcal{D}$; $C, C', \ldots$ are cones in $\mathbb{R}^n$ with the vertex at O ; $T^C = \mathbb{R}^n + iC$ the tube over a cone C ; pr. $C = C \cap S^{n-1}$ (S^{n-1} is the unit spere); $C' \Subset C$ means that $pr\,\overline{C}' \subset pr\,C$; $C^* = [\xi : (\xi, y) \geq 0,\ \forall y \in C]$ is the conjugate cone for the cone C . If int $C^* \neq \emptyset$ the cone C called proper one:

1. The "Edge of the wedge" theorem by Bogoliubov.

In 1956 N.N.Bogoliubov discovered and proved a remarkable theorem called now as Bogoliubov's "edge of the wedge" theorem [1] .

It was firstly reported by N.N.Bogoliubov at the International conference at Seattle U.S.A. in September 1956. This theorem gives a peculiar generalization of the principle of analytical continuation of holomorphic functions.

The Bogoliubov "edge of the wedge" theorem is as follows (the local variant). Let C_+ be a connected open cone in $\mathbb{R}^n_y$, $C_-=-C_+$ and let $\mathcal{O}$ be an open set in $\mathbb{R}^n_x$. Let functions $f_\pm(z)$ be holomorphic in wedgeshaped domains

$$\mathcal{D}_\pm=[z;\ x\in\mathcal{O},\ y\in C_\pm\ ,\ |y|<R]$$

respectively and let their boundary values $bv\, f_\pm(x)$ at the common "edge" $x=\mathcal{O}$, $y=0$ of the domains $\mathcal{D}_\pm$ exist (as distributions from $\mathcal{D}'(\mathcal{O})$) and coincide in $\mathcal{O}$. Then there is exist a single function $f(z)$ holomorphic in a domain $\mathcal{D}_+\cup\mathcal{D}_-\cup\tilde{\mathcal{O}}$ where $\tilde{\mathcal{O}}$ is some (complex) neighborhood of $\mathcal{O}$, and $f(z)$ is equal to $f_\pm(z)$ in $\mathcal{D}_\pm$ respectively.

For case $n=1$ this theorem is classical one and it follows immediately from the Cauchy formula (the distributional boundary values can be overcome without difficulty). For $n\geq 2$ it seemed at first that such theorem can not be true at all because the domains $\mathcal{D}_+$ and $\mathcal{D}_-$ are in contact only along the n-dimensional set $\mathcal{O}$ in the $2n$-dimentional space $\mathbb{C}^n$. But the wedgeshaped structure of the domains $\mathcal{D}_+$ and $\mathcal{D}_-$ contacting along the "edge" $\mathcal{O}$ is here the decisive argument.

Now there are about ten proofs of the "edge of the wedge" theorem and its varions generalizations and sharpenings; there are surveys and books devoted to this problem (see [1 - 15] , [39 - 43] ,[65 - 67],[73 - 76]). Most important generalizations were established for any number of cones (for holomorphic cocycles) and for more general boundary values (up to hyperfunctions) see: Martineau [8,9] , Beurling [11] , Morimoto [10] ,Bros and Lagolnitzer [14] , Epstein [5], Žarinov [65-67],[76] Pinchuk [73], Henkin [75] .

We formulate here one of such generalizations of the Bogoliubov's theorem according to Martineau [8] . Let functions $f_k(z)$, $k=1,\dots,N$, be holomorphic in domains

$$\mathcal{D}_k = [z;\ x\in \mathcal{O},\ y\in C_k,\ |y|<R]$$

respectively where C_k, $k=1,\dots,N$, are mulually d sjoint (connected open) cones in $\mathbb{R}^n$. Let further their boundary values $bv f_k(x)$ exist in $\mathcal{D}'(\mathcal{O})$ and satisfy the relation

$$\sum_{1\le k\le N} bv f_k(x) = 0\ ,\ x\in\mathcal{O}.$$

Then there exist functions $f_{kj}(z), k,j=1,\dots,N, k\neq j$, holomorphic in domains

$$\mathcal{D}_{kj} = [z:\ x\in\mathcal{O},\ y\in ch(C_k\cup C_j)\ ,\ |y|<R]$$

respectively wich satisfy relations

$$f_k(z)=\sum_{1\le j\le N} f_{kj}(z),\ z\in\mathcal{D}_k\ ,\ f_{kj}(z)=-f_{jk}(z)\ ,\ z\in\mathcal{D}_{kj}$$

and possess $bv f_{kj}(x)$ from $\mathcal{D}'(\mathcal{O})$.

Recently Žarinov [65] etablished rather general analogy of the Bogoliubov's "edge of the wedge" theorem: the following sequence of vector spaces

$$0 \to A_N(\mathcal{O};\{C_k\}) \xrightarrow{\delta_N} \cdots \xrightarrow{\delta_2} A_1(\mathcal{O};\{C_k\}) \xrightarrow{\delta_1} A_0(\mathcal{O};\{C_k\}) \to 0$$

is exact. Here $A_p(\mathcal{O};\{C_k\})$, $p=1,\dots,N$, consists of all skew- symmetric elements of direct product of spaces

$$\underset{1\le k_1,\dots,k_p\le N}{\times} bv\, H(\mathcal{O}; ch(C_{k_1}\cup\dots\cup C_{k_p}))$$

and $A_0(\mathcal{O};\{C_k\}) = \mathcal{B}(\mathcal{O}; \mathcal{O}\times pr \bigcup_{1\le k\le N} C_k^*)$ is the space of hyperfunctions in $\mathcal{O}$ which singular spectrum [68] is contained in $\mathcal{O}\times pr \bigcup_{1\le k\le N} C_k^*$; the space $H(\mathcal{O}; C)$ is

$$\operatorname*{lim\,proj}_{C'\Subset C}\ \operatorname*{lim\,ind}_{\tilde{\mathcal{O}}\supset\mathcal{O}} H(\tilde{\mathcal{O}}\cap T^{C'});$$

the operator δ_p is defined by the formula

$$(\delta_p a)_{k_1\dots k_{p-1}} = \sum_{k=1}^{N} a_{kk_1\dots k_{p-1}}.$$

Similar theorem is valid for distributions [78] as well for Fourier -hyperfunctions and for Fourier -ultrahyperfunctions (see Žarinov [66 - 67]).

The exactness in term $A_1(\mathcal{O};\{C_\gamma\})$ is the contents of the "edge of the wedge" theorem for hyperfunctions [10] (similarly - - for distributions [8]).

Remark. The problem of the existence of boundary values of holomorphic functions has been investigated by many authors:

Fatou [52] ($\mathcal{L}^\infty$), F.Riesz [53] ($\mathcal{L}^p$), Köthe [54] ($\mathcal{D}'$), Sato [55] (hyperfunctions), Vladimirov [7], [16], [49], Tillmann [56] ($\mathcal{S}'$), Tillmann [57], Luszczki and Zielezny [58] ($\mathcal{D}'_{\mathcal{L}^p}$), Beurling [11], Komatsu [59] (ultradistributions), Martinean [8], Stein [34], Sato, Kawai, Kashiwara [68], Henkin, Čirka [74].

2. The "C - convex hull" theorem. Domains of the Dyson type $T^{C_+} \cup T^{C_-} \cup \tilde{\mathcal{O}}$ arising in the "edge of the wedge" theorem are not domains of holomorphy. The problem is to construct the envelope of holomorphy

$$\mathcal{H}(T^{C_+} \cup T^{C_-} \cup \tilde{\mathcal{O}}) \tag{1}$$

(the global version of the "edge of the wedge" theorem). For the cones $C_{\pm} = V^{\pm} = [y : \pm y_0 > \sqrt{y_1^2 + y_2^2 + y_3^2}]$ in $\mathbb{R}^4$ (future and past light comes respectively) the global "edge of the wedge" theorem has been solved for some special domains $\mathcal{O}$ by Bogoliubov [1], Bremermann, Oehme and Taylor [2], Jost and Lehman [3], Dyson [39], Bros, Messia and Stora [40], Bros, Itzykson and Pham [41], Seneor [42], Vladimirov [7], [16], Vladimirov and Žarinov [43] (for arbitrary $C_{\pm}$).

For arbitrary cones $C_{\pm}$ and domains $\mathcal{O}$ the envelope of holomorphy (1) has not been yet constructed, Nevertheless it is possible to point out some real points in (1) which are different from the points of $\mathcal{O}$, namely the following "C-convex hull" theorem is valid:

$$ch_C\,\mathcal{O} \subset \mathrm{Re}\,\mathcal{H}(T^{C_-} \cup T^{C_+} \cup \tilde{\mathcal{O}}).$$

Here $ch_C \mathcal{O}$ is the convex hull of $\mathcal{O}$ with respect to C-like curves where $C = C_+ \cup C_-$ (C - like curve is defined similar to a time-like curve when $C = V^+ \cup V^- = V$ is the light cone).

From the "edge of the wedge" and the " C- convex hull" theorems we get following quasianalyticity property of distributions important for applications [7] : if a tempered distribution f vanishes in $\mathcal{O}$ and its Fourier transform vanishes outside $(C_+^* \cup C_-^*) + K$, where K is a compact, then f vanishes in $ch_C \mathcal{O}$ (see also Žarinov [67] for Fourier-hyperfunctions).

The " C -convex hull" theorem was first established in 1960 by Vladimirov [16] . Another proof of this theprem was given one year later by Borchers [17] for the light cone V^+ (and called there as the "double cone" theorem; see also Araki [18]) . Others its generalizations and sharpenings especially for the most general boundary values are in papers by Beurling [11] (ultradistributions) and Morimoto [19], [10] (hyperfunctions).

The theorems "edge of the wedge" and " C - convex hull" have numerous applications in quantum field theory and in partial differential equations (see books by Bogoliubov, Medvedev and Polyvanov [1] , Bogoliubov, Logunov and Todorov [20] , Streater and Wightman [21] , Jost [22] and Vladimirov [7]) .

3. The "finite covariance" theorem. Let V^+ and $V^- = -V^+$ be the future and past cones in $\mathbb{R}^4$ resp. The corresponding tube domains $\tau^{\pm} = T^{V^{\pm}}$ in $\mathbb{C}^4$ are called future and past tubes; the domains $\tau_N^{\pm} = \underbrace{\tau^{\pm} \times \tau^{\pm} \times \ldots \times \tau^{\pm}}_{N \text{times}}$ in $\mathbb{C}^{4N}$ are called N - points future and past tubes.

In 1958 Bogoliubov and Vladimirov [23] etablished the following theorem which is analogous to Liouville's theorem for holomorphic functions (one - point "finite covariance" theorem). Let a function $f(z)$, $z = (z_0, z_1, z_2, z_3)$, be holomorphic in the Dyson domain

$$\tau^+ \cup \tau^- \cup \widetilde{(x^2 < 0)}$$

(here $x^2 = x_0^2 - |\mathbf{x}|^2$, $|\mathbf{x}|^2 = x_1^2 + x_2^2 + x_3^2$, is Lorentz square) and it is tempered in the domains $\tau^{\pm}$ (in other words, $f(z)$ is in $\tau^{\pm}$ the Laplace transform of tempered distributions with support in the cones $\bar{V}^{\pm}$ respectively). Then $f(z)$ is holomorphic in $\mathbb{C}^4$ except the "cut" $z^2 = \rho$, $\rho \geq 0$ and it can be represented in the form

$$f(z) = \sum_{\nu\text{-finite}} \mathcal{P}_\nu(z) f_\nu(z^2)$$

where $\mathcal{P}_\nu$ are polynomials and $f_\nu(\zeta)$ are holomorphic in ζ - plane with the cut: $\zeta = \rho, \rho \geq 0$ and they are tempered in the upper and lower halp-planes.

This theorem has been generalized by Bros, Epstein and Glaser [24] and by Bogoliubov and Vladimirov [25] for the case of N-point ($N+1$ particles) functions under the additional assumption that the N-point extended tube

$$\tau_N' = \bigcup_{\Lambda \in L_+(\mathbb{C})} \Lambda \tau_N^+$$

is a domain of holomorphy; here $L_+(\mathbb{C})$ is the proper complex Lorenz group. (Up to now there is not yet a satisfactory proof that for $N \geq 3$ $\mathcal{T}_N'$ is a domain of holomorphy: it has been proved only that the envelope of holomorphy of the Dyson domain $\mathcal{T}_N^+ \cup \mathcal{T}_N^- \cup \widetilde{J}_N$ contains $\mathcal{T}_N'$ (Streater [37]); here J_N is the Jost set which is the real section of the domain $\mathcal{T}_N'$ (geometrical description of J_N has been done by Jost [63]).

The boundary of $\mathcal{T}_N'$ has been investigated by Källen and Wightman [26-28], Jost [29], Fronsdal [30], Manoharan [31], Möller [32], Zavyalov and Trushin [33]. It was found out that $\mathcal{T}_1' = \mathbb{C}^4 \setminus [z \; z^2 = \rho, \rho \geq 0]$ and $\partial \mathcal{T}_2'$ consists of four pieces of some analitical hypersurfaces. For $N \geq 3$ on $\partial \mathcal{T}_N'$ except the just mentioned simplest analytical hypersurfaces some new surfaces appear which have more complicate structure, so - called DUMUD surfaces. (for $N \geq 7$ new surfaces do not appear at all).

Without assumption that $\mathcal{T}_N'$ is a domain of holomorphy Bogoliubov and Vladimirov [25] proved that each N-point tempered in $\mathcal{T}_N^\pm$ function $f(z^{(1)}, \ldots, z^{(N)})$ holomorphic in $\mathcal{T}_N'$ satisfies the following homogeneous system of partial differential equations

$$A_j (A_j^2 - 1)(A_j^2 - 4) \ldots (A_j^2 - p^2) f = 0 \;, \quad j = 1, 2, 3 \tag{2}$$

for some entire $p = p(f) \geq 0$. Here A_j is the infinitesimal operator for complex Lorentz rotations in the plane (z_0, z_j):

$$A_j = \sum_{1 \leq k \leq N} \left(z_0^{(k)} \frac{\partial}{\partial z_j^{(k)}} + z_j^{(k)} \frac{\partial}{\partial z_0^{(k)}} \right), \quad j = 1, 2, 3.$$

If τ_N' is a domain of holomorphy then general solutions of the systems (2) for all entire $p=0,1,\dots$ in the class of functions which are holomorphic in τ_N' and tempered in $\tau_N^{\pm}$ are given by the formula:

$$f=\sum_{\nu-\text{finite}} \mathcal{P}_\nu f_\nu$$

where $\mathcal{P}_\nu$ are polynomials and f_ν are holomorphic in τ_N', tempered in $\tau_N^{\pm}$ and $L_+(\mathbb{C})$ - invariant functions. It follows from results by Hepp [35-36]. The description of $L_+(\mathbb{C})$ - invariant functions holomorphic in τ_N^+ is contained in the paper by Hall and Wightman [38].

The "finite covariance" and " C - convex hull" theorems indicate on a close connection between the axioms of Lorentz covariance, spectrality and locality in quantum field theory.

4. Holomorphic functions with positive real part. Many problems in Mathematical physics are reduced to one of description of holomorphic functions with positive real part in a tube domain over a (convex) proper cone C (class $H_+(T^C)$).

For $n=1$ (the upper halp-plane) this problem was solved by means of the known Herglotz-Nevanlinna representation:

$$f(z)=\frac{z+i}{\pi i}\int_{-\infty}^{\infty}\frac{d\mu(x')}{(x'+i)(x'-z)}-\frac{1}{\pi}\int_{-\infty}^{\infty}\frac{d\mu(x)}{1+x^2}-iaz+b \tag{3}$$

where a mesure $\mu\geq 0$, numbers: $a\geq 0$ and b is real $(\mu(x)+a\delta(x)\neq 0)$. Moreover

$$\mu=\operatorname{Re}bvf(x),\quad a=\operatorname{Re}f(i)-\frac{1}{\pi}\int_{-\infty}^{\infty}\frac{d\mu(x)}{1+x^2},\quad b=\operatorname{Im}f(i).$$

A priori estimate [62]: if $f \in H_+(T^C)$ then for any cone $C' \Subset C$ there exists a number $M(C')$ such that

$$|f(z)| \leq M(C') \frac{1+|z|^2}{|y|}, \quad z \in T^{C'} \tag{4}$$

From estimate (4) it follows [49] the existence of $bvf(x)$ in $\mathcal{S}'$ and $f(z)$ is the Laplace-transform, $f = L[g]$, of an (unique) spectral function $g = F^{-1}[bvf]$ with support in the cone $C^* = \Gamma$, $g \in \mathcal{S}'(\Gamma)$ (here F is the Fourier transform, F^{-1} is its inverse). Moreover Re[bvf] is a nonnegative tempered mesure μ in $\mathbb{R}^n$ satisfying the estimate [69]

$$\int \mathcal{P}(x-x'; y)\, d\mu(x') \leq \operatorname{Re} f(z), \quad z \in T^C; \tag{5}$$

$\operatorname{Re} f(x+iy)$ takes its boundary value μ in the following precise sense:

$$\int \operatorname{Re} f(x+iy') \mathcal{P}(x,y) \varphi(x)\, dx \to \int \varphi(x) \mathcal{P}(x,y)\, d\mu(x),$$

$$y' \to 0, \; y' \in C', \; \forall C' \Subset C, \; \varphi \in C(\mathbb{R}^n) \cap \mathcal{L}^\infty(\mathbb{R}^n). \tag{6}$$

Here $\mathcal{P}(x,y)$ is the Poisson kernel for the tube domain T^C:

$$\mathcal{P}(x,y) = \frac{|\mathcal{K}(x+iy)|^2}{(2\pi)^n \mathcal{K}(2iy)}, \quad z \in T^C,$$

where $\mathcal{K}(z)$ is the Cauchy kernel for T^C

$$\mathcal{K}(z) = \int_\Gamma e^{i(z,\xi)}\, d\xi = i^n \Gamma(n) \int_{pr\,\Gamma} \frac{d\sigma}{(z,\sigma)^n}, \quad z \in T^C$$

where σ is the Lebesque mesure on the unit sphere $|\sigma| = 1$.

The Schwarz kernel $\mathcal{S}(z; z^0)$ with respect to the point

$z^0 = x^0 + iy^0 \in T^C$ is defined by the formula

$$\mathcal{S}(z;z^0) = \frac{2\mathcal{K}(z)\overline{\mathcal{K}(z^0)}}{(2\pi)^n \mathcal{K}(z-\bar{z}^0)} - \mathcal{P}(x^0;y^0), \quad z \in T^C$$

on the condition that $\mathcal{K}(z) \neq 0$, $z \in T^C$.

A cone C is called regular if $1/\mathcal{K}(z)$ is the Laplace-transform of a tempered distribution from $\mathcal{S}'(\Gamma)$; in particular $\mathcal{K}(z) \neq 0$, $z \in T^C$ if C is a regular cone. (For $n=1,2,3$ all proper cones are regular ones; for $n \geq 4$ it is not so; these results has been proved by Danilov.)

Example 1. $C = R_+^n = [y : y_1 > 0, \ldots, y_n > 0]$ is the positive octant: $(R_+^n)^* = \overline{R_+^n}$,

$$\mathcal{K}(z) = \frac{i^n}{z_1 \ldots z_n}, \quad \mathcal{P}(x,y) = \frac{y_1 \ldots y_n}{\pi^n |z_1|^2 \ldots |z_n|^2}.$$

Example 2. $C = V_n^+ = [y = (y_0, \mathbf{y}) : y_0 > |\mathbf{y}|]$ is the future light cone in $\mathbb{R}^{n+1}$ $(V_3^+ = V^+)$: $(V_n^+)^* = \overline{V_n^+}$,

$$\mathcal{K}(z) = 2^n \pi^{\frac{n-1}{2}} \Gamma\left(\frac{n+1}{2}\right)(-z^2)^{-\frac{n+1}{2}},$$

$$\mathcal{P}(x,y) = 2^n \pi^{-\frac{n+3}{2}} \Gamma\left(\frac{n+1}{2}\right) \frac{(y^2)^{\frac{n+1}{2}}}{|z^2|^{n+1}}.$$

A refined integral estimate [70]: if $f \in H_+(T^C)$ then for all $z^0 \in T^C$ and $y \in C$.

$$\int |\mathcal{K}(x+iy-\bar{z}^0)|^4 |f(x+iy) - \bar{f}(z^0)|^2 dx \leq$$

$$\leq 4(2\pi)^n \mathcal{K}(2iy^0)\mathcal{K}(2iy)\mathcal{K}(2iy^0+2iy) \operatorname{Re} f(z^0) \operatorname{Re} f(z^0+2iy). \tag{7}$$

The smoothness of the spectral function: from estimate (7) it follows that

$$\theta_\Gamma * \theta_\Gamma * g \in \mathcal{L}^2_s(\Gamma), \ s < -\frac{3}{2}n - 1 \tag{8}$$

where θ_Γ is the characteristic function of the cone $\Gamma = C^*$ and $\mathcal{L}^2_s(\Gamma)$ is the set of (measurable) functions v for which $v(\xi)(1+|\xi|)^{s/2} \in \mathcal{L}^2$ and $\operatorname{supp} v \subset \Gamma$.

For a regular cone C we introduce distributions θ^α_Γ, $-\infty < \alpha < \infty$, from $\mathcal{S}'(\Gamma)$ by the formula [49]

$$\mathcal{K}^\alpha(z) = L[\theta^\alpha_\Gamma].$$

They have properties: $\theta^0_\Gamma = \delta$, $\theta^1_\Gamma = \theta_\Gamma$, $\theta^\alpha_\Gamma * \theta^\beta_\Gamma = \theta^{\alpha+\beta}_\Gamma$. (For the cone $\Gamma = [0,\infty)$, $n=1$, θ^α_Γ is the kernel of the operator of fractional integration $(\alpha > 0)$ and differentiation $(\alpha < 0)$).

From (8) it follows: if a cone C is regular then

$$g = \theta^{-2}_\Gamma * g_1, \ g_1 \in \mathcal{L}^2_s(\Gamma), \ s < -\frac{3}{2}n - 1. \tag{9}$$

Examples.

$$\theta^{-2}_{R^n_+} = \frac{\partial^{2n}\delta(\xi)}{\partial\xi_1^2 \ldots \partial\xi_n^2}, \quad \theta^{-2}_{V^+_n} = \frac{\Box^{n+1}\delta(\xi)}{4^n \pi^{n-1} \Gamma^2\left(\frac{n+1}{2}\right)}.$$

The growth indicator of a function $\operatorname{Re} f(z)$, $f \in H_+(T^C)$ [49]:

$$h(\operatorname{Re} f; y) = \lim_{t \to +\infty} \frac{\operatorname{Re} f(z^0 + ity)}{t}, \ z^0 \in \mathbb{C}^n, \ y \in C$$

does not depend on z^0 ; it is a nonnegative concave function homogeneous of degree 1 in the cone C and

$$h(\mathrm{Re} f; y) \leq \mathrm{Re} f(z^0 + iy), \quad z^0 \in T^C, \; y \in C. \tag{10}$$

By means of the results just stated the following main Theorem was peoved.

Main Theorem [69]. If $f \in H_+(T^C)$ where C is a proper (convex) cone, $\mu = \mathrm{Re}\, bv f$ and $h(\mathrm{Re} f; y)$ is the growth indicator. Then the following assertions are equivalent:

(i) The Poisson integral

$$\int \mathcal{P}(x-x'; y)\, d\mu(x') \tag{11}$$

is pluriharmonic function in T^C ;

(ii) The function $\mathrm{Re} f(z)$ can be represented by the Poisson formula

$$\mathrm{Re} f(z) = \int \mathcal{P}(x-x'; y)\, d\mu(x') + (a, y), \quad z \in T^C \tag{12}$$

for some $a \in C^*$;

(iii) For all $z^0 \in T^C$ the function $f(z)$ can be represented by the Schwartz formula

$$f(z) = \int \mathcal{S}(z-x'; z^0-x')\, d\mu(x') - i(z, a) + i b(z^0), \quad z \in T^C \tag{13}$$

where $b(z^0)$ is a real number (on the condition that C is a regular cone).

Moreover $b(z^0) = (a, x^0) + \mathrm{Im} f(z^0)$ and (a, y) is the best linear minoranta of the indicator $h(\mathrm{Re} f; y)$ in the cone C ; $\mu(\xi)+\sum_{1\le j\le n}|a_j|\delta(\xi)\neq 0$.

Corollary. If $f\in H_+(T^C)$ and if the Poisson integral (11) is a pluriharmonic function in T^C then there is exists the best linear minoranta (a,y) of the indicator $h(\operatorname{Re} f;y)$ in the cone C.

Example: $h(\operatorname{Re}\frac{1}{i}\sqrt{z^2};y)=\sqrt{y^2}$, $(a,y)=0$ in V_n^+ .

The proof of the main Theorem uses, in particular, the following

Uniqueness Theorem. If $f\in H_+(T^C)$ and $\mu=\operatorname{Re} bvf=$ $=0$, then $f(z)=-i(z,a)+ib$ where $a\in C^*$ and $\operatorname{Im} b=0$.

Remark. For $n=1$, $C=(0,\infty)$ the main Theorem follows from Herglotz-Nevanlinna representation (3) for $z^0=i$.

Asimptotic behaviour (tauberian theorem). We say that a distribution $g(\xi)$ from $\mathcal{S}'$ has a quasiasymptotics at ∞ of an order α if there exist $g_\infty\in\mathcal{S}'$, $g_\infty\neq 0$ such that

$$\frac{g(\kappa\xi)}{\kappa^\alpha}\to g_\infty(\xi)\ ,\quad \kappa\to\infty \qquad \text{in } \mathcal{S}'.$$

Theorem (Drojjinov [76]). Let $g\in\mathcal{S}'(\Gamma)$ and $f(z)=L[g]$ has a bounded argument in T^C. In order that g has a quasiasymptotics at ∞ of an order α it is necessary:

(i) for any $z\in T^C$ there is exist

$$\lim_{\rho\to+0}\rho^{\alpha+n}f(\rho z)=h(z)\neq 0,$$

(ii) there are exist numbers M, p and q such that

$$|\rho^{\alpha+n}f(\rho z)|\le\frac{M(1+|z|)^q}{[\Delta_C(y)]^p}\ ;\ 0<\rho\le 1,\ z\in T^C,$$

where $\Delta_C(y)$ is the distance from $y \in C$ to ∂C ;
it is sufficient: there is exist a solid cone $C' \subset C$ such that

$$\lim_{\rho \to +0} \rho^{\alpha+n} f(i\rho y) = h(iy), \; y \in C'.$$

In addition

$$h(z) = L[g_\infty].$$

Corollary. If $f(z)$ is holomorphic in T^C , $f(z)$ has a bounded argument in a neighbourhood of 0 and if in some solid cone $C' \subset C$ there is exist

$$\lim_{\rho \to +0} f(i\rho y) = h(iy), \; y \in C'$$

then there is exist an "angular" limit

$$\lim_{\rho \to +0} f(\rho z) = h(z), \; z \in T^C.$$

The question arises of when the Poisson integral (11) is a pluriharmonic function in T^C . This quastion has been studied for two extreme cases: the cones $C = \mathbb{R}_+^n$, $T^{\mathbb{R}_+^n} = T^n$, and $C = V_3^+ = V^+$, $T^{V^+} = \tau^+$.

The cone $\mathbb{R}_+^n$. In this case the condition (i) of the main Theorem holds for any $f \in H_+(T^C)$, so the other assertions (ii)-(iii) of the main Theorem hold ([62],[49]), Therefore the best linear minoranta (a,y) of the indicator $h(\mathrm{Re} f; y)$ in $\mathbb{R}_+^n$ exists and the numbers a_j are defined by the formulas

$$a_j = \lim_{\substack{y \to e_j \\ y \in R_+^n}} h(\operatorname{Re} f; y) = \lim_{\substack{y_j \to +\infty \\ y \in R_+^n}} \frac{\operatorname{Re} f(iy)}{y_j}, \quad j = 1, \dots, n \tag{14}$$

where $\{e_j, j = 1, \dots, n\}$ are the basis (unit) vectors in $\mathbb{R}^n$.

The assertions (i)-(iii) are supplemented by the following: 1) in order that the mesure μ be the real part of the boundary value of some function f of the class $H_+(T^n)$, it is necessary and sufficient that

$$\mu \geq 0, \quad \int \frac{d\mu(x)}{(1+x_1^2) \dots (1+x_n^2)} < \infty,$$

$$\int \prod_{j=1}^{n} \left(\frac{x_j - i}{x_j + i} \right)^{\alpha_j} \frac{1}{1+x_j^2} \, d\mu(x) = 0, \quad 0 \not\leq \alpha \not\geq 0; \tag{15}$$

here $\alpha = (\alpha_1, \dots, \alpha_n)$, $\alpha_j \gtreqless 0$ are integer; $\alpha \geq 0$ iff $\alpha_j \geq 0$, $j = 1, \dots, n$.

2) In order that the function f belongs to $H_+(T^n)$, it is necessary and sufficient that its spectral function has the following properties

a) $g(\xi) + g^*(\xi) \gg 0$ (in Bochner-Schwartz sense) where $g^*(\xi) = \overline{g(-\xi)}$;

b) $$g(\xi) = D_1^2 \dots D_n^2 u(\xi) + (a, D)\delta(\xi) \tag{16}$$

where $a \in \overline{\mathbb{R}_+^n}$ and $u(\xi)$ is a continous function in $\mathbb{R}^n$ with a support in $\overline{\mathbb{R}_+^n}$ satisfying the growth condition

$$|u(\xi)| \le C(1+\xi_1^2)\dots(1+\xi_n^2).$$

In addition $\frac{1}{2}F[g+g^*] = \mathrm{Re}\, bvf = \mu$ and a_j are defined by the Eqs (14); $\mu(\xi) + \sum_{1\le j\le n} a_j\, \delta(\xi) \not\equiv 0$.

Corollary. From representation (16) it follows that the main singularities in the spectral function $g(\xi)$,

$$(a,D)\delta(\xi) = D_1^2 \dots D_n^2 \sum_{1\le j\le n} a_j\, \xi_1 \theta(\xi_1) \dots \theta(\xi_j) \dots \xi_n \theta(\xi_n),$$

are determined by numbers a_j , i.e. the contribution from points at infinity of the space R^n ; here θ is the Heaviside function.

Remark. For $n=1$ the conditions (15) are absent and assertion 1) follows from representation (3); for $n\ge 2$ it was proved by Vladimirov [62]; for $n=1$ the assertion 2) was proved by König and Zemanian [60], for $n\ge 2$ - by Vladimirov ([47], see also [49]).

The cone V^+ . In this case the assertions (i)-(iii) of the main Theorem hold precisely for those $f\in H_+(\tau^+)$ for which the indicator $h(\mathrm{Re} f; y)$ has the properties

$$h(\mathrm{Re} f; y) = h_0(y) + (a,y),\quad h_0(y)\ge 0,\ y\in V^+,\ a\in \bar{V}^+;$$

$$\lim_{|y|\to 1-0} \int_{|\sigma|=1} h_0(1, |y|\sigma)\, d\sigma = 0. \tag{17}$$

If it is the case then the assertions (i)-(iii) are supplemented by the following: in order that the mesure μ be the real part of the bvf for some $f\in H_+(\tau^+)$, it is necessary and sufficient that

$$\mu \geq 0, \quad \int \frac{d\mu(x)}{|(x+i)^2|^4} < \infty, \quad i=(i,0),$$

$$\int \Delta^{jm}_{q_1 q_2}[X(x)] \frac{d\mu(x)}{|(x+i)^2|^4} = 0, \quad 2j=2,3,\ldots, \quad m=-1,-2,\ldots,-2j+1,$$
$$-j \leq q_1, q_2 \leq j \tag{18}$$

where $\Delta^{jm}_{q_1 q_2}(X)$, $m=0,\pm 1,\ldots, 2j=0,1,\ldots, -j \leq q_1, q_2 \leq j$, are spherical functions on the group $U(2)$ and $x \to X(x)$ is the representation of $\mathbb{R}^4$ in $U(2)$, defined by the formulas

$$X(x) = (I - i\underset{\sim}{x})^{-1}(I + i\underset{\sim}{x}), \quad \underset{\sim}{x} = \begin{pmatrix} x_0 + x_3 & x_1 - ix_2 \\ x_1 + ix_2 & x_0 - x_3 \end{pmatrix}. \tag{19}$$

The corresponding complex mapping $z \to w(z)$ maps τ^+ be-holomorphically onto generalized "unit circle" $\tau_0 = [w : ww^* < I]$.

We compactify the space $\mathbb{R}^4$ by adding points at infinity which an image by the mapping (19) is a 3-dimensional manifold

$$S_0 = [X : X \in U(2), \det(I+X) = 0].$$

We shall denote this compactification of $\mathbb{R}^4$ by $\mathcal{M}_4$. The set of points at infinity is topologically equivalent to the 3-dimensional Klein bottle one equator of which is contracted to a point. For details see Uhlmann [72] , Penrose [71] (se also [61]).

The functions

$$V_{q_1 q_2}^{mj}(x) = \Delta_{q_1 q_2}^{mj}[X(x)], \quad m = 0, \pm 1, \ldots, 2j = 0, 1, \ldots, \quad -j \leq q_1, q_2 \leq j \tag{20}$$

are rational and continuous on $\mathcal{M}_4$ and form a complete orthogoman system in the Hilbert space of (mesurable) functions $f(x)$ in $\mathbb{R}^4$ with norm

$$\|f\|^2 = \left(\frac{2}{\pi}\right)^3 \int \frac{|f(x)|^2}{|(x+i)^2|^4} dx.$$

For $m = 0, 1, \ldots$ they are extended holomorphically and boundedly to the domain τ^+ ; when $m = -2j, -2j-1, \ldots$ - to the domain τ^-.

Keeping in mind some applications in Mathematical physics we write out the explicit form of the Poisson formula (12) and Schwartz formula (13) (for $z^0 = i$):

$$\mathrm{Re} f(z) = \left(\frac{2}{\pi}\right)^3 \int \frac{(y^2)^2 d\mu(x')}{|(z-x')^2|^4} + (a, y),$$

$$f(z) = \frac{[(z+i)^2]^2}{\pi^3} \int \frac{d\mu(x')}{[(z-x')^2(x'+i)^2]^2} - \left(\frac{2}{\pi}\right)^3 \int \frac{d\mu(x')}{|(x'+i)^2|^4} -$$

$$-i(z, a) + i \, \mathrm{Im} f(i).$$

The results have applications in the theory of linear systems of convolution equations (see [44 - 51])

$$\mathcal{Z} * u = f \tag{21}$$

(here $Z(\xi)$ is a real $N\times N$ -matrix with components Z_{kj} from $\mathcal{D}'(R^n)$) which are passive with respect to a (convex closed solid) proper cone Γ :

$$Re \int_{-\Gamma} \langle Z * \varphi, \varphi \rangle d\xi \geq 0, \quad \varphi = (\varphi_1, \dots, \varphi_N) \in [\mathcal{D}(R^n)]^{\times N}.$$

Theorem. A matrix $Z(\xi)$ defines a passive operator with respect to a cone Γ iff its impedance $\tilde{Z}(z)$ (the Laplace transform of $Z(\xi)$) is a positive-real matix-function in the tube domain T^C where $C = int\,\Gamma^*$ (that is $\tilde{Z}(z)$ is holomorphic, $Re\,\tilde{Z}(z) \geq 0$ and $\tilde{Z}(z) = \overline{\tilde{Z}}(-\bar{z})$ in T^C).

Remark. For $n=1$ this theorem was proved by Zemenian [64] and for $n \geq 2$ - by Vladimirov [47].

For the system (21) the following problems can be treated by means of the just described techniques: extension of singularities, many-dimensional dispersion relations, existence of inverse operators (admitances), a generalized Cauchy problem, assymptotic behaviors and so on. Many Equations of Mathematical physics are passive with respect to some cone Γ (mainly it is cone $\bar{V}^+$) including equations of Dirac, Maxwell, magnetic hydrodynamics, elastisity, acoustics, electrical, network,... .

Цитированная литература - BIBLIOGRAPHY

I Н.Н.Боголюбов, Б.В.Медведев, и М.К.Поливанов, Вопросы теории дисперсионных соотношений, Физматгиз, I958.

2 H.J.Bremermann, R.Oehme and J.G.Taylor, Proof of dispersion relations in quantized field theories, Phys. Rev., 109(1958), 2178-2190.

3 R.Jost und H.Lehmann, Integral-Darstellung kausaler Kommutatoren, Nuovo Cimento, 5 (1957), 1598-1610.

4 F.J.Dyson, Connection between local commutativity and regularity of Wightman functions, Phys. Rev., 110 (1958), 579-581.

5 H.Epstein, Generalization of the "edge-of-the wedge" theorem, J. Math. Phys., 1 (1960), 524-531.

6 F.E.Browder, On the "edge of the wedge" theorem, Canad. J. of Math., 15 (1963), 125-131.

7 В.С.Владимиров, Методы теории функций многих комплексных переменных, "Наука", I964.

8 A.Martineau, Distributions et valeurs au bord des fonctions holomorphes, Theory distribut. Inst. Gulbenkian cienc. Lisboa, 1964, pp.193-326.

9 A.Martineau, Le "edge of the wedge theorem" en théory des hyperfonctions de Sato, Proc. Intern. Conf. on Functional Analysis and Related Topics, Tokyo, 1969, Univ. Tokyo Press, 1970, pp.95-106.

IO M.Morimoto, Sur la decomposition du faisceau des germes de singularites d'hyperfonctions, J.Fac.Sci.Univ. Tokyo, Sect. I, XVII (1970), 215-239; Edge of the wedge theorem and hyperfunction, Lectures notes in Mathem., 287, Springer, 1973, pp.41-81.

II A.Beurling, Analytic continuation across a linear boundary, Acta Math., 128;3:4(1972), 153-182.

I2 W.Rudin, Lectures on the Edge-of-the-Wedge theorem, Regional Conf.Series in Math., No.6, AMS, 1971.

I3 Å.Kolm and B.Nagel, A generalized edge of the wedge theorem, Commun. math.Phys., 8(1968), 185-203.

I4 J.Bros and D.Iagolnitzer, Causality and local analyticity: mathematical study, Ann.Inst.Henri Poincaré, XVIII, N2(1973), 147-184.

I5 В.С.Владимиров, Теорема об "острие клина" Боголюбова, ее развитие и применения, сб. "Проблемы теоретической физики", "Наука", I969, стр. 6I-67.

I6 В.С.Владимиров, О построении оболочек голоморфности для областей специальности вида, ДАН СССР, I34(I960), 25I-254; О построении оболочек голоморфности для областей специального вида и их применения, Труды Матем.ин-та АН СССР, 60(I96I), IOI-I44.

I7 H.J.Borchers, Über die Vollständigkeit lorentzinvarianter Felder in einer zeitartigen Röhre, Nuovo Cimento, 19(1961), 781-793.

I8 H.Araki, A generalization of Borchers theorem, Helv.Phys. Acta, 36(1963), 132-139.

I9 M.Morimoto, Support et support singulier de l'hyperfonction, Proc.of Japan.Acad., 47(1971), 648-652.

20 Н.Н.Боголюбов, А.А.Логунов и И.Т.Тодоров, Основы аксиоматического подхода в квантовой теории поля, "Наука", I969.

21. R.F.Streater and A.S.Wightman, PCT, Spin and Statistics and all that, Benjamin, 1964.

22 R.Jost, The General Theory of Quantized Fields, AMS, 1965.

23 Н.Н.Боголюбов и В.С.Владимиров, Одна теорема об аналитическом продолжении обобщенных функций, НДВШ, физ.-мат.науки, № 3 (1958), 26-35.

24 J.Bros, H.Epstein and V.Glaser, On the connection between analyticity and Lorentz covariance of Wightman functions, Commun. Math. Phys., 6 (1967), 77-100.

25 Н.Н.Боголюбов и В.С.Владимиров, Представления n-точечных функций, Труды Матем. ин-та им.В.А.Стеклова, 112, 1971, стр.5-21.

26 G.Källén, A.Wightman, The analytic properties of the vacuum expectation value of a product of three scalar local fields, Mat. Fys. Skr.Dan.Vid.Selsk., 1, N 6, 1958.

27 G.Källén, The analyticity domain of the four point function, Nuclear Phys., 25 (1961), 568-603.

28 A.S.Wightman, Quantum field theory and analytic functions of several complex variables, J.Indian Math. Soc., 24 (1960/61), 625-677.

29 R.Jost, Lectures on field theory and the many body problem, Academic press, N.Y., 1961.

30 C.Fronsdal, Analyticity of Wightman functions, J. Math. Phys., 2 (1961), 748-758.

31 A.C.Monoharan, The primitive domains of holomorphy for the 4 and 5 point Wightman functions,J.Math.Phys., 3(1962), 853-859.

32 N.H.Möller, The analyticity domain of the five point functions,Nuclear Phys., 35(1962), 434-450.

33 Б.И.Завьялов, В.Б.Трушин, О расширенной n-точечной трубе, ТМФ,27(I976),3-I5.

34 E.M.Stein, Boundary behavior of holomorphic functions of several complex variables, Mathem.Notes, Princeton Univ.Press,1972.

35 K.Hepp, Klassische komplexe Liesche Gruppen und kovariante analytische Funktionen, Math.Ann., 152(1963),149-158.

36 K.Hepp, Lorentz-kovariante analytische Funktionen,Helv. Phys.Acta,36(1963),355-375.

37 R.F.Streater,Analytic properties of product of field operators, J.Math.Phys.,3(1962), 256-261.

38 D.Hall and A.S.Wightman, A theorem on invariant analytic functions with applications to relativistic quantum field theory,Mat.Fis.Medd.Dan.Vid.Selsk.,31, N5(1957),3-41.

39 F.J.Dyson, Integral representations of causal commutators, Phys. Rev., 110 (1958), 1460-1464.

40 J.Bros, A.Messia, R.Stora, A problem of analytic completion related to the Joest-Lehmann-Dyson formula, J. Math. Phys., 2 (1961), 639-651.

41 J.Bros, C.Itzykson et F.Pham, Representations integrales de fonctions analytiques et formula de Jost-Lehmann-Dyson, Ann. Inst. Henri Poincaré, Sect.A., 5, 1 (1966).

42 R.Seneor, Une generalization de la formula de Dyson, Commun. Math. Phys., 11 (1969), 233-256.

43 B.C.Владимиров и B.B.Жаринов, О представлении типа Йоста-Лемана-Дайсона, ТМФ, 3 (1970), 305-319.

44 A.H.Zemanian, Distribution Theory and Transform Analysis, Mc Graw-Hill, 1965.

45 E.J.Beltrami, M.R.Wohlers, Distributions and the Boundary Values of Analytic Functions, Academic Press,N.Y.-London,1966.

46 H.König, T.Meixner, Lineare systeme und lineare Transformationen, Math. Nachr., 19 (1958), 265-322.

47 B.C.Владимиров, Линейные пассивные системы, ТМФ, 1 (1969), 67-94.

48 B.C.Владимиров, Многомерные линейные пассивные системы,сб. "Механика сплошной среды и родственные проблемы анализа", Москва, 1972, стр.121-134.

49 В.С.Владимиров, Обобщенные функции в математической физике,"Наука", 1979.

50 T.Wu, Some properties of impedance as a causal operator, J.Math.Phys.,3(1962), 262-271.

51 W.Güttinger,Generalized functions in elementary particle physics and passive system theory: recent trends and problems,SIAM J.Appl.Math., 15(1967), 964-1000.

52 P.Fatou,Series trigonometriques et series de Taylor, Acta Mathematica,30(1906),335-400.

53 F.Riesz,Über die Randwerte einer analytischen Functionen, Math.Z., 18(1922),87-95.

54 G.Köthe, Die Randverteilungen analytischer Funktionen, Math.Z., 57(1952), 13-33.

55 M.Sato, On a generalization of the concept of functions, Proc.Japan Acad., 34(1958),126-130, 604-608; Theory of hyperfunctions,J.Fac.Sci., Univ.Tokyo,Sect.I,8(1959/60), 139-193,387-436.

56 H.G.Tillmann,Darstellung der Schwartzschen Distributionen durch analytische Funktionen,Math.Z., 77(1961),106-124.

57 H.G.Tillmann,Distributionen als Randverteilungen analytischer Funktionen,II,Math.Z.,76(1961),5-21.

58 Z.Luszczki and Z.Zielezny,Distributionen der Raume $\mathcal{D}'_{L^p}$ als Ranverteilungen analytischer Funktionen,Colloq.Math., 8(1961),125-131.

59 H.Komatsu, Ultradistributions and Hyperfunctions, Lectures notes in Mathem., 287, Springer, 1973, pp.164-179.

60 H.König, A.H.Zemanian, Necessary and Sufficient Conditions for a Matrix Distribution to have a Positive Real Laplace Transform, SIAM J.Appl.Math., 13 (1965), 1036-1040.

61 В.С.Владимиров, Голоморфные функции с положительной мнимой частью в трубе будущего, Матем.сб., 93 (1974), 3-17; II, 94 (1974), 499-515; IV, 104 (1977), 341-370.

62 В.С.Владимиров, Голоморфные функции с неотрицательной мнимой частью в трубчатой области над конусом, Матем.сб., 79 (1969), 128-152.

63 R.Jost, Eine Bemerkung Zum CTP Theorem, Helv. Phys. Acta, 30 (1957), 409-416.

64 A.H.Zemanian, An N-port Realizability Theory Based on the Theory of Distributions, IEEE Trans. Curcuit Theory, CT-10 (1963), 265-274.

65 В.В.Жаринов, Об одной точной последовательности модулей и теореме "об острие клина" Боголюбова, ДАН СССР, 251 (1980), 19-22.

66 В.В.Жаринов, Фурье-ульрагиперфункции, Изв.АН СССР, сер.матем, 44 (1980), 533-570.

67 В.В.Жаринов, Аналитические представления одного класса аналитических функционалов, содержащего фурье-гиперфункции, Матем.сб., 108(150)(1979), 62-77.

68 M.Sato, T.Kawai and M.Kashiwara, Microfunctions and Pseudo-differential Equations, Lectures notes in Mathem., 287, Springer, 1973, pp.265-529.

69 В.С.Владимиров, Голоморфные функции с неотрицательной мнимой частью в трубчатых областях над конусами, ДАН СССР, 239 (1978), 26-29.

70 В.С.Владимиров, Оценка роста граничных значений неотрицательных плюригармонических функций в трубчатой области над острым конусом, всб. "Комплексный анализ и его приложения", "Наука", 1978, стр.137-148.

71 R.Penrose, The Complex Geometry of the Natural world, Proc. of the ICM, Helsinki, 1978, v.1, 1980, pp.189-194; The apparent shape of a relativistically moving sphere, Proc. Compridge Phylos. Soc., 55 (1959), 137-139.

72 A.Uhlmann, The closure of Minkowski Space, Acta Physica Polonica, XXIV (1963), 295-296.

73 С.И.Пинчук, Теорема Боголюбова об "острие клина" для порождающих многообразий, Мат.сб., 94 (1974), 468-482.

74 Г.М.Хенкин, Е.М.Чирка, Граничные свойства голоморфных функций нескольких комплексных переменных, Современные проблемы математики, т.4, М., 1975, стр.13-142.

75 Г.М.Хенкин, Аналитическое продолжение функций через "острие клина", ДАН СССР,

76 Ю.Н.Дрожжинов, Многомерная тауберова теорема для голоморфных функций с неотрицательной мнимой частью, ДАН СССР, 258, № 3 (1981).

77 P.Lelong, Fonctions plurisousharmoniques et formes differentielles positives, Gordon and Breach, 1968.

78. В.В.Жаринов, Обобщение теоремы об "острие клина" Боголюбова, ДАН СССР, 258, № 4 (1981).

Vol. 759: R. L. Epstein, Degrees of Unsolvability: Structure and Theory. XIV, 216 pages. 1979.

Vol. 760: H.-O. Georgii, Canonical Gibbs Measures. VIII, 190 pages. 1979.

Vol. 761: K. Johannson, Homotopy Equivalences of 3-Manifolds with Boundaries. 2, 303 pages. 1979.

Vol. 762: D. H. Sattinger, Group Theoretic Methods in Bifurcation Theory. V, 241 pages. 1979.

Vol. 763: Algebraic Topology, Aarhus 1978. Proceedings, 1978. Edited by J. L. Dupont and H. Madsen. VI, 695 pages. 1979.

Vol. 764: B. Srinivasan, Representations of Finite Chevalley Groups. XI, 177 pages. 1979.

Vol. 765: Padé Approximation and its Applications. Proceedings, 1979. Edited by L. Wuytack. VI, 392 pages. 1979.

Vol. 766: T. tom Dieck, Transformation Groups and Representation Theory. VIII, 309 pages. 1979.

Vol. 767: M. Namba, Families of Meromorphic Functions on Compact Riemann Surfaces. XII, 284 pages. 1979.

Vol. 768: R. S. Doran and J. Wichmann, Approximate Identities and Factorization in Banach Modules. X, 305 pages. 1979.

Vol. 769: J. Flum, M. Ziegler, Topological Model Theory. X, 151 pages. 1980.

Vol. 770: Séminaire Bourbaki vol. 1978/79 Exposés 525–542. IV, 341 pages. 1980.

Vol. 771: Approximation Methods for Navier-Stokes Problems. Proceedings, 1979. Edited by R. Rautmann. XVI, 581 pages. 1980.

Vol. 772: J. P. Levine, Algebraic Structure of Knot Modules. XI, 104 pages. 1980.

Vol. 773: Numerical Analysis. Proceedings, 1979. Edited by G. A. Watson. X, 184 pages. 1980.

Vol. 774: R. Azencott, Y. Guivarc'h, R. F. Gundy, Ecole d'Eté de Probabilités de Saint-Flour VIII-1978. Edited by P. L. Hennequin. XIII, 334 pages. 1980.

Vol. 775: Geometric Methods in Mathematical Physics. Proceedings, 1979. Edited by G. Kaiser and J. E. Marsden. VII, 257 pages. 1980.

Vol. 776: B. Gross, Arithmetic on Elliptic Curves with Complex Multiplication. V, 95 pages. 1980.

Vol. 777: Séminaire sur les Singularités des Surfaces. Proceedings, 1976-1977. Edited by M. Demazure, H. Pinkham and B. Teissier. IX, 339 pages. 1980.

Vol. 778: SK1 von Schiefkörpern. Proceedings, 1976. Edited by P. Draxl and M. Kneser. II, 124 pages. 1980.

Vol. 779: Euclidean Harmonic Analysis. Proceedings, 1979. Edited by J. J. Benedetto. III, 177 pages. 1980.

Vol. 780: L. Schwartz, Semi-Martingales sur des Variétés, et Martingales Conformes sur des Variétés Analytiques Complexes. XV, 132 pages. 1980.

Vol. 781: Harmonic Analysis Iraklion 1978. Proceedings 1978. Edited by N. Petridis, S. K. Pichorides and N. Varopoulos. V, 213 pages. 1980.

Vol. 782: Bifurcation and Nonlinear Eigenvalue Problems. Proceedings, 1978. Edited by C. Bardos, J. M. Lasry and M. Schatzman. VIII, 296 pages. 1980.

Vol. 783: A. Dinghas, Wertverteilung meromorpher Funktionen in ein- und mehrfach zusammenhängenden Gebieten. Edited by R. Nevanlinna and C. Andreian Cazacu. XIII, 145 pages. 1980.

Vol. 784: Séminaire de Probabilités XIV. Proceedings, 1978/79. Edited by J. Azéma and M. Yor. VIII, 546 pages. 1980.

Vol. 785: W. M. Schmidt, Diophantine Approximation. X, 299 pages. 1980.

Vol. 786: I. J. Maddox, Infinite Matrices of Operators. V, 122 pages. 1980.

Vol. 787: Potential Theory, Copenhagen 1979. Proceedings, 1979. Edited by C. Berg, G. Forst and B. Fuglede. VIII, 319 pages. 1980.

Vol. 788: Topology Symposium, Siegen 1979. Proceedings, 1979. Edited by U. Koschorke and W. D. Neumann. VIII, 495 pages. 1980.

Vol. 789: J. E. Humphreys, Arithmetic Groups. VII, 158 pages. 1980.

Vol. 790: W. Dicks, Groups, Trees and Projective Modules. IX, 127 pages. 1980.

Vol. 791: K. W. Bauer and S. Ruscheweyh, Differential Operators for Partial Differential Equations and Function Theoretic Applications. V, 258 pages. 1980.

Vol. 792: Geometry and Differential Geometry. Proceedings, 1979. Edited by R. Artzy and I. Vaisman. VI, 443 pages. 1980.

Vol. 793: J. Renault, A Groupoid Approach to C*-Algebras. III, 160 pages. 1980.

Vol. 794: Measure Theory, Oberwolfach 1979. Proceedings 1979. Edited by D. Kölzow. XV, 573 pages. 1980.

Vol. 795: Séminaire d'Algèbre Paul Dubreil et Marie-Paule Malliavin. Proceedings 1979. Edited by M. P. Malliavin. V, 433 pages. 1980.

Vol. 796: C. Constantinescu, Duality in Measure Theory. IV, 197 pages. 1980.

Vol. 797: S. Mäki, The Determination of Units in Real Cyclic Sextic Fields. III, 198 pages. 1980.

Vol. 798: Analytic Functions, Kozubnik 1979. Proceedings. Edited by J. Ławrynowicz. X, 476 pages. 1980.

Vol. 799: Functional Differential Equations and Bifurcation. Proceedings 1979. Edited by A. F. Izé. XXII, 409 pages. 1980.

Vol. 800: M.-F. Vignéras, Arithmétique des Algèbres de Quaternions. VII, 169 pages. 1980.

Vol. 801: K. Floret, Weakly Compact Sets. VII, 123 pages. 1980.

Vol. 802: J. Bair, R. Fourneau, Etude Géometrique des Espaces Vectoriels II. VII, 283 pages. 1980.

Vol. 803: F.-Y. Maeda, Dirichlet Integrals on Harmonic Spaces. X, 180 pages. 1980.

Vol. 804: M. Matsuda, First Order Algebraic Differential Equations. VII, 111 pages. 1980.

Vol. 805: O. Kowalski, Generalized Symmetric Spaces. XII, 187 pages. 1980.

Vol. 806: Burnside Groups. Proceedings, 1977. Edited by J. L. Mennicke. V, 274 pages. 1980.

Vol. 807: Fonctions de Plusieurs Variables Complexes IV. Proceedings, 1979. Edited by F. Norguet. IX, 198 pages. 1980.

Vol. 808: G. Maury et J. Raynaud, Ordres Maximaux au Sens de K. Asano. VIII, 192 pages. 1980.

Vol. 809: I. Gumowski and Ch. Mira, Recurences and Discrete Dynamic Systems. VI, 272 pages. 1980.

Vol. 810: Geometrical Approaches to Differential Equations. Proceedings 1979. Edited by R. Martini. VII, 339 pages. 1980.

Vol. 811: D. Normann, Recursion on the Countable Functionals. VIII, 191 pages. 1980.

Vol. 812: Y. Namikawa, Toroidal Compactification of Siegel Spaces. VIII, 162 pages. 1980.

Vol. 813: A. Campillo, Algebroid Curves in Positive Characteristic. V, 168 pages. 1980.

Vol. 814: Séminaire de Théorie du Potentiel, Paris, No. 5. Proceedings. Edited by F. Hirsch et G. Mokobodzki. IV, 239 pages. 1980.

Vol. 815: P. J. Slodowy, Simple Singularities and Simple Algebraic Groups. XI, 175 pages. 1980.

Vol. 816: L. Stoica, Local Operators and Markov Processes. VIII, 104 pages. 1980.

Vol. 817: L. Gerritzen, M. van der Put, Schottky Groups and Mumford Curves. VIII, 317 pages. 1980.

Vol. 818: S. Montgomery, Fixed Rings of Finite Automorphism Groups of Associative Rings. VII, 126 pages. 1980.

Vol. 819: Global Theory of Dynamical Systems. Proceedings, 1979. Edited by Z. Nitecki and C. Robinson. IX, 499 pages. 1980.

Vol. 820: W. Abikoff, The Real Analytic Theory of Teichmüller Space. VII, 144 pages. 1980.

Vol. 821: Statistique non Paramétrique Asymptotique. Proceedings, 1979. Edited by J.-P. Raoult. VII, 175 pages. 1980.

Vol. 822: Séminaire Pierre Lelong–Henri Skoda, (Analyse) Années 1978/79. Proceedings. Edited by P. Lelong et H. Skoda. VIII, 356 pages, 1980.

Vol. 823: J. Král, Integral Operators in Potential Theory. III, 171 pages. 1980.

Vol. 824: D. Frank Hsu, Cyclic Neofields and Combinatorial Designs. VI, 230 pages. 1980.

Vol. 825: Ring Theory, Antwerp 1980. Proceedings. Edited by F. van Oystaeyen. VII, 209 pages. 1980.

Vol. 826: Ph. G. Ciarlet et P. Rabier, Les Equations de von Kármán. VI, 181 pages. 1980.

Vol. 827: Ordinary and Partial Differential Equations. Proceedings, 1978. Edited by W. N. Everitt. XVI, 271 pages. 1980.

Vol. 828: Probability Theory on Vector Spaces II. Proceedings, 1979. Edited by A. Weron. XIII, 324 pages. 1980.

Vol. 829: Combinatorial Mathematics VII. Proceedings, 1979. Edited by R. W. Robinson et al.. X, 256 pages. 1980.

Vol. 830: J. A. Green, Polynomial Representations of GL_n. VI, 118 pages. 1980.

Vol. 831: Representation Theory I. Proceedings, 1979. Edited by V. Dlab and P. Gabriel. XIV, 373 pages. 1980.

Vol. 832: Representation Theory II. Proceedings, 1979. Edited by V. Dlab and P. Gabriel. XIV, 673 pages. 1980.

Vol. 833: Th. Jeulin, Semi-Martingales et Grossissement d'une Filtration. IX, 142 Seiten. 1980.

Vol. 834: Model Theory of Algebra and Arithmetic. Proceedings, 1979. Edited by L. Pacholski, J. Wierzejewski, and A. J. Wilkie. VI, 410 pages. 1980.

Vol. 835: H. Zieschang, E. Vogt and H.-D. Coldewey, Surfaces and Planar Discontinuous Groups. X, 334 pages. 1980.

Vol. 836: Differential Geometrical Methods in Mathematical Physics. Proceedings, 1979. Edited by P. L. García, A. Pérez-Rendón, and J. M. Souriau. XII, 538 pages. 1980.

Vol. 837: J. Meixner, F. W. Schäfke and G. Wolf, Mathieu Functions and Spheroidal Functions and their Mathematical Foundations Further Studies. VII, 126 pages. 1980.

Vol. 838: Global Differential Geometry and Global Analysis. Proceedings 1979. Edited by D. Ferus et al. XI, 299 pages. 1981.

Vol. 839: Cabal Seminar 77 – 79. Proceedings. Edited by A. S. Kechris, D. A. Martin and Y. N. Moschovakis. V, 274 pages. 1981.

Vol. 840: D. Henry, Geometric Theory of Semilinear Parabolic Equations. IV, 348 pages. 1981.

Vol. 841: A. Haraux, Nonlinear Evolution Equations- Global Behaviour of Solutions. XII, 313 pages. 1981.

Vol. 842: Séminaire Bourbaki vol. 1979/80. Exposés 543–560. IV, 317 pages. 1981.

Vol. 843: Functional Analysis, Holomorphy, and Approximation Theory. Proceedings. Edited by S. Machado. VI, 636 pages. 1981.

Vol. 844: Groupe de Brauer. Proceedings. Edited by M. Kervaire and M. Ojanguren. VII, 274 pages. 1981.

Vol. 845: A. Tannenbaum, Invariance and System Theory: Algebraic and Geometric Aspects. X, 161 pages. 1981.

Vol. 846: Ordinary and Partial Differential Equations, Proceedings. Edited by W. N. Everitt and B. D. Sleeman. XIV, 384 pages. 1981.

Vol. 847: U. Koschorke, Vector Fields and Other Vector Bundle Morphisms – A Singularity Approach. IV, 304 pages. 1981.

Vol. 848: Algebra, Carbondale 1980. Proceedings. Ed. by R. K. Amayo. VI, 298 pages. 1981.

Vol. 849: P. Major, Multiple Wiener-Itô Integrals. VII, 127 pages. 1981.

Vol. 850: Séminaire de Probabilités XV. 1979/80. Avec table générale des exposés de 1966/67 à 1978/79. Edited by J. Azéma and M. Yor. IV, 704 pages. 1981.

Vol. 851: Stochastic Integrals. Proceedings, 1980. Edited by D. Williams. IX, 540 pages. 1981.

Vol. 852: L. Schwartz, Geometry and Probability in Banach Spaces. X, 101 pages. 1981.

Vol. 853: N. Boboc, G. Bucur, A. Cornea, Order and Convexity in Potential Theory: H-Cones. IV, 286 pages. 1981.

Vol. 854: Algebraic K-Theory. Evanston 1980. Proceedings. Edited by E. M. Friedlander and M. R. Stein. V, 517 pages. 1981.

Vol. 855: Semigroups. Proceedings 1978. Edited by H. Jürgensen, M. Petrich and H. J. Weinert. V, 221 pages. 1981.

Vol. 856: R. Lascar, Propagation des Singularités des Solutions d'Equations Pseudo-Différentielles à Caractéristiques de Multiplicités Variables. VIII, 237 pages. 1981.

Vol. 857: M. Miyanishi. Non-complete Algebraic Surfaces. XVIII, 244 pages. 1981.

Vol. 858: E. A. Coddington, H. S. V. de Snoo: Regular Boundary Value Problems Associated with Pairs of Ordinary Differential Expressions. V, 225 pages. 1981.

Vol. 859: Logic Year 1979–80. Proceedings. Edited by M. Lerman, J. Schmerl and R. Soare. VIII, 326 pages. 1981.

Vol. 860: Probability in Banach Spaces III. Proceedings, 1980. Edited by A. Beck. VI, 329 pages. 1981.

Vol. 861: Analytical Methods in Probability Theory. Proceedings 1980. Edited by D. Dugué, E. Lukacs, V. K. Rohatgi. X, 183 pages. 1981.

Vol. 862: Algebraic Geometry. Proceedings 1980. Edited by A. Libgober and P. Wagreich. V, 281 pages. 1981.

Vol. 863: Processus Aléatoires à Deux Indices. Proceedings, 1980. Edited by H. Korezlioglu, G. Mazziotto and J. Szpirglas. V, 274 pages. 1981.

Vol. 864: Complex Analysis and Spectral Theory. Proceedings, 1979/80. Edited by V. P. Havin and N. K. Nikol'skii, VI, 480 pages. 1981.

Vol. 865: R. W. Bruggeman, Fourier Coefficients of Automorphic Forms. III, 201 pages. 1981.

Vol. 866: J.-M. Bismut, Mécanique Aléatoire. XVI, 563 pages. 1981.

Vol. 867: Séminaire d'Algèbre Paul Dubreil et Marie-Paule Malliavin. Proceedings, 1980. Edited by M.-P. Malliavin. V, 476 pages. 1981.

Vol. 868: Surfaces Algébriques. Proceedings 1976–78. Edited by J. Giraud, L. Illusie et M. Raynaud. V, 314 pages. 1981.

Vol. 869: A. V. Zelevinsky, Representations of Finite Classical Groups. IV, 184 pages. 1981.

Vol. 870: Shape Theory and Geometric Topology. Proceedings, 1981. Edited by S. Mardešić and J. Segal. V, 265 pages. 1981.

Vol. 871: Continuous Lattices. Proceedings, 1979. Edited by B. Banaschewski and R.-E. Hoffmann. X, 413 pages. 1981.

Vol. 872: Set Theory and Model Theory. Proceedings, 1979. Edited by R. B. Jensen and A. Prestel. V, 174 pages. 1981.